POROUS SILICON
SCIENCE AND TECHNOLOGY

Winter School

Les Houches, 8 to 12 February 1994

Editors

Jean-Claude Vial, CNRS-Université Joseph Fourier

Jacques Derrien, CNRS-Université Aix-Marseille 2

Springer-Verlag Berlin Heidelberg GmbH

ISBN 978-3-540-58936-5   ISBN 978-3-662-03120-9 (eBook)
DOI 10.1007/978-3-662-03120-9

The international winter school LUMINESCENCE OF POROUS SILICON AND SILICON NANOSTRUCTURES (Les Houches 8 to 12 February 1994) was organized by :

- Le Groupement de Recherche du CNRS (GDR LUSIL)

It was supported by :

- Université J. Fourier de Grenoble
- Centre National de la Recherche Scientifique (CNRS-formation permanente)
- Programme interdisciplinaire de recherche sur les techniques poussées à leur limite (Ultimatech)
- Ministère de l'Enseignement Supérieur
- Direction des Recherches et Etudes Techniques (DGA-DRET)
- France Télécom-CNET
- Institut des Etudes Scientifiques Avancées de Grenoble
- International Science Fondation (ISF)
- Laboratoire de Spectrométrie Physique de Grenoble
- Centre de Recherche sur la Croissance Cristalline (CRMC2) de Marseille

FOREWORD

The discovery of bright visible light emission from porous silicon has opened the door to various nanometer sized silicon structures where the confinement of carriers gives rise to interesting physical properties. While the high efficiency of the light emission in the visible range is the common and the most prominent feature, their structures display similar properties with other highly divided materials (even non semiconductors), and then justify a multidisciplinary approach. This along with potential applications has attracted a large number of researchers followed by students to be trained. Until now international conferences have provided the exchange of information but have remained highly specialised so it was time to give thought to the organisation of topical and advanced lectures where the multidisciplinarity and the didactic approach are paramount. L'école des Houches was ideally devoted to that purpose. The meeting : " Luminescence of porous silicon and silicon nanostructures" was the first international school on this topic but some aspects in the organisation and the attendance have given an international workshop flavor to it.

The school by itself has trained 82 « students », most of them were students starting their Ph. D thesis. 50% were French citizens and the other represented countries were Germany, England, USA, Czechoslovakia, The Netherlands, Italy, Japan, Poland, Spain, Canada, Brazil, India and Russia.

The originality of this school was to give an important place to disciplines not usually taught to physicists and thanks to lectures on the electrochemistry of semiconductors (7 hours with L.M. Peter, J.N. Chazalviel, R. Hérino and A. Halimaoui) or through a course on the basic techniques used in the characterisation of highly divided or porous material (7 hours with W. Theiss, I. Berbezier, H.J. von Bardeleben, B. Champagnon, A. Naudon, H. Muender and C. Ortega). A good overview on theoretical models has been particularly appreciated during the 4 hour course given by M. Hybertsen, R. Tsu and M. Lannoo. More prospectively, the aspects related to optoelectronic devices based on porous silicon were presented by W. Lang, A. Bsiesy and N. Koshida. The optical properties of porous silicon were indeed not forgotten (P. Calcott, J.C. Vial and F. Koch) but they were presented along with what was well known and well established in other systems (R. Ferreira for II-VI and III-V nanostructures, E. Bustarret for the amorphous silicon). In addition we had reserved one hour a day for the evening seminars where hot topics were discussed (P. Fauchet, P.A. Badoz, H. Muender, et C. Lévy-Clément). The school was introduced by a general review on porous silicon given by L.T. Canham.

Jacques Derrien, CRMC2-CNRS, Université Aix -Marseille 2
Jean-Claude Vial, Laboratoire de Spectrométrie Physique, Université J. Fourier.

CONTENTS

LECTURE 3

Porous silicon: material processing, properties and applications
by A. Halimaoui

LECTURE 4

Luminescence of porous silicon after electrochemical oxidation
by R. Herino

LECTURE 5

Mechanism for light emission from nanoscale silicon
by M.S. Hybertsen

LECTURE 6

Theory of silicon crystallites. Part II
by C. Delerue, G. Allan, E. Martin and M. Lannoo

x

LECTURE 10

Ion beam analysis of thin films. Applications to porous silicon
by C. Ortega, A. Grosman and V. Morazzani

LECTURE 11

IR spectroscopy of porous silicon
by W. Theiss

LECTURE 12

Nano characterization of porous silicon by transmission electron microscopy
by I. Berbezier

LECTURE 13

Electron paramagnetic resonance spectroscopy: Defect and structural analysis of solids
by H.J. von Bardeleben and M. Schoisswohl

LECTURE 14

Raman scattering in silicon nanostructures
by B. Champagnon, I. Gregora, Y. Monin, E. Duval, L. Saviot

LECTURE 15

Scattering of X-rays
by A. Naudon, P. Goudeau and V. Vezin

LECTURE 16

X-ray photoemission spectroscopy
by H. Münder

LECTURE 17

Optoelectronic properties of porous silicon. The electroluminescent devices
by W. Lang, P. Steiner, F. Kozlowski

LECTURE 18

Porous silicon luminescence under cathodic polarisation conditions
by A. Bsiesy

LECTURE 19

Interrelation between electrical properties and visible luminescence of porous silicon
by N. Koshida

LECTURE 20

Characteristics of porous n-type silicon obtained by photoelectrochemical etching
by C. Levy-Clement

LECTURE 21

Porous Si: From single porous layers to porosity superlattices
by M.G. Berger, St. Frohnhoff, W. Theiss, U. Rossow
and H. Münder

Fundamental aspects of the semiconductor-solution interface

L.M. Peter

School of Chemistry, University of Bath,
Bath BA2 7AY, U.K.

1. ELECTRONS IN SOLIDS

1.1 The Band Model and Conductivity

Let us begin with a brief summary of the solid state physics of semiconductors. The electronic properties of solids are determined by their *band structure*. The electron energy levels are grouped into energy bands separated by regions of forbidden energies, and the filling of the available energy levels with electrons determines whether a material is a *metal* or an *insulator*. In the case of semiconductors such as silicon, the *intrinsic* conductivity of the pure material is very low at room temperature, since the density of mobile carriers created by thermal excitation is small. In most practical applications of semiconductors, the conductivity is *extrinsic*, i.e. it is determined by the concentration of dopants (electron donors in the case of n-type material and acceptors in the case of p-type material). Doping densities vary over a very wide range from 10^{15} to 10^{19} cm^{-3}.

The population of available electron energy levels is controlled by *Fermi-Dirac* statistics. The Fermi-Dirac distribution function is

$$f_{FD}(E) = \frac{1}{1 + \exp[(E - E_F)/kT]} \qquad 1)$$

$f_{FD}(E)$ is the probability that an electronic level of energy E is occupied and E_F is the *Fermi energy*. As we shall see, E_F plays an important role in discussions of *electronic equilibrium* at semiconductor junctions; E_F is related to the *free energy* of electrons. The density of electrons and holes is determined by the product of the Fermi-Dirac function and the density of states function. For an insulator or semiconductor, the product of the hole (*p*) and electron (*n*) densities is given by

$$np = N_C N_V \exp(-E_G/kT) \qquad 2)$$

where E_G is the bandgap and N_C, N_V are the density of states in the conduction and valence bands. In the case of intrinsic semiconductors or insulators, the Fermi energy is located near

L.M. Peter

the centre of the energy gap. For doped semiconductors, E_F is close to the valence band (p-type) or conduction band (n-type). If E_F is more than a few kT below the conduction band (or above the valence band), the *majority carrier densities* are given by

$$n = N_C \exp[-\frac{(E_C-E_F)}{kT}] \qquad\qquad 3a)$$

$$p = N_V \exp[-\frac{(E_F-E_V)}{kT}] \qquad\qquad 3b)$$

Figure 1 illustrates the band diagrams and Fermi energies for metals and for intrinsic, n- and p-type semiconductors.

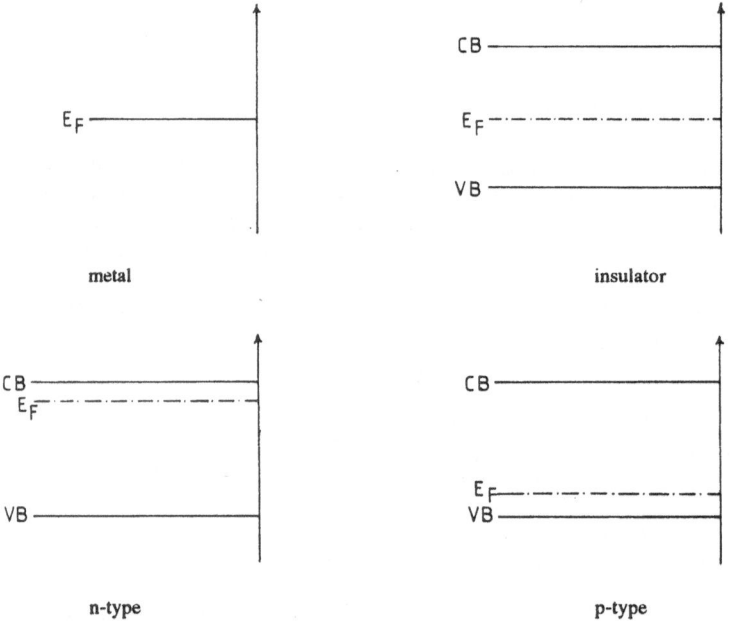

Fig. 1. Simplified band diagrams for metals and semiconductors.

The energy of electrons in a semiconductor can also be related to the *work function*, Φ, the *ionization energy I* and the *electron affinity A*. Here the arbitrary zero of the electron energy scale is taken as the *vacuum level*. Figure 2 shows how Φ, I, A and E_G are related.

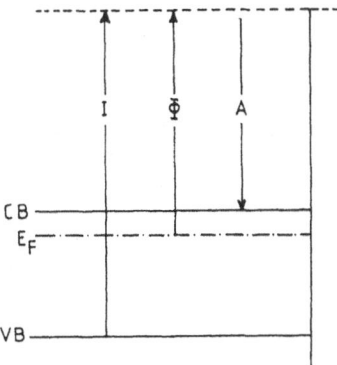

Fig. 2. Ionization energy (I), electron affinity (A) and work function (Φ)
defined for an n-type semiconductor.

2. SOLID STATE SEMICONDUCTOR JUNCTIONS.

2.1 The Contact Potential

When two electronically conducting phases are brought into contact, equilibrium is established by the transfer of electrons from one phase to the other, giving rise to a *contact potential*. The formation of semiconductor junctions is illustrated by the formation of an ideal *Schottky barrier (blocking contact)*. If, for example, an n-type semiconductor is contacted with a metal with a higher work function, then electrons will tend to move from the semiconductor into the metal. This results in a charged region at the interface and a corresponding contact potential given by

$$\Delta V_{M\text{-}S} = \frac{\Phi_{SC} - \Phi_M}{q} \qquad\qquad 4)$$

2.2 Band Bending

The movement of electrons across the metal-semiconductor junction leaves ionized donor atoms fixed in the lattice that form a *space charge*, whereas the mobile electrons on the metal side form a *surface charge*. The width of the space charge region, W_{SC}, depends on the magnitude of the contact potential and on the doping density; typical values are in the range 10^{-5} to 10^{-4} cm. The distribution of charge in the semiconductor leads to a parabolic dependence of electron energy on distance which is referred to as *band bending*. The magnitude of the band bending is simply $q\Delta_{M\text{-}Sc}$ as shown in Figure 3 for a Schottky junction at an n-type material.

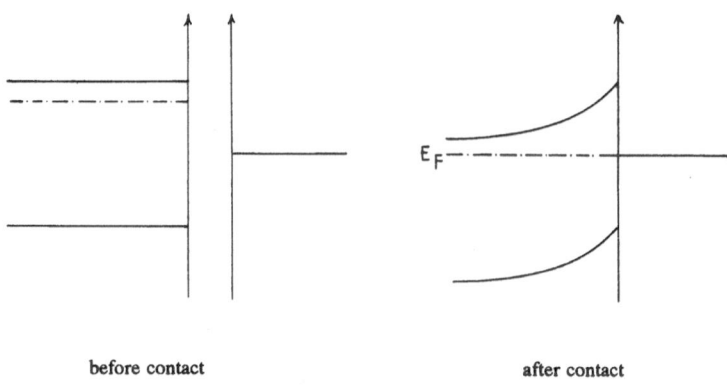

before contact   after contact

Fig. 3. Formation of a Schottky barrier at the metal-semiconductor junction

If the work function of the metal is *lower* than that of the n-type semiconductor, electrons flow the other way, creating a contact potential of opposite sign. Now the charge in the semiconductor is due to electrons in the conduction band, so that there is no space charge region and no large barrier to electron flow. The junction then forms an *ohmic contact*. Analogous arguments with inverted sign apply to contacts between metals and p-type semiconductors. Junctions between n- and p-type semiconductors of the same material (*homojunctions*) as well as different materials (*heterojunctions*) are also described in this way. As we shall see in the next section, these concepts can be carried over to the semiconductor-electrolyte contact.

3. ELECTROLYTE SOLUTIONS

3.1 Standard Electrode Potentials and the Vacuum Energy Scale.

Electrolyte solutions are *ionic conductors*. There are no delocalized electron states of the kind associated with the band structure of solids; instead electrons are localized on *molecular orbitals* of ionic species. It is common practice to distinguish between *inert electrolytes* which do not participate in electron transfer reactions at the semiconductor-solution interface, and *redox electrolytes* which can accept or donate electrons to the semiconductor. In practice, of course, any electrolyte and the solvent itself can take part in electron transfer processes if the 'oxidizing' or 'reducing' power of the solid phase is large enough.

In general, redox reactions can be written as

$$O + ne^- \rightleftarrows R \qquad\qquad 5)$$

The properties of redox systems are usually discussed in terms of their *standard reduction potentials, E^o*. These potentials are related by definition to the reaction

$$O + n/2\ H_2 \rightleftarrows R + n\ H^+ \qquad (\Delta G^o) \qquad\qquad 6)$$

by $$\Delta G^o = - nFE^o \qquad\qquad 7)$$

where ΔG^O is the standard Gibbs free energy change (all components in their standard states).

The concentration (or activity) dependence of the equilibrium reduction potential is described by the *Nernst equation*

$$E_{EQ} = E^O + \frac{RT}{nF} \ln[\frac{C_O}{C_R}]$$

8)

It is clear that the standard potentials must be related to the ionization energy or electron affinity of the redox species. In order to establish a relationship between the free energies of electrons in solids and in redox equilibria, it is necessary to know the *work function of the standard hydrogen electrode*. This corresponds to the process

$$1/2 \, H_2 \; \rightarrow \; H^+ + e^-$$

9)

There are several ways of obtaining a good estimate of the energy associated with this process; recent calculations suggest a value of 4.7 eV. This allows us to convert the *electrochemical scale* of standard reduction potentials to the *vacuum scale*. This conversion is valuable when considering the relative energy levels of electrons in the solution and in the solid.

In the case of a metal electrode, the feasibility of a particular reaction is determined by the *electrode potential*. The reduction reaction in eqn 5 can occur under standard conditions, provided that the electrode potential is more negative than $E°$. If, on the other hand, the electrode potential is more positive than $E°$, the reaction proceeds from right to left, i.e. oxidation of R occurs.

The feasibility of reactions at semiconductor electrodes is a more complex matter. In principle, we can distinguish between conduction band and valence band reactions. The reaction rate is determined in part by the availability of electrons (or holes) in the solid; this contrasts with a metal, which acts as an effective source or sink for electrons.

3.2 Energy Fluctuations in Solution Redox Species.

The localized electrons on individual redox ions in solution are associated with chemically distinguishable species. Consider, for example, the redox reaction

$$Fe^{3+}_{AQ} + e_{VAC}^- \; \rightleftarrows \; Fe^{2+}_{AQ}$$

10)

The charge on the two ions is different, and as a consequence the interactions with the solvent (ion-dipole interactions) are not the same. Let us assume that the Fe^{3+} ion is surrounded by an 'equilibrium' configuration of water molecules. Now the electron is transferred from vacuum to the Fe^{3+} ion, creating a Fe^{2+} ion. Electron transfer is subject to the *Franck Condon principle,* which in effect means that the solvent molecules remain 'frozen'. However, after electron transfer has occurred, the ion has a lower positive charge, and the ion-dipole interaction is correspondingly weaker. The solvation shell will therefore relax to a new 'equilibrium' distribution. As a consequence, the ionization energy of Fe^{2+} is not the same magnitude as the electron affinity of Fe^{3+}; the polarization energy arising from interaction with the solvent effectively splits the electron energy levels of the oxidized and reduced species apart.

A second complication is that the polarization experienced by an electron on a redox ion is not constant. Instead it fluctuates continuously as a result of rotations and vibrations of the solvent molecules. The electron energy levels are therefore described in terms of *probability distributions*. Although the energy of the thermal fluctuations is small (of the order of kT), the resulting changes in electron energy are large, typically of the order of electron volts. The *temporal distributions* of electron energy levels in a redox system are therefore described by the 'Gaussian' *W(E)* curves shown in Figure 4. The work function of the redox system is determined by the point of intersection of the two curves as shown in the figure. This energy has also been referred to as the *redox Fermi energy*, $E_{F,REDOX}$ although the term can lead to confusion since it encourages analogies with the distribution of energies in solids.

In many cases, redox ions are metal ligand complexes such as $Fe(CN)_6^{4-}$ or $Ru(bpy)_3^{3+}$ for example. Here it is useful to distinguish between the *inner sphere* of ligands and the *outer sphere* of solvent molecules. Changes in the inner sphere bond energies and fluctuations in bond lengths and angles also contribute to the spread of electron energy levels shown in Figure 4.

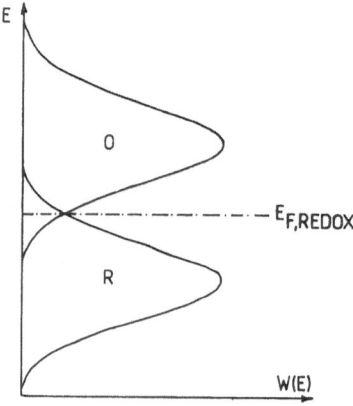

Fig. 4.   Probability Distributions for Electron Levels on Redox Ions O and R

4. THE SEMICONDUCTOR-SOLUTION JUNCTION

4.1 Electron Transfer at the Metal Solution Interface.

Electronic equilibrium between a redox system and a metal is established by *electron transfer*. The situation is exactly analogous to a metal-semiconductor contact. At equilibrium $E_{F,METAL}$ = $E_{F,REDOX}$, and the electron energy levels in the solid and solution phases align as shown in Figure 5a.

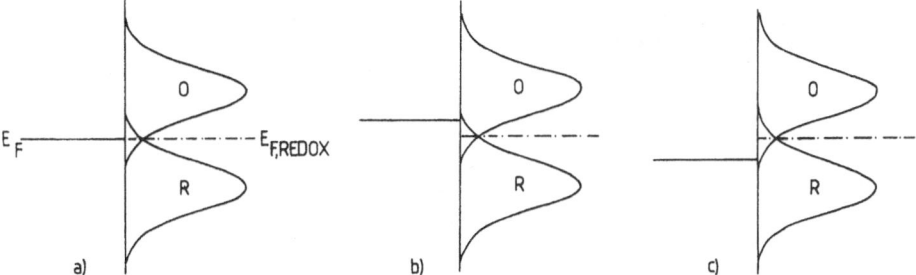

Fig. 5. The metal-redox electrolyte interface:
(a) at equilibrium, (b) $E < E°$, (c) $E > E_o$

Figure 5 shows that the overlap between vacant states in the metal and filled states in the solution (R) is restricted to a region close to E_F. The same is true of the overlap between O states and filled metal states. It follows that electron transfer at metal electrodes takes place predominantly near E_F. Since the 'most probable energies' of O and R are typically 1 eV or more distant from E_F, electron transfer is an *activated process*; the width of the distribution functions determines the so-called *reorganization energy*.

Since the metal and electrolyte phases are highly conducting, changes in electrode potential alter the potential difference in a narrow region at the surface, so that the O/R curves are displaced relative to the metal states, allowing either reduction (5b) or oxidation (5c). In effect, the *activation energy* for electron transfer at metals is potential dependent. As a consequence, the rate constants for electrode reactions at metals have the form

$$k_O = k° \exp[- \alpha nF(E - E°)/RT] \qquad \textit{anodic} \qquad \text{11a)}$$

$$k_R = k° \exp[(1-\alpha)nF(E - E°)/RT] \qquad \textit{cathodic} \qquad \text{11b)}$$

4.2 Electron Transfer at the Semiconductor Electrolyte Interface.

Consider the case of an n-type semiconductor in contact with a redox electrolyte which has a redox potential that lies below the Fermi energy of the semiconductor. Electrons (majority carriers) will initially move across the interface, reducing O ions. As this takes place, the ionized donor species remaining in the semiconductor form the space charge layer. The potential difference that results from the formation of this layer compensates the original tendency of electrons to escape from the semiconductor, and the Fermi levels on both sides of the interface become equal. As before, the space charge region extends over distances of the order of 10^{-6} to 10^{-4} cm, depending on doping density. Figure 6 illustrates the energy

L.M. Peter

diagrams for junctions to n- and p-type semiconductors for redox systems which form a
depletion layer in the semiconductors.

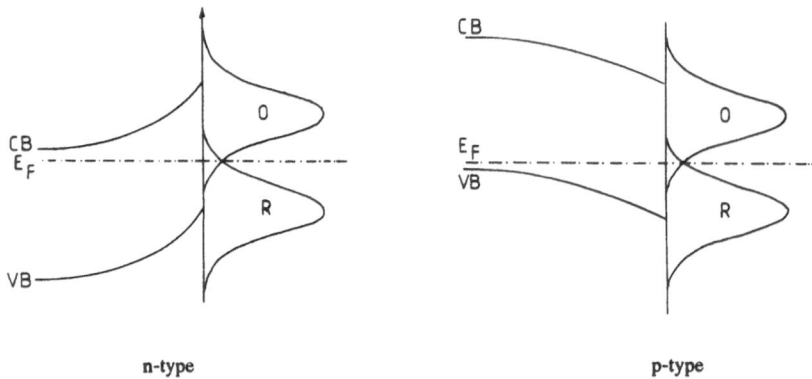

Fig. 6. Energy Diagrams for n- and p-type Semiconductors
in Contact with Redox Electrolytes

Figure 6 shows that electron exchange at semiconductors does not involve states near the
Fermi energy if a Schottky barrier is formed, since E_F is located in the bandgap. Electron
exchange with redox species may involve either the conduction or valence bands or both.
The overlap between the energy bands and redox states determines the relative contributions
of the bands.

An important difference between metal electrodes and semiconductor electrodes under
depletion conditions is the location of the potential drop. For metal-solution junctions, it is
restricted to a narrow region of a few Ångstroms in the solution that is referred to as the
Helmholtz layer. For semiconductors, on the other hand, the potential drop under depletion
conditions is almost entirely located in the solid phase so that changes in electrode potential
do not influence the activation energy for electron transfer. However, the surface
concentrations of charge carriers do change with potential; under ideal depletion conditions
by a factor of 10 for a change in potential of 0.059 V (see eqn 3).

4.3 Space Charge Capacitance: the Mott Schottky Equation.

The potential difference between the semiconductor and solution phases can be varied using
an external bias. If no redox species are present, $E_{F,REDOX}$ is undefined, and the band bending
can be varied over a wide range without current flow. In electrochemistry, the potential of
the solid phase is referred to that of a second *reference electrode* immersed in the solution.
This reference potential is related to the standard hydrogen potential, which is taken as an
arbitrary zero (see 3.1). If the potential is changed by an external source, the junction can
be varied from *accumulation* through *flatband* to *depletion* and finally *inversion*. In practice,
the potential range under accumulation conditions can be limited by solvent reactions, e.g.

hydrogen evolution in water, or by reactions leading to decomposition of the semiconductor. For highly doped or narrow band semiconductors, inversion conditions may also lead to current flow associated with the same reactions.

The changes in band bending are shown in Figure 7 for the case of an n-type semiconductor (band bending for p-type semiconductors has opposite sign, because the space charge consists of negatively charged acceptors).

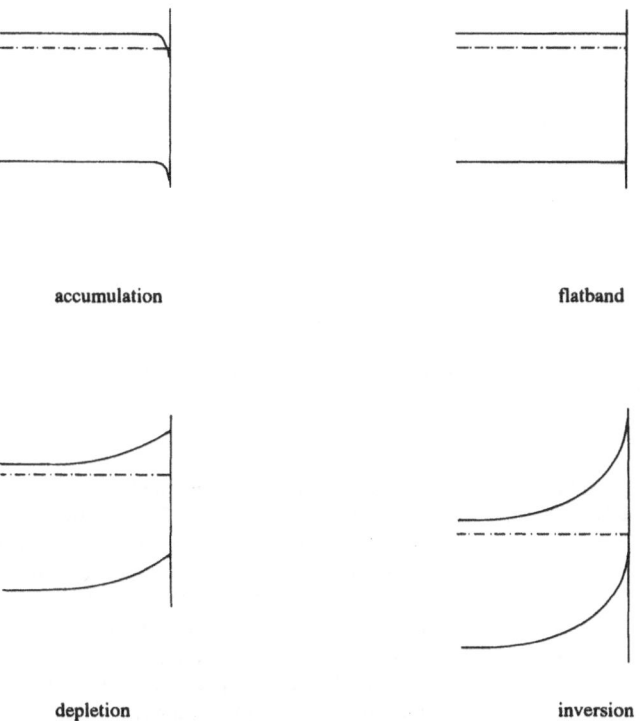

accumulation   flatband

depletion   inversion

Fig. 7. Changes of band bending with applied potential.

The space charge in the semiconductor is balanced by an equivalent charge of opposite sign in the electrolyte. This arises from an excess of ions of one sign near the electrode, and in concentrated solutions this region can be as thin as 10^{-7} cm since the density of ionic charges is high (1 mol dm^{-3} is roughly equivalent to 10^{21} cm^{-3}). The charge in the solid and solution phases varies with potential, and under depletion conditions the corresponding *space charge capacitance* is described by the *Mott Schottky equation*.

$$C_{SC}^{-2} = (\frac{2}{qN\epsilon\epsilon_o})(U - U_{FB} - \frac{kT}{q})$$

12)

The electrode potential has been given the symbol U rather than E here to avoid confusion with energy E. A plot of C_{SC}^{-2} vs electrode potential has a slope inversely proportional to the doping density N and intercept close to U_{FB}, the *flatband potential*. An example is shown in Figure 8.

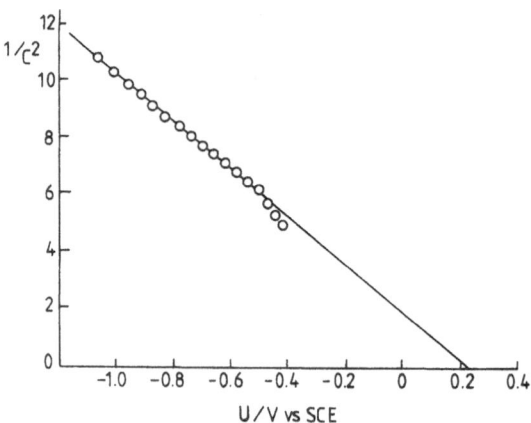

Fig. 8. Mott Schottky plot for p-GaAs in H_2SO_4

The flatband potential is an important quantity in semiconductor electrochemistry, since the band bending is defined by $U - U_{FB}$.   The value of U_{FB} is influenced by the fact that the potential difference at the interface is determined not only on the electronic charges, but also by adsorbed ionic charges and oriented dipoles that give rise to a *surface dipole potential*. This potential difference can arise from termination of the semiconductor surface by -H or -OH groups. The surface dipole potential therefore depends on solution composition, and it is commonly observed that the flatband potential changes with pH. Oxide semiconductors generally exhibit a variation in U_{FB} of -0.059 V per pH unit, corresponding to ideal 'Nernstian' behaviour. Other ions may influence U_{FB}; for example, the flatband potential of CdS is displaced to more negative potentials by sulphide ions.

The measured capacitance may differ from the predictions of the Mott Schottky equation for several reasons. The first is that there is also a capacitance associated with the finite width of the charge distribution in the electrolyte. This *Helmholtz capacitance*, C_H appears in series with C_{SC}, and usually it is much larger than C_{SC}. Only in the case of highly doped semiconductors ($N > 10^{19}$ cm^{-3}) is it necessary to take C_H into account. A second, and potentially more complex, reason for deviations from the Mott Schottky equation is that there may be electronic states on the surface of the semiconductor that can store charge. These *surface states* are specifically excluded from the derivation of C_{SC}. Their inclusion leads to an additional potential dependent capacitance C_{SS}, and Figure 9 is an example of the equivalent circuits used to describe the more complicated impedance response.

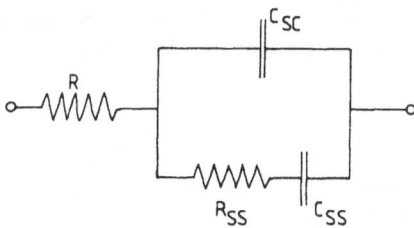

Fig. 9. Equivalent circuit including elements to describe surface states

5. THE ILLUMINATED SEMICONDUCTOR-ELECTROLYTE JUNCTION.

5.1 The Gärtner equation.

Illumination of the semiconductor with light of energy $h\upsilon > E_G$ excites electrons from the valence band to the conduction band. The subsequent reactions of the *photogenerated minority carriers* form the basis of *photoelectrochemistry*. Under depletion conditions, minority carriers generated within the space charge region are swept rapidly to the surface. Carriers generated outside the space charge region move by diffusion, and if they reach the edge of the space charge region, they are collected and transported to the surface. The generation collection problem can be described conveniently in terms of three *characteristic lengths:*

W_{SC} - the width of the space charge region
$1/\alpha$ - the penetration depth of the light
L_{MIN} - the diffusion length of minority carriers.

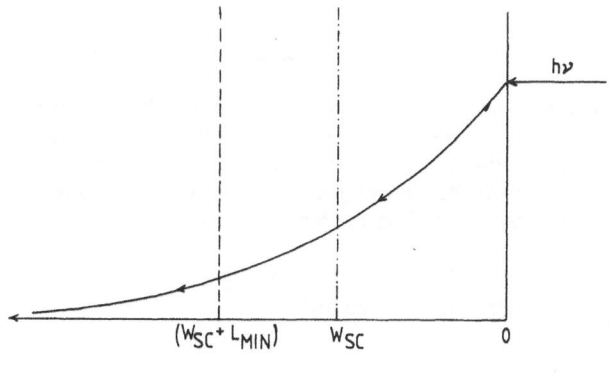

semiconductor   solution

Fig. 10. Characteristic lengths used to derive photocurrent response

If we assume that all carriers reaching the surface are transferred to redox species in solution, then the *photocurrent* is given by the *Gärtner equation*.

$$\frac{j_{PHOTO}}{I_o} = 1 - \frac{(\exp - \alpha W_{SC})}{(1 + \alpha L_{MIN})} \qquad 13)$$

where I_o is the incident light flux corrected for reflection. It can be seen that the photocurrent is a linear function of light intensity.

In principle, eqn 13 can be used to derive values of the absorption coefficient and the minority carrier diffusion length L_{MIN}. However, real semiconductor-electrolyte junctions deviate substantially from ideal behaviour, particularly close to the flatband potential. The existence of *surface recombination centres* allows holes and electrons to recombine, and the photocurrent potential curve deviates from eqn 12 as shown in Figure 11. The photocurrent onset potential is often used to estimate U_{FB}, but this method is unreliable, since surface recombination can lead to a significant displacement of the photocurrent potential curve.

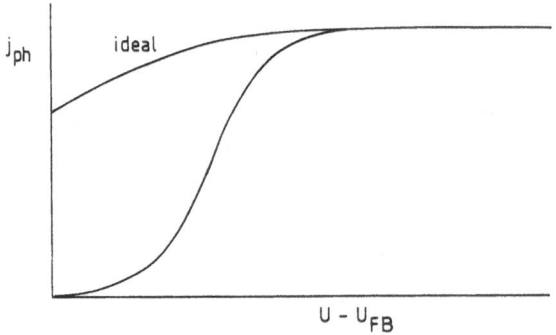

Fig. 11. Ideal and real photocurrent potential characteristics

The derivation of the Gärtner equation is based on the assumption that minority carriers reaching the semiconductor-solution interface are consumed rapidly in an electron transfer reaction. The simplest type of reaction involves one electron transfer to or from solution redox species, e.g. at n-silicon in methanol, photogenerated holes can react with ferrocene $Fe(cp)_2$ in a fast outer sphere reaction to produce the ferrocinium cation.

$$h\upsilon \rightarrow h^+ + e^- \qquad 14a)$$
$$h^+ + Fe(cp)_2 \rightarrow Fe(cp)_2^+ \qquad 14b)$$

5.2 Photocorrosion Reactions.

The situation is more complicated, however, in the absence of a suitable redox species. Photogenerated holes or electrons reaching the surface can react with the semiconductor lattice, leading to *Photocorrosion*. Examples of reactions of this kind are *n-CdS*

$$h\upsilon \rightarrow h^+ + e^- \tag{15a}$$
$$CdS + 2 h^+ \rightarrow Cd^{2+} + S \tag{15b}$$

p-CdTe

$$h\upsilon \rightarrow h^+ + e^- \tag{16a}$$
$$CdTe + 2 H^+ + 2 e^- \rightarrow Cd + H_2Te \tag{16b}$$

The Photocorrosion of n-Si normally results in the formation of a surface layer of SiO_2. In acidic fluoride solutions, however, dissolution occurs. The reactions are complex and involve not only hole capture but also *electron injection into the conduction band* from reaction intermediates. At low light intensities the quantum efficiency for dissolution can reach 4. This corresponds to the overall reaction

$$h\upsilon \rightarrow h^+ + e^- \tag{17a}$$

$$Si + h \rightarrow Si(I) \qquad \text{\textit{hole capture}} \tag{17b}$$

$$Si(I) \rightarrow Si(IV) + 3 e^- \qquad \text{\textit{electron injection}} \tag{17c}$$

In this scheme, Si(I) is a reactive intermediate that can inject electrons into the conduction band until finally the silicon atom leaves the lattice as SiF_6^{2-}. Other examples of *photocurrent multiplication* are found during photooxidation of organic species such as formic acid at n-type semiconductors (ZnO, TiO_2 and CdS) as well as during photoreduction of oxygen and hydrogen peroxide at p-type semiconductors (p-GaP and p-GaAs).

5.3 Photosensitization.

Photocurrents can also result from the photoexcitation of an adsorbed dye. This process is known as *photosensitization*. In this case, photoexcitation of the dye is followed by *electro injection* into the conduction band of the semiconductor as shown in Figure 12. Efficient sensitization is limited to adsorbed layers of monolayer dimensions, but substantial effects can be observed with high surface area porous electrodes.

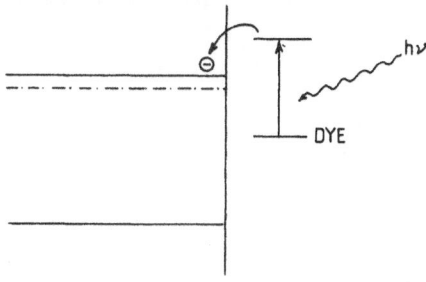

Fig. 12. Photosensitization by electron injection from an adsorbed dye.

Since electron injection results in oxidation of the adsorbed dye molecules, it is necessary to add a *supersensitizer*, which is a reducing species that can donate an electron to fill the vacancy in the LUMO orbital of the dye.

6. EXPERIMENTAL ASPECTS OF SEMICONDUCTOR ELECTROCHEMISTRY

Electrochemical experiments are usually carried out using a *three electrode cell* that incorporates working, secondary and reference electrodes. Glass is a suitable material for non-agressive electrolytes; it can be cleaned easily, and optical windows for photocurrent measurements are easily built in. Glass is not suitable for agressive media such as acidic fluoride solutions or strongly alkaline electrolytes; PTFE or polypropylene are used instead. Platinum is often used for the secondary electrode, and in order to minimize contamination of the semiconductor by traces of platinum, it may be necessary to use a two compartment cell, separated by a frit. This arrangement may not be suitable for experiments involving high current densities since it leads to a non-uniform current distribution at the working electrode. Suitable reference electrodes are palladium-hydrogen (PdH), saturated Ag/AgCl, and saturated calomel, Hg/Hg_2Cl_2. Contamination with traces of Ag or Hg as well as Cl⁻ can be a problem with the last two electrodes, but they have the advantage of being available commercially. PdH electrodes can be made by charging palladium with hydrogen electrochemically. Good performance is obtained if the palladium is present as a high surface area electrodeposit ('palladium black'). Connection to the cell is best made through a tap or frit to prevent contamination. The potential electrode should monitor the potential as close to the working electrode as possible to minimize the 'iR drop' error caused by current flow in the electrolyte. A Luggin capillary, in the form of a fine open ended tube is used to connect the reference electrode compartment to the main cell body. The tip of the capillary is placed a few mm from the surface of the working electrode. In PTFE cells, a thin PTFE tube can be used for the same purpose.

Potential and current can be controlled using a *three electrode potentiostat*. In a potentiostatic experiment, the *working electrode* is connected to the virtual ground input of a current-voltage converter amplifier. The *secondary* electrode is driven by the output of the potentiostat control amplifier, and the *reference electrode* provides the necessary negative feedback. It is necessary to use a buffer amplifier (impedance converter) for the reference electrode to prevent significant currents flowing. Low cost potentiostats are easily constructed from operational amplifiers; a typical arrangement is shown in Figure 13.

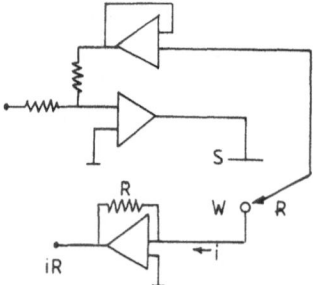

Fig. 13. Potentiostat design based on operational amplifiers

A potentiostat can also be modified to provide a constant current source. The connections are shown in Figure 14. The current is controlled by the series resistor and the input voltage. Note that measurement of the electrode potential requires a differential instrumentation amplifier since the working electrode is no longer grounded in this configuration.

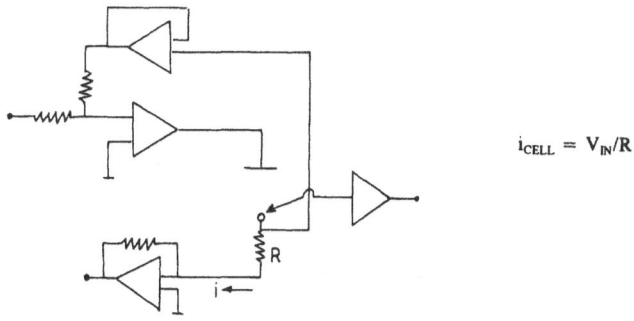

$$i_{CELL} = V_{IN}/R$$

Fig. 14. Constant current circuit based on a potentiostat.

7. ACKNOWLEDGMENTS.

I should like to take this opportunity to express my gratitude to numerous colleagues who have worked with me in the field of semiconductor electrochemistry. Financial support has been provided by the UK Science and Engineering Council and the University of Bath.

8. FURTHER READING

This overview has touched briefly on many aspects of semiconductor electrochemistry. Further details can be found in the following books and reviews.

1.   S.R. Morrison. *Electrochemistry at Semiconductor Electrodes*.
        Plenum, New York (1980).

2.   Yu.V. Pleskov and Yu. Ya. Gurevich. *Semiconductor Electrochemistry*.
        Consultants Bureau, New York (1986).

3.   A. Hamnett in *Comprehensive Chemical Kinetics*. Editor R.G. Compton.
        Vol. 27, pp 61-246. Elsevier, Amsterdam (1987).

4.   V.A. Myamlin and Yu. V. Pleskov, *Semiconductor Electrochemistry*.
        Plenum, New York (1967).

5.   S.R. Morrison, *The Chemical Physics of Surfaces*.
        Plenum, New York (1977).

6.   A. Many, Y. Goldstein and N.B. Grover, *Semiconductor Surfaces*.
        North Holland, Amsterdam (1965).

7.   R. Memming in Topics in Surface Chemistry, Editors E. Katz and P.S. Bagus
        Plenum , New York (1978).

LECTURE 2

The silicon/electrolyte interface

J.-N. Chazalviel

*Laboratoire de Physique de la Matière Condensée,
Ecole Polytechnique, 91128 Palaiseau, France*

---

1. INTRODUCTION

Silicon is known as a highly oxidizable material. Stability of silicon electrodes has been a major problem to electrochemists. Fortunately, the ability of this material to oxidize or dissolve may be taken as an advantage, for example for the formation of anodic oxides or the generation of porous silicon. The fundamental aspects of the latter phenomena are nonetheless complex, and we will adopt a presentation going from the simpler systems to the more complex ones. The most nearly ideal and simple silicon/electrolyte interfaces may be found with certain organic solvents. We will recall typical results obtained at n-Si/organic-electrolyte interfaces and from that we will go to a presentation of the present ideas about surface chemistry of silicon in ambient and in contact with wet, fluoride or alkaline media. Then we will turn to the increasingly complex situations encountered in aqueous electrochemistry. We will successively consider the interfaces under cathodic polarization, the interfaces under anodic polarization in non-fluoride electrolytes, which lead to anodic oxide formation, and the interfaces under anodic polarization in fluoride media, which may lead either to electropolishing or to porous silicon formation.

2. STABLE SILICON/ELECTROLYTE INTERFACES: THE CASE OF ORGANIC ELECTROLYTES

The anodic oxidation reaction of silicon in aqueous media is characterized by a rather negative potential (~−0.8 V relative to NHE at pH=0). Despite the recent effort on semiconductor photoelectrochemistry,[1,2] this trend toward easy oxidation of silicon in the presence of water has somewhat discouraged and delayed electrochemical investigations on this material. According to common sense, chemical oxidation will tend to form a silica layer at the silicon surface, and this layer should be of considerable hindrance as silica is known as an especially "blocking" oxide. Things get even worse if one intends to build a photoelectrochemical system based upon an n-Si photoanode, because the possibility of anodic oxidation adds up to that of chemical oxidation. The first attempts of working with silicon in non-aqueous media were themselves rather discouraging: for example, in acetonitrile, one might hope to slow down electrode oxidation upon thorough elimination of water from the organic solvent, and to protect it further by incorporating a fast-transfer oxidizable species, such as ferrocene, into the electrolyte. This strategy has been used with success on many semiconducting photoanodes.[3] Yet, in the case of silicon, even for water contents as small as a few p.p.m., an n-Si photoanode hardly works more than a few minutes under a typical illumination:[4] water tends to physisorb preferentially at the electrode surface;[5] when a hole reaches the surface, the oxidation reaction unavoidably takes place, and the formed oxide rapidly passivates the electrode. The lifetime of such electrodes is significantly higher when they are left in the dark, and fundamental studies have been feasible in such conditions.

2.1 Interface energetics

The silicon/acetonitrile-electrolyte interface has been studied with special care by a number of investigators. The results, somewhat disperse, indicate a flatband potential at $\approx$ –0.2 to –0.5 V (versus Ag/Ag+ reference in acetonitrile) for n-Si,[6-9] which means that the conduction-band potential is at $\approx$ –0.5 to –0.75 V and the valence-band potential at $\approx$ 0.35 to 0.6 V, relative to the same reference (if aqueous SCE is taken as the reference instead of Ag/Ag+ in acetonitrile, all these numbers become $\approx$ 0.3 V more positive). This potential becomes slightly more negative (by ~ 100 to 150 mV) upon letting the electrode age in the electrolyte.[9] This evolution, which takes place on a time scale of hours, is just an early manifestation of oxidation by residual water in the electrolyte.

Another early manifestation of electrode oxidation is the phenomenon of Fermi-level pinning.[10,11,6] A fresh (i.e., HF rinsed) electrode in acetonitrile presents a nearly ideal behaviour: when redox couples are incorporated into the electrolyte, the flatband potential stays about unchanged, and the electrical characteristics of the junction are just as one may expect from the difference between the flatband potential and the potential of the redox system (in other words, this is the electrochemical analog of an "ideal" Schottky diode whose barrier height would depend on metal workfunction with nearly unit slope). After ageing for some hours in the electrolyte, this ideal behaviour disappears. The flatband potential becomes increasingly dependent upon the redox couple added to the system, and the electrical behaviours of the junctions realized with various redox couples become rather similar (Schottky-barrier height $\approx$ 0.7 eV). This change in interface energetics arises from an increase in surface-state density upon slow oxidation of silicon by residual water in the organic electrolyte.[6] In-situ, potential-modulated infrared data provide some evidence for a correlation between the surface states and the SiOH groups formed on the surface.[12,5]

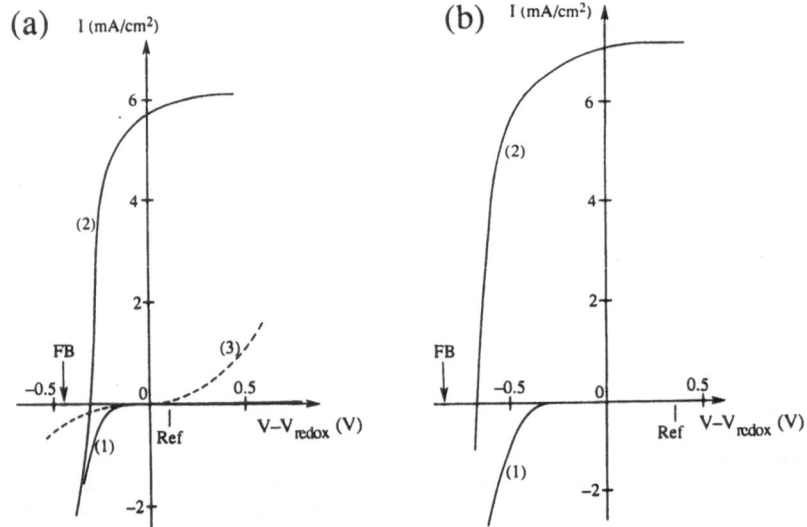

Figure 1: Current/potential curves: (a) n-Si/acetonitrile + LiClO$_4$ + dimethylferrocene: (1) dark, (2) illuminated, (3) illuminated, oxidized surface. (b) n-Si/methanol + LiClO$_4$ + dimethylferrocene: (1) dark, (2) illuminated. "FB" = flatband, "Ref" = Ag/Ag+ reference. After ref.[8].

2.2 Methanol-based regenerative solar cells

A breakthrough was performed in 1984, when the group of Lewis started working with methanol as a solvent. In electrolytes such as [methanol + $LiClO_4$ + dimethylferrocene / dimethylferricinium], an n-Si photoanode may allow one realizing a photoelectrochemical solar cell with an energetic yield of 14%, almost comparable to those obtained with solid-state *pin* junctions, and stable on a scale of several months.[13] Though such a duration (actually limited by reliability of the packaging) is still far from the required performance for a device, the considerable stability improvement relative to other comparable systems suggests the existence of an effect specific to the silicon/methanol interface. We will come again to this point later on.

These cells exhibit a quantum yield almost constant up to the blue spectral range, which is the fingerprint of a very low surface recombination velocity. Finally, using polycrystalline silicon results in a drop of the quantum yield to the still respectable value of 7.3%, a rather remarkable performance as compared to solid-state devices.[14]

3.   SILICON   SURFACE   CHEMISTRY   UNDER   OPEN-CIRCUIT CONDITIONS

The behaviour of Si/organic-electrolyte interfaces leads one naturally to wondering about the initial state of the surface, i.e., to questions about surface preparation and chemical etching.

3.1 Fluoride media

Hydrofluoric acid, one of the few media where silicon dioxide is readily dissolved, has long been used for preparing silicon surfaces.[15] Upon rinsing in HF, a silicon sample comes out with a hydrophobic surface. Such a surface has been reported to exhibit the lowest surface recombination velocity ever measured at a semiconductor surface, viz. 0.25 cm/s.[16] In the early 70's, using infrared spectroscopy, Beckmann had shown that this surface appeared covered with Si-H bonds, an observation that he interpreted in terms of a polymer-like layer ("polymerized silene").[17] This rather counter-intuitive result had stayed poorly known, and has become widely accepted only in the late eighties, after having received several confirmations from independent studies.[16,18-20] Actually, the detailed studies from the groups of Burrows and Chabal show that the surface of silicon rinsed in concentrated HF is completely saturated with a *monolayer* of hydrogen, but is atomically rough, as indicated by the presence of SiH, SiH_2 and SiH_3 groups.[20] These groups are best identified through their infrared vibrational spectra (2080-2150 cm^{-1} range). The fine structure exhibited by these spectra gives further information on the local environment of the Si atom of the group.[20]

On the opposite, if a [111] silicon surface is rinsed in a fluoride solution of a higher pH (for example, ammonium fluoride solution of pH≈10), the resulting surface appears unreconstructed, hydrogenated and atomically flat, as can be deduced from the observation of a single SiH vibrational peak, which is extraordinarily sharp (half width at mid height smaller than 1 cm^{-1} at room temperature).[21] This provides the most nearly ideal silicon surfaces that have ever been prepared. They have been used as a model system by surface physicists, and have been the subject of many fundamental studies (LEED, STM, EELS, and detailed vibrational investigations: IR, Raman, sum-frequency generation, time-resolved studies leading to the determination of dephasing time and vibrational relaxation, ...).[22-25]

It is remarkable that the chemical dissolution rate of silicon in fluoride media, very weak in concentrated HF, tends to increase with pH.[26] This suggests a dissolution mechanism catalyzed by the OH$^-$ ions, and which is probably anisotropic. The obtention of atomically flat surfaces would result from the anisotropy of this mechanism.

The hydrogenated silicon surface appears astonishingly stable in the presence of water. For example, in organic electrolyte (acetonitrile + supporting salt + 10 ppm H_2O), an *in-situ* infrared study has shown that from the beginning water from the electrolyte tends to become physisorbed at the interface (characteristic scissor-mode deformation of the water molecule at 1650 cm^{-1}), but, in the dark and in the absence of applied potential, the SiH groups survive on a 10-hour time scale.[5,27] The infrared spectra indicate the progressive appearance of SiOH and SiOSi peaks, together with an SiH component shifted toward higher energy (2270 cm^{-1}), characteristic of SiH groups backbonded to oxygen atoms (O_3SiH).[12] Also, this study shows that spontaneous oxidation of a [111] surface in such a medium proceeds via the formation of oxide islands, of typical lateral dimensions 20 Å.[5] The atomically flat NH_4F-rinsed surface has been found to exhibit even better stability.[28,29] This suggests that silicon oxidation is initiated through an attack of the multiply hydrogenated Si sites, which are absent on the atomically flat surface. Water treatments have even been used as a selective attack of the multiply hydrogenated silicon groups, thereby providing more nearly perfect [111] surfaces.[28,30]

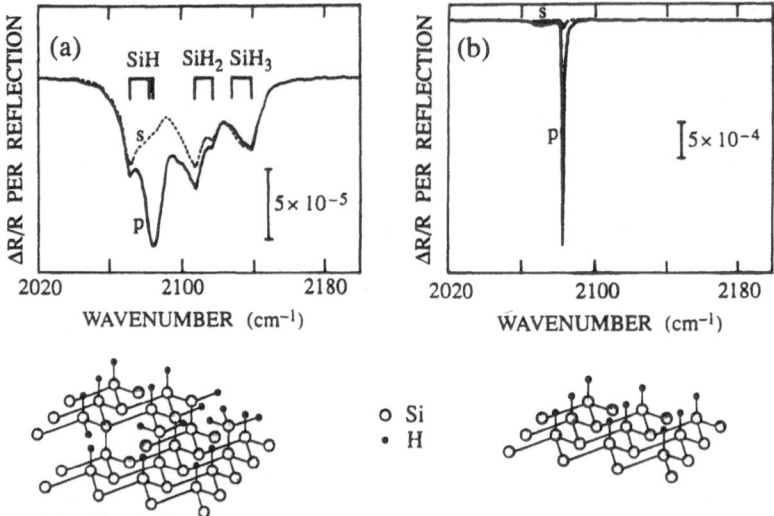

Figure 2: Infrared spectra (in air) of a [111] Si surface rinsed in (a) 50% HF, (b) NH_4F (pH≈10). Notice the several SiH_x groups in (a), and the single narrow peak in (b). After refs.[20-21].

3.2 Alkaline media

Silicon dissolution in alkaline media has long been known and used for its remarkable anisotropic properties.[31] The dissolution speed of [111] faces is typically two orders of magnitude lower than that of other orientations. This property is routinely used in technology for micromachining. The used solutions are often ethylenediamine/pyrocatechol/water ("EPW") mixtures. The essential role of water and of OH^- ions in the dissolution process has been evidenced[32] and similar anisotropic effects are indeed observed in simple aqueous solutions of sodium hydroxide or potassium hydroxide.[33,34] The anisotropic effect is undoubtedly related to that found in alkaline fluoride solutions.

As a matter of fact, recent studies have shown that the dissolution mechanism appears very similar whether or not fluoride ions are present in the medium. In either case, the surface of [111] terraces appears unreconstructed and covered with hydrogen.[35-37] This can be interpreted quite simply in the framework of a mechanism where the limiting step is breaking of a surface SiH bond (catalyzed by the OH^- ions); the resulting SiOH group will polarize the Si-Si backbonds, which can then be easily broken by fast reaction with water. Once the silicon atom has gone into solution, the hydrogenated surface is regenerated. Microscopic investigations using *in-situ* STM have shown that the chemical dissolution is slowed down under cathodic polarization. The anisotropy is then reinforced, the dissolution of a [111] face proceeding then exclusively through attack at steps. At intermediate potentials, dissolution may also proceed by pitting on the terraces. These observations have allowed an attempt for elaborating a detailed mechanism for the chemical and electrochemical dissolution of silicon in alkaline media.[35,36] Both mechanisms would proceed in the same way (OH attachment followed by backbond breaking). Both are anisotropic, though the anisotropy of the electrochemical mechanism appears comparatively weaker.

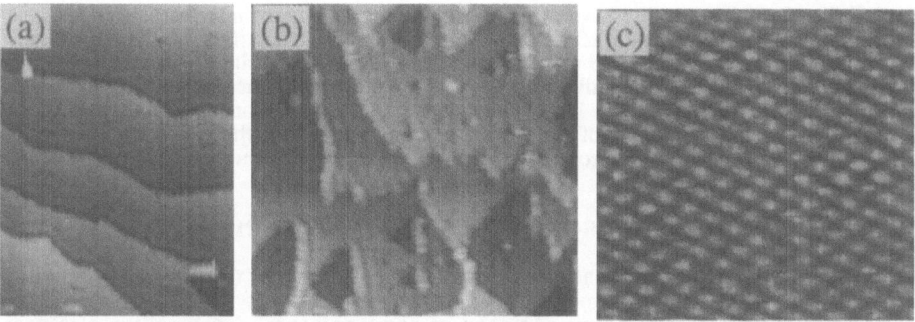

Figure 3: In-situ STM of a [111] n-Si surface in 2M NaOH, under cathodic bias. (a) $J=-150$ μA/cm^2, (b) same after zero-current polarization (notice the pitting), (c) same as (a) with atomic resolution of the 1×1 structure. Sizes (a) 1270 Å × 935 Å, (b) 1100 Å × 1100 Å, (c) 50 Å × 50 Å. Courtesy of P. Allongue [35-36].

3.3 Surface modification by organic reagents

Surface SiOH (silanol) groups are well known to provide a means for chemically grafting organic groups on the surface of silica.[38] In the case of silicon, it has been demonstrated that, in the same way as for silica, surface exposure to vapors of chlorotrimethylsilane or hexamethyldisilazane leads to chemical grafting of $-OSi(CH_3)_3$ surface groups.[39,9] This possibility has been considered, either as a means for stabilizing the silicon surface against oxidation,[39,9] or as a way for grafting electroactive species.[40] These graftings have been made by using an organosilicon compound, which leads to a grafting through a SiOSi bridge.

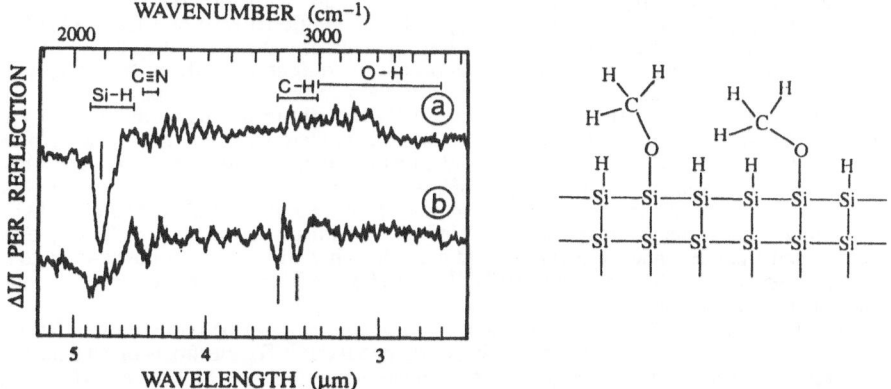

Figure 4: In-situ infrared (electromodulated) spectra of a n-Si/(acetonitrile + 0.1M LiClO$_4$) interface. (a) HF-rinsed surface (see SiH peak), (b) surface pretreated in methanol vapor. On the right, schematic representation of the modified surface. After ref. [9].

The surprizing stability and performance of silicon as a photoanode in methanol electrolyte have also led to wondering about the chemical nature of the silicon surface in such a medium. An infrared study has shown that a silicon surface exposed to methanol (either liquid or vapor) becomes partly covered with SiOCH$_3$ groups.[9] The obtained surface is found to be more stable than the hydrogenated surface: after such a pretreatment, a silicon sample can be exposed to acetonitrile electrolyte during several days without any change in interface characteristics (especially, weak density of surface states).[9] Furthermore, such a surface can undergo water rinses, in either acidic or alkaline medium, without being affected significantly. Finally, the modified surface exhibits a negative shift of the flatband potential in acetonitrile electrolyte, which is clearly related to the large photopotentials obtained on n-Si photoanodes in methanol.[9,41] Yet, in the case of the experiments in methanol, the role of a very slow dissolution of the silicon electrode (with formation of tetramethoxysilane) may not be completely rejected.

4. THE SILICON/AQUEOUS-ELECTROLYTE INTERFACE

The electrochemistry of silicon in aqueous electrolytes is hampered by the blocking character of silicon oxide. In non-fluoride media, HF-rinsed silicon surfaces will hardly keep their hydrogen coating for a long time. As soon as some oxidation develops, the electrical characteristics of the interface are deeply affected, so that obtaining reproducible results requires much care. In this section, we will summarize the main general results, and especially those obtained in non-fluoride media, and go more specifically to fluoride media in the last section.

4.1 Interface energetics: effect of pH

The flatband potential of silicon/aqueous-electrolyte interfaces has been the subject of several studies, in non-fluoride as well as in fluoride electrolytes.[42,43] The results obtained in both kinds of media are rather similar, and are summarized by the relation $V_{fb} = a + b \times pH$, with $b \approx -0.03$ V per pH unit, $a = -0.3$ to -0.4 V vs. SCE (n-type, $\sim 10^{15}$ cm^{-3}) and $+0.2$ to $+0.4$ V (p-type). The pH dependence of the flatband potential is a phenomenon familiar to semiconductor electrochemists. Typically, a shift of the flatband potential may occur if a charged species can get adsorbed at the interface, thereby modifying the electrostatic balance. One expects the adsorption of a positive (resp. negative) species to lead to a positive (resp. negative) shift of the flatband, and the shift will increase with the concentration of the corresponding species in the electrolyte.[44] Typically, an additional shift of $kT/e = 25$ mV is expected each time the concentration is increased by a factor e (here e is the base of natural logarithms), which means a slope of 60 mV per decade of concentration. Beautiful examples of such effects have been given in the case of S^{2-} adsorption on II-VI semiconductors.[45]

The case of pH-induced flatband shifts is usually thought to belong to this category of effects: if there is some oxide at a semiconductor surface, there will be in general adsorption sites for either OH$^-$ or H$^+$ ions, or both (in a chemist language, acidic or basic surface sites). In any case, a dependence of flatband upon pH, with a typical slope of 60 mV per pH unit, is to be expected.[44] The silicon surface in non-fluoride electrolytes is probably contaminated with oxide, which would tend to suggest that this kind of mechanism may apply here. The problem is that the same pH dependence seems to hold in media where the surface is now known to be fully hydrogenated, such as aqueous HF[42,43] or NaOH.[36] This fact stands rather strongly against the adsorption mechanism.

An alternate mechanism has been proposed and proved to describe correctly the flatband shift in the case of III-V or II-VI semiconductor compounds.[46] The electrochemical oxidation of any material involves pH-dependent steps, typically $M + OH^- \rightarrow MOH + e^-$. The redox potential associated with such a step depends upon pH with a slope of 60 mV per pH unit. If there are any surface states inducing Fermi-level pinning, the flatband potential is expected to follow the redox potential, hence a shift of V_{fb} with pH. It is most plausible that the observed shift in the case of silicon stems from such a kinetic mechanism.

4.2 Cathodic polarization: hydrogen evolution and hydrogen penetration

Cathodic polarization of silicon might be expected to be innocuitous toward the material. Upon cathodic polarization, hydrogen evolution takes place. Yet, in the case of silicon as well as for other electrode materials, the hydrogen evolution reaction takes place through two successive steps.[47] The first step is proton discharge and formation of an adsorbed hydrogen species

$$H^+ + e^- \rightarrow H_{ads} \tag{1}$$

The second step may be either a chemical recombination step or an electrochemical desorption step

$$H_{ads} + H_{ads} \rightarrow H_2 \tag{2a}$$

$$H_{ads} + H^+ + e^- \rightarrow H_2 \tag{2b}$$

In the case of silicon, it has been found that hydrogen adsorbed at the surface may penetrate into the electrode material, where its diffusion coefficient is non-negligible, even at room temperature. A classical electrochemical study on germanium,[48] confirmed by later infrared investigations on Ge and Si,[49] has shown that for both semiconductors a thin layer (~100 Å), disordered and highly concentrated in hydrogen, is formed below the electrode surface. Furthermore, in the case of p-type silicon, fast hydrogen diffusion leads to neutralization of the acceptors over thicknesses of ~ 10^3-10^4 Å.[50] This may occur under cathodic polarization but also possibly during chemical treatments generating hydrogen at the surface, such as some chemo-mechanical polishing treatments.[51] This "undoping" is of considerable hindrance if one wants to determine the flatband potential through surface-capacitance measurements, because the Mott-Schottky plots then become unreliable.[52]

4.3 Anodic polarization and SiO$_2$ formation

Silicon oxidation is an essential phenomenon for silicon technology. Silicon dioxide may be used either as the insulator in field-effect devices (MOS structures), or as a protecting (passivating) layer on integrated circuit chips. Anodic silica had been studied from the early ages of silicon technology.[53] Organic electrolytes with controlled amounts of water are generally used.[54] Applying a constant potential leads to an initial current burst followed by a decay. When the current has decayed to a small value, the obtained oxide thickness is essentially a function of applied potential (typically, thickness is 4 Å per applied volt). Potentials up to a few hundred volts may typically be applied before dielectric breakdown of the oxide occurs, hence thicknesses in the 1000 Å range may be obtained. In practice, one usually prefers applying a constant current rather than constant potential (this avoids the initial current burst), and monitoring the increasing potential. The experiment is stopped when the desired potential is attained. A curing stage at the final potential may also be added.

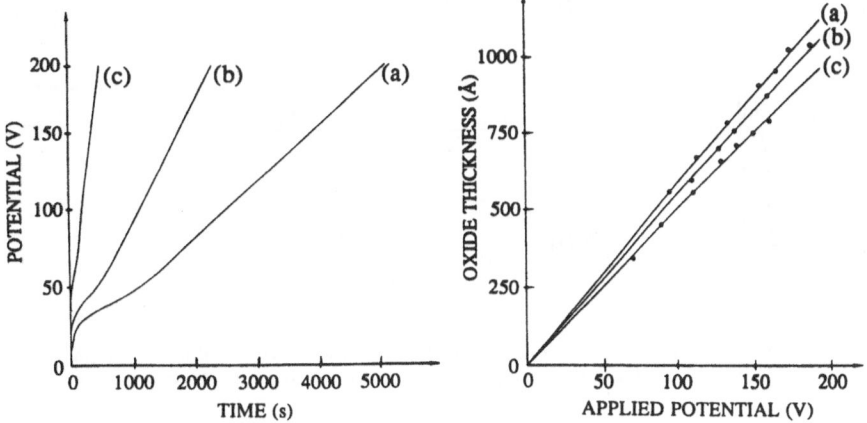

Figure 5: Anodic oxidation of silicon: p-Si/acetonitrile + 10^{-3}M KNO$_3$ + 200 ppm H$_2$O. Left: potential versus time for a constant applied current density of (a) 1 mA/cm^2, (b) 2 mA/cm^2, (c) 7 mA/cm^2. Right: corresponding oxide thicknesses as a function of potential. After ref.[58].

Such oxides have yet been found to be unappropriate for MOS structures, firstly because of practical considerations, and secondly because the first attempts had shown that the obtained oxides are far from being perfect, especially because they exhibit high densities of electronic defects.[55] Such defects are clearly due to ions from the electrolyte. Although the mechanism of electrochemical oxidation is not completely clear, the general belief is that it takes place through inwards migration of oxygen species.[56] Ions from the electrolyte may also be incorporated into the layer. They affect the electronic quality of the interfaces by creating fixed charges and inducing electronic states in the forbidden gap, which makes these oxides unusable for manufacturing MOS structures. As a matter of fact, thermal oxidation has become the exclusive tool for the manufacturing of MOS devices. Anodic films may yet be of interest as a low temperature passivation process.[54]

Furthermore, recent studies have shown that it is possible to realize "good" anodic oxides, by using as the electrolyte ultra-pure water, in the absence of any added ions.[57] The obtained oxides exhibit electronic quality comparable to that of the best thermal oxides. Furthermore, the oxidation current critically depends upon the thickness of the already-formed oxide. This allows an improved control of the layer thickness and an exceptional homogeneity. This technique might be advantageous for realizing ultra-thin oxide layers (below 100 Å).[57]

5. ANODIC DISSOLUTION OF SILICON IN FLUORIDE MEDIA

5.1 General behaviour: n and p type, dissolution valence

Typical voltammograms of n- and p-Si in a fluoride electrolyte are shown in Figure 6.[59-63] As expected for a semiconducting electrode, anodic dissolution takes place in the dark in the case of p-type material, but illumination is necessary in the case of n-type, unless high doping is used (n+).[64] The voltammograms of silicon in HF and fluoride electrolytes are qualitatively similar for a wide range of fluoride concentrations and pH. Yet the currents may change by several orders of magnitude, depending on these parameters.[63] Especially, increasing the fluoride ion concentration will increase the currents. The effect of pH is far from being negligible.[64]

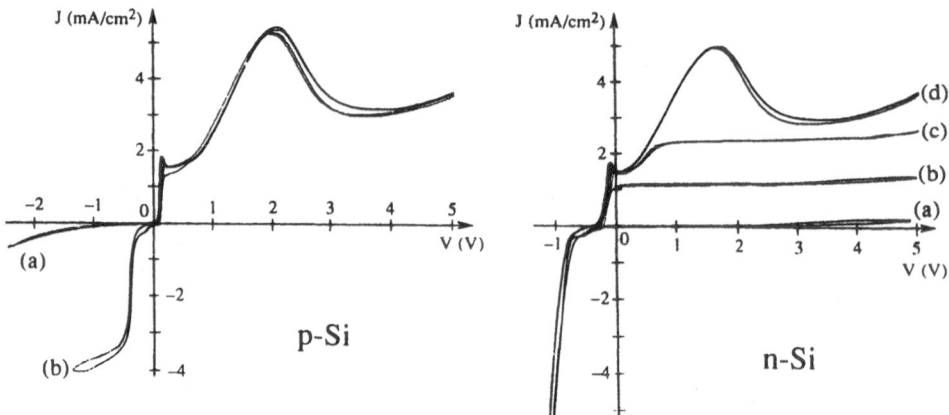

Figure 6: Typical current /potential curves at a Si/fluoride interface (here 0.2M HF). Left: p-Si (a) in the dark, (b) illuminated. Right: n-Si (a) in the dark, (b-d) under increasing illumination. Reference is Ag/AgCl in saturated KCl. Current increases at higher polarizations (breakdown).

The voltammograms of p-Si in HF are rather similar to those of typical metallic electrodes. At moderate potentials, there is a region where current rises sharply with increasing potential.

This is the region where direct dissolution of the electrode material occurs. At higher potentials, the current becomes a smoother function of potential. This is the classical electropolishing regime, where a layer of weakly soluble oxidation products is present at the interface. Finally, at still higher potentials, the current rises again, corresponding to dielectric breakdown of the oxide layer associated to a strong oxygen evolution. This scheme is obeyed in the case of p-Si: the fast-rising part, associated to direct dissolution, occurs near $0\ V_{SCE}$. This is the region where porous silicon formation occurs.[65-68] [We will not discuss here the various porosities and morphologies which may be found depending upon Si doping, electrolyte composition and current density, since these matters are treated by A. Halimaoui in another chapter]. The electropolishing "plateau" extends from ~ 0.1 to $\sim 5\ V_{SCE}$. It is reached only at very high current densities ($\sim 1\ A/cm^2$) in the concentrated fluoride solutions generally used for porous-silicon fabrication. The reaction becomes limited by mass transport for typical fluoride concentrations above ~ 0.1 M [1% aqueous HF corresponds to ≈ 0.55 M], so that the voltammograms in this composition range become dependent upon stirring of the electrolyte.[69] In-situ infrared spectroscopy shows that the surface is covered with hydrogen in the first region, and with an oxide/hydroxide layer in the second region.[70]

The voltammograms for n-Si depend upon the illumination intensity. At high illumination levels, the voltammograms are similar to those on p-Si, except for a ~ -0.5 V shift in the potential scale (photopotential).[71] At lower illuminations, the photocurrent may become limited by hole generation, and a perfectly constant, light-limited, plateau is obtained instead of the polishing regime. The anodic currents at p+-Si and n+-Si electrodes are essentially the same as for p-Si. The absence of a dark-current limitation at an n+-Si electrode is associated to reverse breakdown of the thin space-charge layer (avalanche and/or tunneling).[64]

An intriguing observation, from the pioneering times, was that a gas is evolved at the electrode surface in the region of porous-silicon formation. This gas was found to be hydrogen. Subsequent studies (e.g., coulometric or rotating-ring-disk-electrode studies[72]) have shown that this hydrogen evolution, —a surprizing effect for an anodic reaction,— is associated to *divalent dissolution* of the silicon electrode: in principle, silicon dissolution is expected to be tetravalent, i.e., four electric charges should be consumed per dissolved silicon atom, according to the scheme

$$Si + 6\ F^- \xrightarrow{+4e} SiF_6^{2-} \tag{3}$$

This scheme is indeed observed to take place in the electropolishing range, where a dissolution valence between 3.6 and 4 has been determined by several authors.[59,62,72,73] The divalent dissolution in the current-rising region means that only two charges are consumed per dissolved silicon atom, which is typically accounted for in a scheme such as

$$Si + 6\ F^- + 2H^+ \xrightarrow{+2e} SiF_6^{2-} + H_2 \tag{4}$$

This may be understood as a divalent electrochemical step, followed by a chemical step:

$$Si + 2\ F^- \xrightarrow{+2e} SiF_2 \tag{5a}$$

$$SiF_2 + 2H^+ + 4\ F^- \rightarrow SiF_6^{2-} + H_2 \text{ (chemical)} \tag{5b}$$

Step (5b) accounts for the hydrogen evolution. However, an alternate chemical step may be

$$2SiF_2 + 2F^- \rightarrow SiF_6^{2-} + Si \tag{5c}$$

("dismutation" reaction). Such a reaction path would not result in hydrogen evolution, but instead amorphous silicon would be redeposited at the surface. This second reaction path seems at present rather unlikely.[72] Yet it might exist as a small percentage of the total reaction. Let us emphasize that here the species SiF_2 has been taken as a schematic illustration of a typical possible intermediate. Though the presence of a *divalent* intermediate species is necessary, there is at present no clear information on the nature of this species (molecular species such as SiF_2 or $SiHF_3$, surface-attached species ?)

5.2 Porous silicon regime: electron/hole participation in the electrochemical process

A second question is about the nature of the carriers (electrons or holes) involved in the electrochemical reaction. A hole is certainly necessary for initiating the process, since only a very small anodic dark current is observed at n-Si electrodes.[42] Yet electron injection may occur at later stages, if highly oxidizable intermediates are involved in the reaction mechanism.[74] Such electron injection steps can be evidenced by measuring the photocurrent at an n-Si photoanode. If a complex reaction, involving (x+y) elementary charges, proceeds through x holes and y injected electrons, the quantum yield will be (1+y/x), which may be higher than unity.[74] [Alternately, equivalent information may be obtained from an analysis of the voltammograms in oxidizing media, i.e., electrolytes containing hole-injecting species[75,76]]. Such measurements have been carried out in fluoride media. The quantum yield is unity in the electropolishing regime, —as well as in non-fluoride media—. However, in the regime of porous-silicon formation, it goes from 2 under typical illumination levels to nearly 4 under very low illumination.[77,78,72,73] (photocurrent < 10 $\mu A/cm^2$). At the same time, hydrogen evolution goes from one H_2 molecule per dissolved silicon atom to about zero.[72] This gives the proof that highly oxidizable species are involved in the reaction mechanism, and that these species may become oxidized through either electron injection or hydrogen evolution. The change of behaviour induced by a change in illumination intensity has been taken as an indication that the hydrogen evolution mechanism exhibits second-order kinetics with respect to the divalent intermediate species.[72]

Further information on the kinetics may be obtained from modulated photocurrent measurements: the electron injection steps will usually take place some time after the injecting species is formed (the more stable the intermediate species, the longer the time before electron injection). If the photoexcitation light beam is modulated at a frequency larger than the inverse electron injection rate, the fraction of the photocurrent associated with the corresponding injection step will not follow the modulation. More precisely, the photocurrent response can be taken as a complex impedance, whose quantitative analysis allows one to reach the characteristic kinetic constants of the system. Such an analysis has allowed Peter et al. to give a picture with characteristic rate constants for the three injection steps of 2×10^4, 500, and 0.5 s^{-1}.[79] Yet the interpretation of such data becomes difficult if second-order mechanisms are present. Another indication may be obtained from the electrical behaviour of the p-Si/fluoride-electrolyte interface: the low-frequency impedance of this system exhibits an inductive behaviour in the regime of porous-silicon formation.[80,81] Although the interpretation of this behaviour is not unique (one may think, for example, of a relaxation of the specific surface area[81]), it may be regarded as a further indication for delayed electron injection.[82]

Another information on highly oxidizable species may be obtained from observing the current during the chemical dissolution of an oxide layer on an n-Si anode in the dark.[77,83] In the dark and while the oxide layer is present, no significant current flows through the interface. When the oxide layer is fully dissolved, there is a hydrogenated surface under reverse polarization, and no significant current is observed either. However, in the short time when oxide dissolution is not fully completed, a sizable anodic current is observed, proving that electron-injecting species are formed at the surface during *chemical* dissolution of the oxide layer.[77,83] This may suggest that such injecting species are surface-attached silicon fluoride groups. Yet the observation that photocurrent doubling may be transiently observed when an HF-rinsed silicon surface is immersed in a non-fluoride electrolyte shows that the fluoride species are not a necessary ingredient for electron injection.[84] The relevant electron-injecting species might then just consist of an intermediate in the oxidation of a surface SiH bond, whether the oxidation takes place in fluoride or non-fluoride medium.

In spite of these various indications, the overall reaction mechanism of the anodic dissolution of silicon in fluoride media is still uncompletely understood. During about twenty years, there has been many proposals, all ignoring that the surface is essentially in a hydrogenated state,[59,77] a fact which is now well established. Among recent studies taking this fact into account, is the reaction scheme proposed by Lehmann and Gösele.[85] This mechanism, when starting from a SiH_2 surface step site, can be schematized by

$$=SiH_2 + F^- + h^+ \rightarrow =SiFH_2{}^{\cdot} \text{ (hole capture)} \tag{6}$$

$$=SiFH_2{}^{\cdot} + F^- \rightarrow =SiF_2 + H_2 + e^- \text{ (electron injection)} \tag{7}$$

$$\begin{matrix} \equiv Si & & \equiv SiH \\ & SiF_2 + HF \rightarrow & + & \text{(chemical step)} \\ \equiv Si & & \equiv SiSiF_3 \end{matrix} \tag{8}$$

$$\equiv SiSiF_3 + HF \rightarrow \equiv SiH + SiF_4 \text{ (chemical step)} \tag{9}$$

$$SiF_4 + 2HF \rightarrow 2H^+ + SiF_6{}^{2-} \text{ (chemical step)} \tag{10}$$

Another scheme has recently been proposed by Gerischer et al.[86]

$$=SiH_2 + F^- + h^+ \rightarrow =SiH{}^{\cdot} + HF \text{ (hole capture)} \tag{11}$$

$$=SiH{}^{\cdot} + F^- \rightarrow =SiHF + e^- \text{ (electron injection)} \tag{12}$$

$$\begin{matrix} \equiv Si & & \equiv SiH \\ & SiHF + HF \rightarrow & + & \text{(chemical step)} \\ \equiv Si & & \equiv SiSiHF_2 \end{matrix} \tag{13}$$

$$\equiv SiSiHF_2 + HF \rightarrow \equiv SiH + SiHF_3 \text{ (chemical step)} \tag{14}$$

$$SiHF_3 + 3HF \rightarrow 2H^+ + SiF_6{}^{2-} + H_2 \text{ (chemical step in solution)} \tag{15}$$

The second mechanism differs from the first one in that the initial hole transfer results in H (rather than Si) oxidation, the first chemical step is located on a Si backbond [see difference between (7) and (12-13)], and a dissolved intermediate $SiHF_3$ is involved, whereas the first mechanism proceeds entirely at the surface. Though they are both plausible, these schemes are only two choices among many, and may appear somewhat speculative. They do not account for the possibility of tetravalent dissolution of n-Si at very weak illuminations. It is therefore highly desirable to design appropriate experiments for identifying the reaction intermediates involved in the reaction.

5.3 Porous silicon formation: possible mechanisms

Early investigators had tried to associate porous silicon formation to the specificities of the above electrochemical mechanism.[68] For example, redeposition of amorphous silicon upon reaction step (5c) had been considered.[87] Although the mechanism of porous silicon formation is still unclear, the present ideas rather tend to favour physical mechanisms.[68] Since there is increasing evidence for distinct mechanisms in the cases of n- and p-Si substrates, we will first discuss the case of p-Si, and mention shortly the case of n-Si afterwards.

Porous silicon is known to be a high-resistivity material. This experimental fact may be ascribed, for example, to an increase of the donor binding energy due to quantum-confinement effects, or else to depletion due to the small size of the porous-Si wires and Fermi-level pinning at the surface [a clear indication for the existence of surface states during porous-Si formation is given by impedance measurements[80,81,88]]. Whatever the origin of the high resistivity may be, this results in the electrode material being much more resistive than the electrolyte. In a naive electrical approach, this tends to concentrate the electric field and current flow near the tips of the more conducting medium, i.e., near the bottom of the pores. A serious approach must yet consider the interface in more detail. According to Beale et al.,[89] and in agreement with flatband potential measurements,[80,81,88,90] under typical dissolution conditions, a depletion layer is present at the interface, and the electrochemical transfer is limited by thermionic hole emission over the potential barrier [or by hole tunneling in the case of degenerately doped silicon substrates]. A barrier lowering is then to be expected in the presence of the interface electric field. As indeed shown by Gaspard et al.,[90] the electrochemical current density is proportional to the electric field at the interface, which is consistent with a potential drop in the Helmholtz layer, resulting in a barrier lowering proportional to the interface electric field (this incidentally accounts for the fact that porous silicon is formed at less positive potentials on more highly doped p-type substrates). Now, solving Poisson equation in 3D space around the bottom of a pore is expected to lead to a thinning of the depletion layer, with enhanced electric field at the pore tip. The resulting barrier lowering should lead to enhanced current flow, hence to preferential dissolution at the bottom of the pores.[89]

Alternately, Smith and Collins have suggested that a diffusion-limited mechanism may lead to ramified growth,[91] as it is well known from the physics of Diffusion-Limited-Aggregation (DLA).[92] If carrier transport to the interface is ruled by a diffusion process, interface roughening and pore growth are to be expected. Computer simulations show that this mechanism may lead to morphologies highly suggestive of porous silicon.[68] Actually, there is little difference between the morphologies that the two classes of mechanisms would predict. The essential difference is that the main characteristic length in the first class of models is a screening length, whereas it is a diffusion length in the second class. The first class of models seems more appropriate for porous silicon formation on p-type material (where the process is controlled by majority carriers), whereas the second class of models might be more relevant in the case of porous silicon on n-type material (where the process is controlled by light-induced photocarriers). The case of n-Si is yet probably more complex (see below). In either case, it is not clear what would fix the scale of the smallest structures observed in porous silicon.

A third kind of idea has been introduced by Lehmann and Gösele.[85] Since very-small-scale structures are present in porous silicon, quantum-confinement effects (QCE) are expected to affect the electron states at the growing interface, and possibly to govern porous-silicon morphology, the smallest-scale structures becoming frozen due to carrier depletion. This effect should be sensitive to illumination. For example, a very thin porous-silicon wire will possess electron states only above an effective bandgap, higher than the bulk-silicon gap, due to QCE. Photocarriers will be created in such a wire if the light energy is taken higher than this effective bandgap, hence higher-energy illumination should allow one to create smaller-scale structures. Lehmann and Gösele have provided some experimental support in favour of this model.[93]

The case of porous silicon on n-Si appears rather different: its morphology exhibits macropores,[94,95] whose characteristic size and spacing appear governed by the screening length in the substrate material.[94,96-99] Here one must emphasize the distinct role of the space charge for n-Si and p-Si: in the case of p-Si, it acts as a barrier for the holes, and only near the pore tips is the field sufficiently large for opening gaps at the top of the barrier, whereas in the case of n-Si, it rather acts as a sink channeling the holes to the pore tips. This different energetic configuration may be related to the different morphology. The bottom of the pores is thought to be under local electropolishing conditions.[94] Yet the porous layer on n-Si may be formed under front or back illumination, or just under strong anodic polarization (avalanche breakdown) in the dark. In the last case, the possibility of electron tunneling at the interface may also be considered.[97,98] There is also some evidence that initiation of the macropores may be favoured in the vicinity of ionized donors.[94,96] This kind of effect is familiar to the world of photoelectrochemical etching of semiconductors.[100] However, nanostructures are also observed besides the macropores, especially if front-side illumination is used.[94,95] The origin of these nanostructures may be relevant to the same ideas as for the case of p-Si substrates.

Surprisingly, the *quantitative* understanding of the formation of porous silicon appears rather more advanced for the macropores on n-Si than for p-Si. In the latter case, one may yet notice that the existing conflicting models are not mutually exclusive. It seems reasonable to think that the basic ideas are here and a full understanding of the mechanism(s) of porous-silicon formation mainly depends on the quantitative development of the above models.

5.4 Electropolishing regime: interfacial film, ionic transport, resonant behaviour

An intriguing feature in the voltammograms is the current maximum between 1 and 2 V_{SCE}, which separates the electropolishing "plateau" into two regions. In-situ infrared measurements[70] have confirmed the conjecture, based on coulometric and other transient measurements,[101] that this behaviour is due to the presence of two kinds of oxides: when the potential is increased above the porous-silicon-formation region, first the hydrogenated coating disappears and is replaced by a wet oxide or hydroxide layer (mostly SiOH groups); then a true oxide appears (SiOSi groups), but this takes place only above the broad current maximum. The picture of the interface in the high-potential region is then in terms of a dry oxide layer lying immediately against the silicon surface, and a wet hydroxide overlayer in contact with the electrolyte. The total thickness of the oxide/hydroxide layer increases monotonically with potential and is of the order of 100 Å at 5 V_{SCE} potential.[101]

Anodic dissolution in the presence of an oxide layer requires ionic transport through the layer. The mechanism of ionic transport in the anodic oxidation of silicon has been rather disputed.[102,103,56] Especially, it was not clear whether the migrating species is silicon (moving outwards) or oxygen (moving inwards), although the last mechanism appears generally favoured (see above Sec.4-C). Recent transient experiments have provided evidence for a positive charge stored in the dry oxide.[101] This suggests that the migrating species might be positively charged oxygen "semivacancies" (O_3Si^+ centers), which are also a well known defect in thermal oxides.[104]

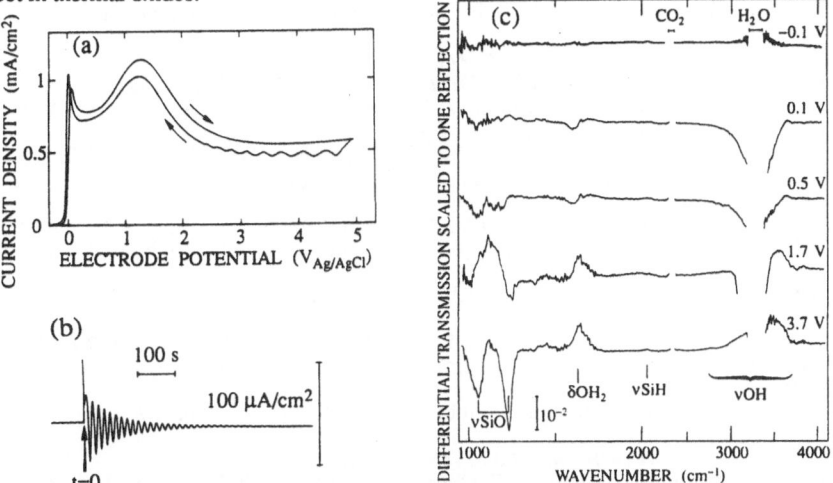

Figure 7: p-Si electrode in the electropolishing regime. (a) voltammogram (electrolyte 1M NH_4Cl + 0.025M NH_4F + 0.025M HF). Notice the damped oscillations triggered by scan reversal at +5 V. (b) Current transient upon stepping the potential from +3.0 to +3.1 V. (c) Differential infrared spectra as a function of potential (reference state at –0.5 V). After refs.[70,108].

A conspicuous effect in the high potential region is the tendency of the current to exhibiting damped oscillations upon applying a small potential perturbation to the interface.[105] Such a tendency is seen in the voltammogram of Fig. 7. Here the damped oscillation is triggered by reversal of the potential sweep at the maximum potential. The impedance of the interface in this regime exhibits a resonant behaviour of the admittance at the natural oscillation frequency and at its overtones. This resonant behaviour can be understood as due to an oscillation of the current density and oxide thickness (mainly of the dry oxide, as indicated by infrared data) on the microscopic scale: the oscillation is not correlated between the various locations on the surface, so that a stable macroscopic current is observed. However, a potential perturbation, by inducing partial synchronization of the various "domains", makes the hidden oscillation appear on the macroscopic scale. After the potential excitation is turned off, the domains slowly return to incoherency, and the macroscopic oscillation is damped.[81,106] From noise measurements, an estimate of ~ 10^3 Å for the characteristic domain size has been obtained.[107] Occasionnally, the series resistance of the electrolyte may be large enough to synchronize the whole interface together, and a spontaneous oscillation is then observed.[108] The microscopic mechanism responsible for the oscillation is still to be elucidated.

6. CONCLUSION

Although silicon is a simple elemental semiconductor, its electrochemistry still counceals many mysteries. The understanding of its surface chemistry has progressed significantly these last ten years. It is to be hoped that the fundamental understanding of more complex phenomena, such as porous-silicon formation, anodic oxidation, or electropolishing mechanism, will also see progress in the future. The need for appropriate in-situ microscopic tools is certainly a problem in this respect.

References

1. See, e.g., *Semiconductor Electrodes*, edited by H.O. Finklea (Elsevier, Amsterdam, 1988).
2. See, e.g., H. Gerischer, *Electrochim. Acta* **35**, 1677 (1990).
3. See, e.g., D. Vanmaekelbergh, W.P. Gomes and F. Cardon, *J. Electrochem. Soc.* **129**, 546 (1982).
4. D. Laser and A.J. Bard, *J. Phys. Chem.* **80**, 459 (1976).
5. F. Ozanam and J.-N. Chazalviel, *J. Electroanal. Chem.* **269**, 251 (1989).
6. J.-N. Chazalviel and T.B. Truong, *J. Electroanal. Chem.* **114**, 299 (1980); *J. Am. Chem. Soc.* **103**, 7447 (1981).
7. H.J. Byker, V.E. Wood and A.E. Austin, *J. Electrochem. Soc.* **129**, 1982 (1982).
8. N. Gabouze, Thèse de l'Université Paris VI (1988); N. Gabouze and O. Gorochov (unpublished).
9. J.-N. Chazalviel, *J. Electroanal. Chem.* **233**, 37 (1987).
10. A.J. Bard, A.B. Bocarsly, F.R.F. Fan, E.G. Walton and M.S. Wrighton, *J. Am. Chem. Soc.* **102**, 3671 (1980).
11. A.B. Bocarsly, D.C. Bookbinder, R.N. Dominey, N.S. Lewis and M.S. Wrighton, *J. Am. Chem. Soc.* **102**, 3683 (1980).
12. A. Venkateswara Rao and J.-N. Chazalviel, *J. Electrochem. Soc.* **134**, 2777 (1987).
13. J.F. Gibbons, G.W. Cogan, C.M. Gronet and N.S. Lewis, *Appl. Phys. Lett.* **45**, 1095 (1984).
14. G.W. Cogan, C.M. Gronet, J.F. Gibbons and N.S. Lewis, *Appl. Phys. Lett.* **44**, 539 (1984).
15. W. Kern, *J. Electrochem. Soc.* **137**, 1887 (1990).
16. E. Yablonovitch, D.L. Allara, C.C. Chang, T. Gmitter and T.B. Bright, *Phys. Rev. Lett.* **57**, 249 (1986).
17. N.J. Harrick and K.H. Beckmann, in *Characterization of Solid Surfaces*, edited by P.F. Kane and G.R. Larrabee (Plenum, New York, 1974), p. 243.
18. H. Ubara, T. Imara and A. Hiraki, *Solid State Comm.* **50**, 673 (1984).
19. A. Tardella and J.-N. Chazalviel, *Appl. Phys. Lett.* **47**, 334 (1985).
20. V.A. Burrows, Y.J. Chabal, G.S. Higashi, K. Raghavachari and S.B. Christman, *Appl. Phys. Lett.* **53**, 998 (1988).
21. G.S. Higashi, Y.J. Chabal, G.W. Trucks and K. Raghavachari, *Appl. Phys. Lett.* **56**, 656 (1990).
22. G.S. Higashi, R.S. Becker, Y.J. Chabal and A.J. Becker, *Appl. Phys. Lett.* **58**, 1656 (1991).
23. P. Dumas and Y.J. Chabal, *Chem. Phys. Lett.* **181**, 537 (1991).
24. P. Guyot-Sionnest, *Phys. Rev. Lett.* **66**, 1489 (1991).
25. P. Guyot-Sionnest, P. Dumas, Y.J. Chabal and G.S. Higashi, *Phys. Rev. Lett.* **64**, 2156 (1990).

26. See, e.g., P. Allongue, V. Costa-Kieling and H. Gerischer (to be published).
27. A. Venkateswara Rao, J.-N. Chazalviel and F. Ozanam, *J. Appl. Phys.* **60**, 696 (1986).
28. P. Jakob and Y.J. Chabal, *J. Chem. Phys.* **95**, 2897 (1991).
29. E.P. Boonekamp (private communication and to be published).
30. S. Watanabe, M. Shigeno, N. Nakayama and T. Ito, *Jap. J. Appl. Phys.* **30**, 3575 (1991).
31. See, e.g. H. Seidel, L. Csepregi, A. Heuberger and H. Baumgärtel, *J. Electrochem. Soc.* **137**, 3612 (1990) and references therein.
32. S.A. Campbell, D.J. Schiffrin and P.J. Tufton, *J. Electroanal. Chem.* **344**, 211 (1993).
33. E.D. Palik, V.M. Bermudez and O.J. Glembocki, *J. Electrochem. Soc.* **132**, 871 (1985).
34. R.L. Smith, B. Kloeck, N. de Rooij and S.D. Collins, *J. Electroanal. Chem.* **238**, 103 (1987).
35. P. Allongue, H. Brune and H. Gerischer, *Surf. Sci.* **275**, 414 (1992).
36. P. Allongue, V. Costa-Kieling and H. Gerischer, *J. Electrochem. Soc.* **140**, 1009 (1993); ibid. **140**, 1018 (1993).
37. J. Rappich, H.J. Lewerenz and H. Gerischer, *J. Electrochem. Soc.* **140**, L187 (1993).
38. R.K. Iler, *The Chemistry of Silica* (Wiley, New York, 1979).
39. H. Yoneyama, Y. Murao and H. Tamura, *J. Electroanal. Chem.* **108**, 87 (1980).
40. M.S. Wrighton, R.G. Austin, A.B. Bocarsly, J.M. Bolts, O. Haas, K.D. Legg, L. Nadjo and M.C. Palazzotto, *J. Am. Chem. Soc.* **100**, 1602 (1978).
41. M.L. Rosenbluth and N.S. Lewis, *J. Am. Chem. Soc.* **108**, 4689 (1986).
42. J.-N. Chazalviel, *Surf. Sci.* **88**, 204 (1979).
43. M.J. Madou, B.H. Loo, K.W. Frese and S.R. Morrison, *Surf. Sci.* **108**, 135 (1981).
44. S.R. Morrison, *Electrochemistry at Semiconductor and Oxidized Metal Electrodes* (Plenum, New York, 1980), Chap. 5, pp. 153-156.
45. D. Lincot and J. Vedel, *J. Phys. Chem.* **92**, 4103 (1988).
46. D. Lincot and J. Vedel, *J. Electroanal. Chem.* **220**, 179 (1987).
47. See, e.g., J. O'M. Bockris and A.K.N. Reddy, *Modern Electrochemistry* Vol.2 (Plenum-Rosetta, New York, 1970) Chap. 10.3, pp. 1231-1251.
48. H. Gerischer, *Anales de Física y Química* B**56**, 535 (1960).
49. K.C. Mandal, F. Ozanam and J.-N. Chazalviel, *Appl. Phys. Lett.* **57**, 2788 (1990).
50. P. de Mierry, A. Etcheberry and M. Aucouturier, *J. Appl. Phys.* **69**, 1099 (1991).
51. A. Schnegg, H. Prigge, M. Grundner, P.O. Hahn and H. Jacob, *MRS Symp. Proc.* **104**, 291 (1988).
52. P. de Mierry, D. Ballutaud, M. Aucouturier and A. Etcheberry, *J. Electrochem. Soc.* **137**, 2966 (1990).
53. P.F. Schmidt and W. Michel, *J. Electrochem. Soc.* **104**, 230 (1957).
54. See, e.g., G. Mende, H. Flietner and M. Deutscher, *J. Electrochem. Soc.* **140**, 188 (1993) and references therein.
55. M.J. Madou, S.R. Morrison and V.P. Bondarenko, *J. Electrochem. Soc.* **135**, 229 (1988).
56. See, e.g., K. Ghowsi and R.J. Gale, *J. Electrochem. Soc.* **136**, 867 (1989) and references therein.
57. F. Gaspard, A. Halimaoui and G. Sarrabayrouse, *Revue Phys. Appl. (Orsay)* **22**, 65 (1987).
58. A. Halimaoui, Thèse de Doctorat d'Etat, Université Joseph Fourier, Grenoble (1991).
59. R. Memming and G. Schwandt, *Surf. Sci.* **4**, 109 (1966).
60. X.G. Zhang, S.D. Collins and R.L. Smith, *J. Electrochem. Soc.* **136**, 1561 (1989).
61. A.E. Gershinskii and L.V. Mironova, *Sov. Electrochem.* **25**, 1224 (1990).
62. H. Gerischer and M. Lübke, *Ber. Bunsenges. Phys. Chem.* **92**, 573 (1988).
63. J.-N. Chazalviel, M. Etman and F. Ozanam, *J. Electroanal. Chem.* **297**, 533 (1991).
64. R.L. Meek, *Surf. Sci.* **25**, 526 (1971); *J. Electrochem. Soc.* **118**, 437 (1971).
65. A. Uhlir, *Bell Syst. Tech. J.* **35**, 333 (1956).
66. D.R. Turner, *J. Electrochem. Soc.* **105**, 402 (1958).
67. G. Bomchil, A. Halimaoui, I. Sagnes, P.A. Badoz, I. Berbezier, P. Perret, B. Lambert, G. Vincent, L. Garchery and J.L. Regolini, *Appl. Surf. Sci.* **65/66**, 394 (1993).
68. R.L. Smith and S.D. Collins, *J. Appl. Phys.* **71**, R1 (1992).

69. M. Etman, M. Neumann-Spallart, J.-N. Chazalviel and F. Ozanam, *J. Electroanal. Chem.* **301**, 259 (1991).
70. A. Venkateswara Rao, F. Ozanam and J.-N. Chazalviel, *J. Electrochem. Soc.* **138**, 153 (1991); F. Ozanam and J.-N. Chazalviel, *J. Electron Spec.* **64/65**, 395 (1993).
71. M.J. Eddowes, *J. Electroanal. Chem.* **280**, 297 (1990).
72. J. Stumper and L.M. Peter, *J. Electroanal. Chem.* **309**, 325 (1991).
73. E. Peiner and A. Schlachetzki, *J. Electrochem. Soc.* **139**, 552 (1992).
74. S.R. Morrison, *The Chemical Physics of Surfaces* (Plenum, New York, 1977) Chap. 8, pp. 280-283.
75. See, e.g., D. Vanmaekelbergh and J.J. Kelly, *J. Electrochem. Soc.* **136**, 108 (1989).
76. H. Gerischer and M. Lübke, *J. Electrochem. Soc.* **135**, 2782 (1988).
77. M. Matsumura and S.R. Morrison, *J. Electroanal. Chem.* **147**, 157 (1983).
78. D.J. Blackwood, A. Borazio, R. Greef, L.M. Peter and J. Stumper, *Electrochim. Acta* **37**, 889 (1992).
79. H.J. Lewerenz, J. Stumper and L.M. Peter, *Phys. Rev. Lett.* **61**, 1989 (1988).
80. P.C. Searson and X.G. Zhang, *J. Electrochem. Soc.* **137**, 2539 (1990); *Electrochim. Acta* **36**, 499 (1991).
81. F. Ozanam, J.-N. Chazalviel, A. Radi and M. Etman, *J. Electrochem. Soc.* **139**, 2491 (1992).
82. D. Vanmaekelbergh and P.C. Searson, *J. Electrochem. Soc.* **141**, 697 (1994).
83. H.J. Lewerenz, *Electrochim. Acta* **37**, 847 (1992).
84. J.-N. Chazalviel and F. Ozanam, *MRS Symp. Proc.* **283**, 359 (1993).
85. V. Lehmann and U. Gösele, *Appl. Phys. Lett.* **58**, 856 (1991).
86. H. Gerischer, P. Allongue and V. Costa Kieling, *Ber. Bunsenges. Phys. Chem.* **97**, 753 (1993).
87. T. Unagami, *J. Electrochem. Soc.* **127**, 476 (1980).
88. J.D. L'Ecuyer and J.P.G. Farr, in *4th Intl. Symp. on Silicon-on-Insulator Tech. and Devices*, ed. by D.N. Schmidt (The Electrochem. Soc. Softbound Proc. Series, Pennington, NJ, 1990) PV 90-6, p.375.
89. M.I.J. Beale, N.G. Chew, M.J. Uren, A.G. Cullis and J.D. Benjamin, *Appl. Phys. Lett.* **46**, 86 (1985); M.I.J. Beale, J.D. Benjamin, M.J. Uren, N.G. Chew and A.G. Cullis, *J. Cryst. Growth* **73**, 622 (1985).
90. F. Gaspard, A. Bsiesy, M. Ligeon, F. Muller and R. Herino, *J. Electrochem. Soc.* **136**, 3043 (1989); I. Ronga, A. Bsiesy, F. Gaspard, R. Hérino, M. Ligeon, F. Muller and A. Halimaoui, *J. Electrochem. Soc.* **138**, 1403 (1991).
91. R.L. Smith and S.D. Collins, *Phys. Rev.* A **39**, 5409 (1989).
92. T.A. Witten, Jr. and L.M. Sander, *Phys. Rev. Lett.* **47**, 1400 (1981).
93. V. Lehmann and U. Gösele, *Adv. Mater.* **4**, 114 (1992).
94. V. Lehmann and H. Föll, *J. Electrochem. Soc.* **137**, 653 (1990); V. Lehmann, *J. Electrochem. Soc.* **140**, 2836 (1993).
95. C. Lévy-Clément, A. Lagoubi, D. Ballutaud, F. Ozanam, J.-N. Chazalviel and M. Neumann-Spallart, *Appl. Surf. Sci.* **65/66**, 408 (1993).
96. V.M. Dubin, *Surf. Sci.* **274**, 82 (1992).
97. X.G. Zhang, *J. Electrochem. Soc.* **138**, 3750 (1991).
98. P.C. Searson, J.M. Macaulay and F.M. Ross, *J. Appl. Phys.* **72**, 253 (1992).
99. Y. Kang and J. Jorné, *Appl. Phys. Lett.* **62**, 2224 (1993).
100. R. Tenne, V. Marcu and N. Yellin, *Appl. Phys. Lett.* **45**, 1219 (1984).
101. J.-N. Chazalviel, *Electrochim. Acta* **37**, 865 (1992).
102. W.D. Mackintosh and H.H. Plattner, *J. Electrochem. Soc.* **124**, 396 (1977).
103. G. Mende, *J. Electrochem. Soc.* **127**, 2085 (1980).
104. W.L. Warren, E.H. Poindexter, M. Offenberg and W. Müller-Warmuth, *J. Electrochem. Soc.* **139**, 872 (1992).
105. F. Ozanam, J.-N. Chazalviel, A. Radi and M. Etman, *Ber. Bunsenges. Phys. Chem.* **95**, 98 (1991).
106. J.-N. Chazalviel and F. Ozanam, *J. Electrochem. Soc.* **139**, 2501 (1992).
107. F. Ozanam, N. Blanchard and J.-N. Chazalviel, *Electrochim. Acta* **38**, 1627 (1993).
108. J.-N. Chazalviel, F. Ozanam, M. Etman, F. Paolucci, L.M. Peter and J. Stumper, *J. Electroanal. Chem.* **327**, 343 (1992).

Porous silicon: material processing, properties and applications

A. Halimaoui

*France Telecom - CNET, BP. 98,
38243 Meylan cedex, France*

1. POROUS SILICON FORMATION: ANODIZATION CELLS:

Porous silicon (PS) is known to form during electrochemical dissolution of silicon in HF-based solutions. This dissolution is obtained by monitoring either the anodic current or potential. In general, constant current is preferable, as it allows a better control of both the porosity and thickness and a good reproducibility from run to run. The simplest cell which can be used to anodize silicon is shown in Figure 1. The silicon wafer serves as the anode. The cathode is made of platinum or any HF-resistant and conducting material. The cell body itself is in general made of highly acid-resistant polymer such as Teflon. Since the entire silicon wafer serves as the anode, PS is formed on any wafer surface in contact with the HF solution, including the cleaved edges. The advantage of such equipment is its simplicity and ability to anodize Silicon-On-Insulator structures. Its drawback is the non uniformity in both the porosity and thickness of the resulting layer. This inhomogeneity is mainly due to a lateral potential drop. In fact, since the current flows laterally along the bulk of the silicon wafer, there is a difference in potential between the top (point A in Fig. 1) and the bottom (point B in Fig. 1). The potential drop across the wafer leads to different values of the local current density which induce porosity and thickness gradients.

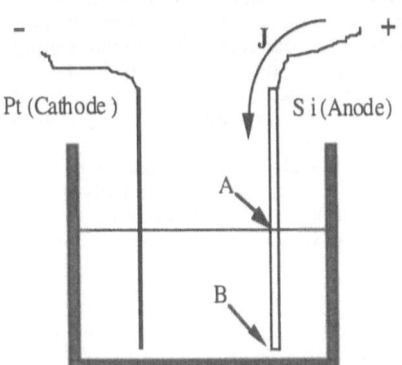

Figure 1: Schematic of lateral anodization cell

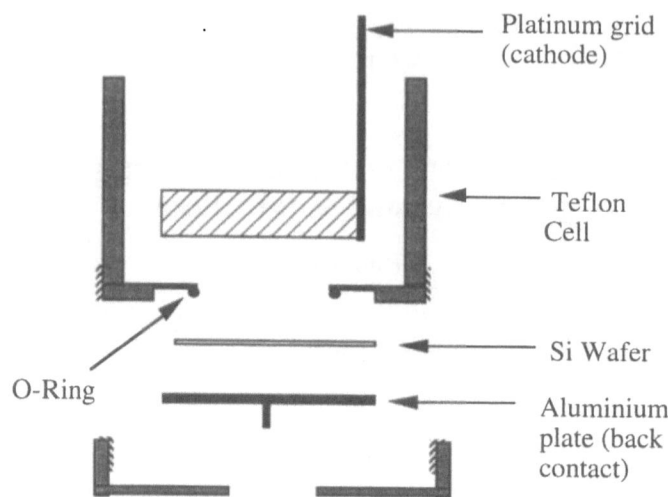

Figure 2: Schematic of a conventional single-tank cell

The second type of anodization cell is the single-cell approach using a back-side contact. In this second type of cell, schematically shown in Fig. 2, a metal contact is made to the back-side of the wafer and sealed so that only the front side of the sample is exposed to the anodizing electrolyte. It should be noticed that for a silicon wafer with low resistivity (typically < few mΩ.cm) a good uniformity is obtained without need of the metallic contact. However, for highly resistive silicon wafers (typically > few Ω.cm) a high-dose implantation (~10^{15} cm^{-2} at 80 Kev) of boron (for p-type) or phosphorus (for n type), on the back-side, is required for a good uniformity. This implantation step is followed by a drive-in at 1000°c for 30 minutes, an aluminium deposition and a thermal annealing at 450°C for 30 minutes. Under these conditions, the resulting layers are uniform, except very close (~2mm) to the O-ring. This type of cell is the most commonly used, leads to PS layers of good uniformity, simplifies the interpretation of the current-voltage characteristic and offers a good control of both thickness and porosity.

The third type of anodization cell is the double-tank cell using an electrolytic back-side contact. This type of equipment (Fig. 3) consists of two half-cells in which Pt electrodes are immersed and the silicon wafer is used to separate and isolate the two half-cells. HF is used for both anodization of the polished side and as a back contact. The electrolyte is circulated by chemical pumps. This circulation removes the gas bubbles generated during the anodic reaction and avoids any decrease in the local concentration of electro-active species such as HF. A good uniformity is obtained by using symmetrical and large Pt plates as the cathode and the anode. The two Pt electrodes are connected to a power supply and the current flows from one half-cell to the other through the silicon wafer. The back-side of the Si wafer acts as a secondary cathode where the proton reduction takes place leading to hydrogen evolution. The front side of the wafer acts as a secondary anode where porous silicon is formed.

Since the back-side contact is made electrolytically, no metallization of the back-side is required. Consequently, the as anodized wafer can be heated or chemically treated without any risk of metallic contamination. However, a high-dose implantation on the back-side is necessary, especially for highly resistive silicon wafers. Under these conditions, The uniformity of the layers obtained with this system is sufficiently good and comparable to that obtained with a conventional single-tank cell.

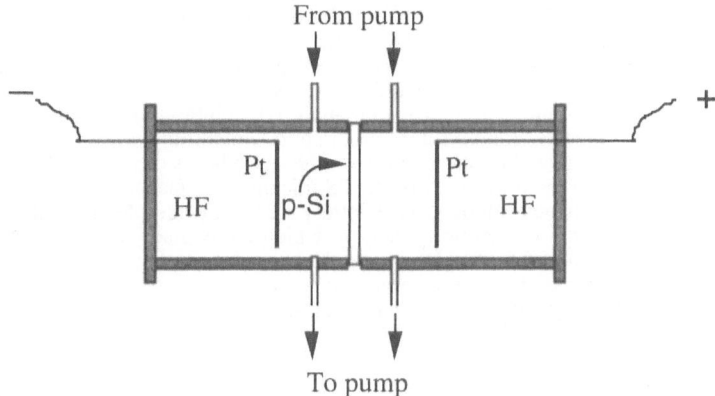

Figure 3: Schematic of a double-tank cell.

2. DISSOLUTION CHEMISTRIES

The exact dissolution chemistries of silicon are still unclear and different mechanisms have been proposed. Turner, Memming and Schwandt have proposed the following overall reaction for the dissolution of silicon [1, 2]:

$$Si + 2HF + \lambda\,h^+ \quad \text{-------->} \quad SiF_2 + 2H^+ + (2-\lambda)e^- \quad (1)$$
$$SiF_2 + 2HF \quad \text{-------->} \quad SiF_4 + H_2 \quad (2)$$
$$SiF_4 + 2HF \quad \text{--------->} \quad H_2SiF_4 \quad (3)$$

where h^+ and e^- are the exchanged hole and electron, respectively, and l is the number of charges exchanged during the elementary step.

More recently, Lehman and Gösele [3] have proposed another variant for the dissolution mechanism based on a surface bound oxidation scheme, with hole capture and subsequent electron injection, which leads to the divalent silicon oxidation state shown in Fig.4 given below:

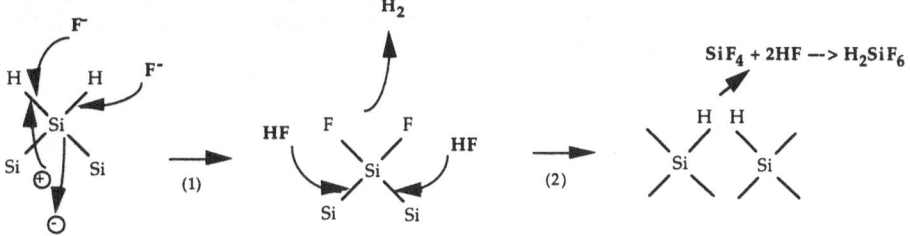

Figure 4: Dissolution mechanism of silicon in HF after Lehmann and Gösele[3].

From the chemical reactions given above, we should note that (i) silicon dissolution requires HF and holes(h^+); and (ii) during this chemical dissolution, there is hydrogen evolution. When purely aqueous HF solutions are used for the PS formation, the hydrogen bubbles stick to the surface and induce lateral and in-depth inhomogeneity. To improve the PS layer uniformity, these bubbles must be readily eliminated. One of the most appropriate means to overcome this problem is to add a surfactant agent to the HF solution. The most widely used surfactant in the case of PS formation is absolute ethanol. For an efficient bubble elimination, the ethanol concentration should not be less than 15%.

We have recently demonstrated that ethanoic HF solution completely infiltrates the pores while a purely aqueous solution does not, due to wettability and capillary phenomena [4, 5]. These phenomena play an important role in the smoothness of the interface between Si and porous Si and thus in the uniformity of the PS thickness. In fact, when the anodization is performed in a purely aqueous solution, the electrolyte does not completely infiltrate the pore and the propagation of the dissolution reaction, which takes place at the interface, is not uniform, thus leading to interface roughness and thickness inhomogeneity. Consequently, the role of ethanol is to improve the PS layer uniformity by elimination of the hydrogen bubbles and to improve the electrolyte penetration in the pores.

3. CHARGE EXCHANGE MECHANISM AND REACTION SELECTIVITY

Figure 5 shows typical current-voltage (i-v) characteristics for p-type silicon wafers of different dopant concentrations in an HF solution. These characteristics, corresponding to porous silicon formation, are plotted after correction of the ohmic potential-drop across the system. The i-v curves are found to be independent of the thickness of the PS layer. Furthermore successive potential sweeps on the same silicon electrode lead exactly to the same i-v characteristics. However, a difference is always found between the first sweep and the following ones. The plot of these curves in a logarithmic representation leads to a linear variation with a slope 59 mV/decade, indicative of a thermionic emission mechanism for the charge exchange during PS formation [6]. An important feature of these characteristics is the shift towards cathodic potential when the dopant concentration is increased. This shift has been quantitatively modelled by Gaspard *et al* [6] and was attributed to an increase in the potential drop across the Helmholtz layer when the dopant concentration is increased. This dependency of the i-v characteristic on the doping level, can be used for the selective formation of porous silicon. In fact, if one anodizes a silicon wafer with a dopant concentration of $\sim 2 \times 10^{15}$ cm^{-3} where heavily doped (10^{19} cm^{-3}) regions are present, PS formation can only take place only in the heavily doped region according to Fig. 5. For example, a potential close to zero (Fig. 5) gives current densities of ~ 50 mA/cm2 and zero in the heavily and lightly doped regions, respectively, thus leading to a reaction selectivity.

For n-type silicon, typical measured anodic i-v characteristics are given in Fig.6 and show a potential shift when the dopant concentration is changed. This potential shift can be exploited to selectively form porous silicon in differently doped regions. To our knowledge, the shape of these curves have not yet been quantitatively modelled. However, a qualitative model has already been proposed by Beale et al [7].

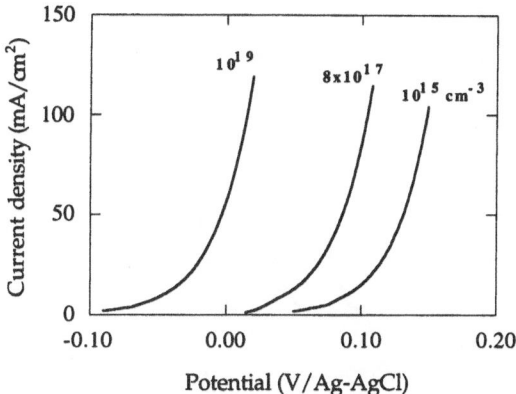

Figure 5: Anodic current-voltage characteristics of p-type silicon in a 35% HF solution and different substrate doping concentrations.

From the i-v characteristic discussed above and summarized in Fig. 7, one can predict that preferential porous silicon formation occurs for different dopant types. For example, p+ can be preferentially anodized over p or n, and n+ can be preferentially anodized over p+, p or n, etc. The different possibilities of selectivity are schematically listed in Fig. 8. For the first three cases (Figs. 8a, b, and c) porous silicon is preferentially formed on p+, p and n+ regions, respectively. However, for the last case (Fig. 8d) there is no PS formation even on the p region, because under anodic bias the p/n junction, which results from the formation of the p region on an n-type substrate, is reverse biased and thus no significant current flow is possible.

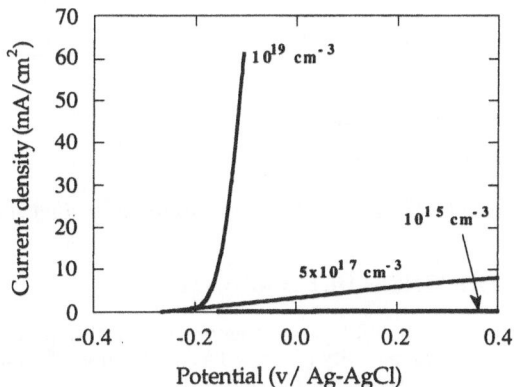

Figure 6: Anodic current-voltage characteristics of n-type silicon in a 35% HF solution and different substrate doping concentrations.

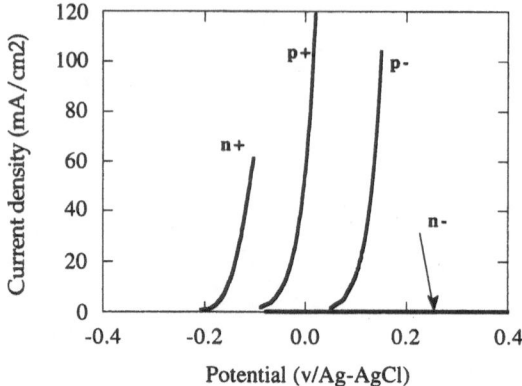

Figure 7: Comparison of the anodic current-voltage characteristics obtained from p and n-type silicon in a 35% HF solution and different doping levels (n+ and p+: 10^{19} cm^{-3}, n$^-$ and p$^-$: 10^{15} cm^{-3}).

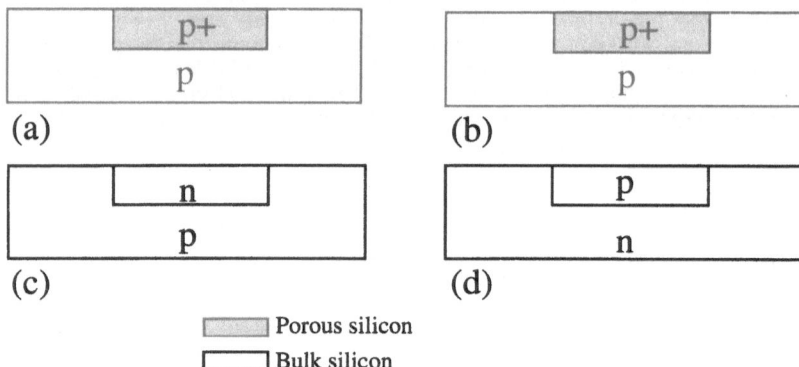

Figure 8: Schematic of different structures for selective formation of porous silicon.

4. POROSITY AND THICKNESS DETERMINATION

The porosity and thickness of the PS layers are among the most important parameters which characterize porous silicon. The porosity is defined as the fraction of void within the PS layer and can be easily determined by weight measurements. The virgin wafer is first weighed before anodization (m_1), then just after anodization (m_2) and finally after dissolution of the whole porous layer in a molar NaOH aqueous solution (m_3). Uniform and rapid stripping in the NaOH solution is obtained when the PS layer is covered with a small amount of ethanol which improve the infiltration of the aqueous NaOH in the pores. The porosity is simply given by the following equation:

$$P(\%) = (m_1-m_2)/(m_1-m_3) \qquad (1)$$

From these measured masses, it is also possible to determine the thickness of the layer according to the following formula

$$W = (m_1-m_3)/(Sxd) \qquad (2)$$

where d is the density of bulk silicon and S the wafer area exposed to HF during anodization.

The thickness can also be directly determined by scanning electron microscopy (SEM) or by step measurement after a part of the layer has been completely dissolved to generate a step corresponding to the layer thickness. However, the step measurement is not well suited to high-porosity (> 80%). In fact, such material is fragile and can be scratched by the tip during the step measurement. In general, this scratch results in to a measured thickness smaller than the actual one. One technique to overcome this problem is to metallize (aluminium) the sample surface before the step measurement. Non destructive optical-techniques such as ellipsometry can be used to determine both the thickness and porosity. However, these methods are model dependent and further work is needed to fully understand the ellipsometric data [8, 9].

5. VARIATION IN POROSITY AND THICKNESS AS A FUNCTION OF ANODIZATION CONDITIONS

All the properties of a PS layer, such as porosity, thickness, pore diameter, microstructure are strongly dependent on the anodization conditions. These conditions include HF concentration, pH of the solution and its chemical composition, current density, wafer type and resistivity, crystallographic orientation, temperature, anodization duration, stirring conditions, illumination (or not) during anodization. Optimum control of the fabrication and the result reproducibility are only possible if all the parameters listed above are taken into account.

The data given here are obtained by anodizing the wafer in the dark, at room temperature (~ 23°C), without stirring the solution and using (100)-oriented wafers. The solutions were prepared from a 50 wt. % HF solution, by dilution in absolute ethanol. The HF concentration of the anodizing solution often given in the literature should be treated with care. This concentration, in percent, corresponds neither to a percentage in weight nor in volume. It is a mixture of the two! For example, an ethanoic solution of 35% HF is obtained by 3 volumes of ethanol and 7 volumes of 50% wt. HF (the HF concentration in the ethanoic solution is given by $(3x50\%)/(3+7) = 35\%$). Consequently, it is preferable and more useful to give the exact composition of the solution (see Table I).

HF "concentration"	volume of 50% wt. HF	Volume of added ethanol
15%	3 volumes	7 volumes
25%	1 volume	1 volume
35%	7 volumes	3 volumes

Table I: Composition of the ethanoic solutions of HF.

The measured porosity as a function of current density and HF concentration for lightly doped (~ 1Ω.cm) p-type silicon is shown in Fig. 9. It appears that for a given HF concentration, the porosity increases with increasing current density and for a fixed current density the porosity decreases with increasing HF concentration. For given anodization conditions (current density, HF concentration) the porosity is much higher for the thicker layer. Such an effect is shown in Fig.10 which represents the variations in porosity as a function of current density for two different thicknesses. The thickest layer is more porous due to the extra chemical dissolution of the PS layer in HF. In fact, the thicker the layer, the longer the anodization time and thus the much higher the mass of chemically dissolved silicon (from the porous layer). The chemical dissolution of porous silicon is discussed in some detail below.

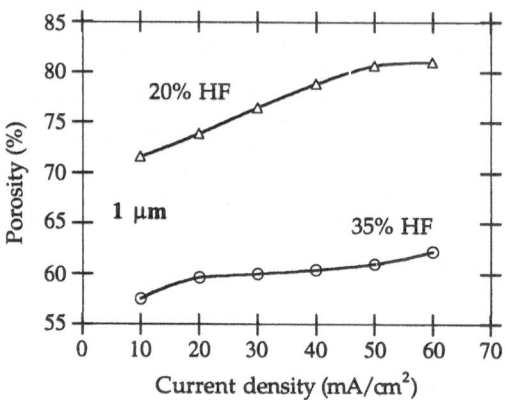

Figure 9: Porosity as a function of current density for two different HF concentrations and p-substrate (~ 1Ω.cm). Thickness of the porous layer=1μm.

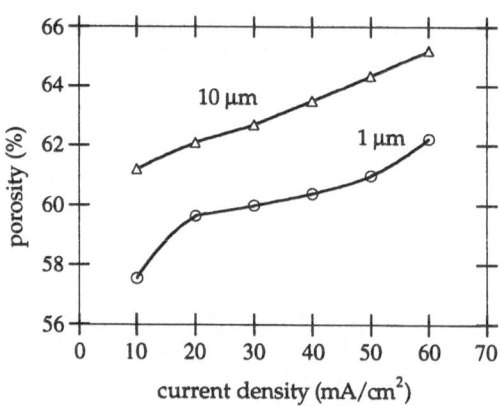

Figure 10: Porosity as a function of current density for two different thicknesses (1 and 10μm) . HF concentration = 35%, p- substrate (1Ω.cm).

The thickness of the layer as determined by gravimetric measurements linearly increases with increasing anodization time, as shown in Fig. 11.

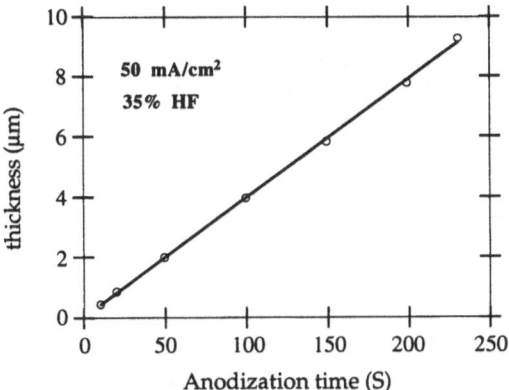

Figure 11: Thickness of the porous silicon layer as a function of anodization time for p-substrate (1Ω.cm) and a fixed current density of 50 mA/cm². HF concentration=35%.

Similarly to the lightly doped silicon substrate, the porosity of the layers obtained from heavily doped (0.01Ω.cm) wafers increases with increasing current density (Fig. 12) and with decreasing HF concentration. For the heavily doped substrate, the effect of the chemical dissolution on the porosity is negligible and not measurable due to the low specific surface area (200 m²/cm³) [10] of the material compared to that of the layers obtained from low resistivity silicon (~600 m²/cm³). The thickness of the layer obtained for p+ material also varies linearly with the anodization time as shown in Fig. 13.

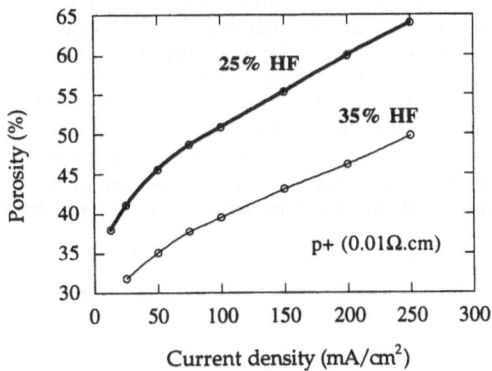

Figure 12: Porosity as a function of current density for two different HF concentrations and p+ substrate (0.01Ω.cm).

A. Halimaoui

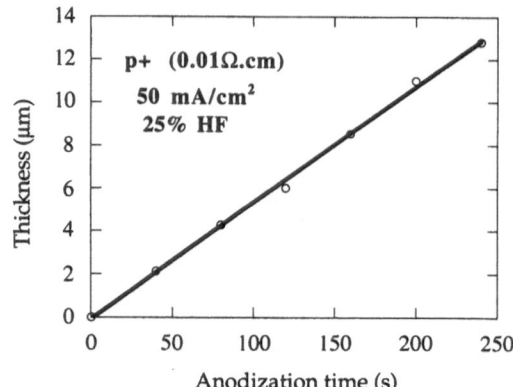

Figure 13: Thickness of the porous silicon layer as a function of anodization time for a p+ substrate (0.01Ω.cm) and a fixed current density of 50 mA/cm². HF concentration=25%.

The measured variation in porosity as a function of current density for heavily doped n-type Si (n+) is shown in Fig. 14. The shape of the obtained curve is different from that obtained with a p-type substrate and exhibits a marked minimum around 20 mA/cm2. For higher current densities (> 20mA/cm²), the porosity increases with increasing current density, similarly to the p-type substrates. However, at lower current densities (< 20mA/cm²), a sharp increase in porosity is observed when the current density is decreased. In a first attempt, this sharp increase in porosity has been attributed to the extra chemical dissolution of the layer, as the time required to form a given thickness is much higher at low current density. However, we have found that such a large increase in porosity (from 15 up to 40%, when the current density is decreased from 20 to 5 mA/cm²) is not consistent with the weight loss in the electrolyte. In fact, at 5 mA/cm², the anodization time is only 100 seconds longer than at 20 mA/cm². Such a small difference in anodization time should lead to only a difference of a few percent in porosity and could not explain the large difference observed in the porosity. We believe that the high value of porosity obtained at low current density is due to a difference in the microstructure. To support this hypothesis, we have measured the specific surface area of two layers obtained at high (> 20 mA/cm²) and low (< 20mA/cm²) current densities. The measured specific surface areas are of the order of 50 and 250 m²/cm³ for current densities of 100 and 5 mA/cm², respectively, indicative of a different microstructure. Furthermore, the layers obtained at low current density are more luminescent as their structure is much finer.

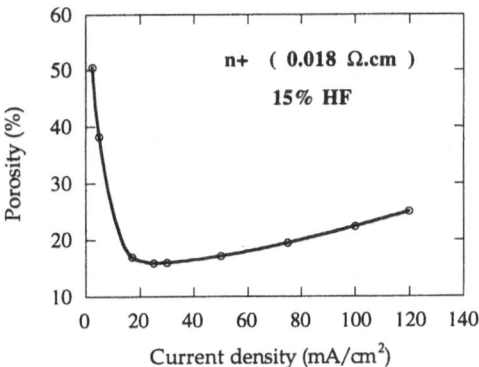

Figure 14: Porosity as a function of current density for an n+ substrate (0.018 Ω.cm) and HF concentration of 15%.

6. MATERIAL PROPERTIES

6.1 Microstructure of the material

The microstructure of porous silicon layers has been extensively studied by microscopy by many groups. Using transmission electron microscopy (TEM), Beale *et al* [7] have characterized porous silicon layers obtained from different substrates and using a wide range of anodization conditions. The layers obtained from heavily doped silicon (resistivity lower than ~0.05 Ω.cm), either p or n-type, consisted of long voids (diameter ~100 Å) running perpendicular to the surface. For lightly doped silicon, the situation is somewhat different: for n-type obtained under illumination and p-type, the material consists of apparently random arrays of very small holes (~ 30 Å). For the n-type obtained in the dark, the pores look like cylinders parallel to each other, with a small branching density and the obtained porosities remain very small (below 10%) [11]. Recently, the microstructure of porous silicon has been studied by high resolution scanning microscopy (HRSEM) [12]. An example of the results obtained using this technique is shown in Fig. 15. The microstructure of the p+ material is anisotropic while that of lightly doped material seems to be isotropic. The layers have the same porosity of 60%, but the pore and crystallite dimensions are quite different. The crystallite dimensions are around ~100Å and 30Å for p+ and p material, respectively. Although TEM and HRSEM are powerful techniques, it is rather difficult to extract an exact size distribution for the crystallites and pores from these observations. The other technique which has been extensively used for the characterization of porous silicon films is based on the analysis of gas adsorption isotherms. This technique allows determination of the pore sizes and the specific surface area of the material [10]. Porous silicon layers formed on n+ and p+ substrates exhibit a pore diameter of ~100Å and a specific surface area in the range 200 - 250 m2/cm3, whereas the films obtained on lightly doped p-type Si have a pore diameter smaller than 40Å and a specific surface area of the order of 600m2/cm3, attesting to the much finer texture of this material. As a rule of thumb, an increase in porosity leads to a pore enlargement. A new and simple technique which allows determination of the specific surface area of PS layers has been recently developed[5]. This technique, based on the etch rate of PS in HF solutions, is discussed in some detail below.

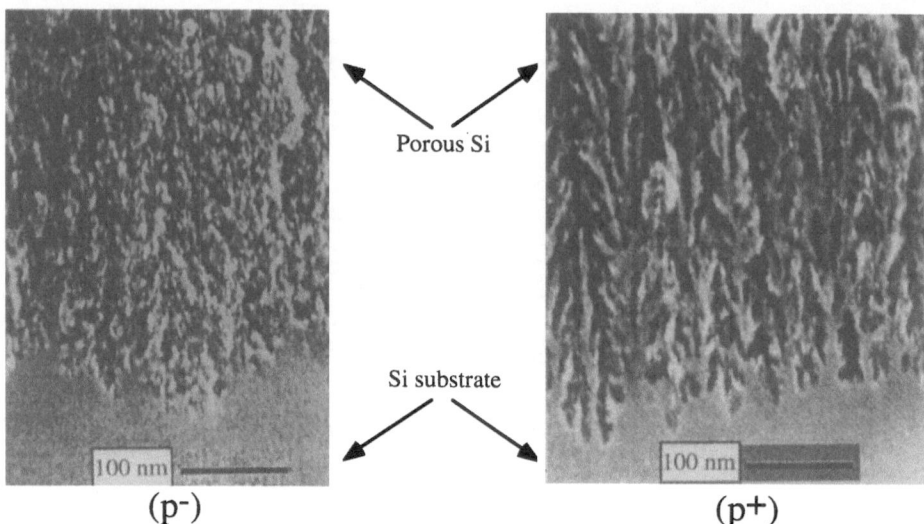

Figure 15: Cross-sectional micrographs (high resolution scanning electron microscope) of PS layers obtained from lightly doped (p-, 1Ω.cm) and heavily doped (p+, 0.01Ω.cm) silicon substrates. Layer porosity ~ 60%.

6.2 X-ray diffraction

Concerning the crystallinity of the material, PS layers with porosities as high as 72% for p+ samples and 56% for p- samples show, in double crystal X-ray experiments, a narrow Bragg peak well separated from the silicon substrate peak. Measured variations in the lattice parameter relative to the substrate increase with increasing porosity, resulting in maximum values of $\Delta a/a = 10^{-3}$ and 4×10^{-3} for the p+ and p- samples, respectively, and the above-indicated porosity [11]. Thus, even for these porosities, PS maitains the monocrystalline nature of the silicon substrate and a quite similar lattice parameter. The origin of the lattice parameter expansion has been attributed to the silicon-hydrogen bonds at the inner surface of the porous layer [13].

To support this hypothesis, Sugiyama *et al* [13] have measured the porous silicon strain after annealing in vacuum. These authors found a correlation between the strain and hydrogen desorption during annealing. After annealing at below 350°C, the lattice parameter of PS is in expansion. Above 350°C, there is a significant hydrogen desorption and a sharp change in strain corresponding to a contraction of the lattice parameter.

6.3 Chemical composition

Due to its large specific surface area, porous silicon was considered for a long time to contain a large amount of contaminant impurities. However, several studies have shown that when porous silicon is prepared under clean conditions and rinsed in pure water, the material consists of hydride termination and occasionally fluorine [11]. The other elements such as carbon, oxygen, nitrogen are all minor impurities in freshly prepared layers. However, this chemical composition changes dramatically with time during ambient air exposure. Storage in air, changes the hydride surface into a contaminated native oxide [14].

6.4 Thermal behavior and oxidation of porous silicon

The most important technological application of porous silicon is the dielectric isolation of integrated circuits. This application is based on the transformation of PS into homogeneous film of silicon dioxide (SiO_2). At first sight, oxidation of PS may appear to be an easy process. The material is expected to thermally oxidize throughout its entire volume due to its open porosity and the easy penetration of oxygen gas into the pore. This oxidation should occur in a relatively short time, as the silicon thickness to be oxidized does not exceed the pore wall dimension (about a hundred angströms). However, due to the thermal instabilities of the material, its oxidation requires some precautions. In fact, when a PS layer is annealed under vacuum or in a non-oxidizing gas, coarsening of the pore texture occurs at temperatures above 400°C, and increases with increasing temperature. After heating at ~900°C [15, 16], large voids of about 1μm are formed within the material, surrounded by thick silicon walls. It is clear that a complete oxidation of such a coarsened structure is difficult to achieve, as this restructuration results in a collapsing of the pores. A quantitative measurement of the coarsening can be obtained from determination of the specific surface area change induced by the thermal annealing under vacuum. Figure 16 shows a sharp decrease in the specific surface area, for an annealing temperature in the 400 - 450°C range. The coarsening of the PS texture is attributed to surface diffusion of silicon atoms along the pore walls, which tends to minimize the high surface energy. Why does this restructuration occur above ~400°C? The answer can probably be found in the experiments which show that during annealing, there is hydrogen desorption from the PS surface [17]. The hydrogen coverage seems to be stable up to ~300°C. For higher temperatures, the partial hydrogen pressure increases, reaching a maximum around 400 °C and total hydrogen desorption is complete at ~ 600°C. Silicon-hydrogen bonds appear, therefore, as a stabilizing factor which prevents the surface diffusion of silicon atoms below ~300 °C. For higher temperatures, the microstructure can be stabilized, against any coarsening, by covering the whole internal surface with a thin SiO_2 layer which is thermally

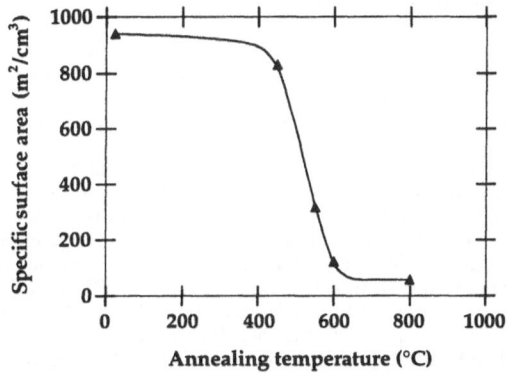

Figure 16: Variation in the specific surface area as a function of annealing temperature.
Sample thickness and porosity are 1μm and 52%, respectively. P⁻ silicon substrate (1Ω.cm).

stable. Such a stabilizing oxide layer is obtained by a low temperature (300°C) oxidation for ~2 hours. The thin oxide layer formed during this "preoxidation" step stabilizes the structure against heat treatment at higher temperatures. As long as no reducing atmosphere is used, heating of the layer can be performed for hours up to 900 °C without significant change in the porous texture (Fig. 17). The complete oxidation of such a stabilized PS layer is obtained around 700-800°C. However, the resulting oxide layers are porous and a densification step is required. This densification is achieved by a viscous flow of silica and requires temperatures higher than 1000°C.

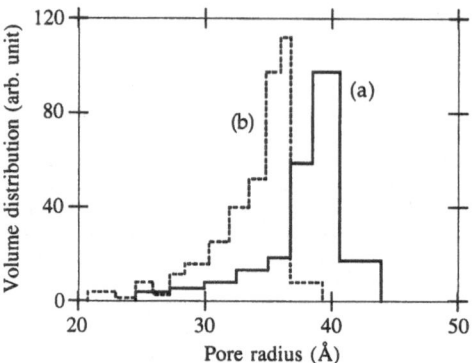

Figure 17: Pore size distribution of PS layers: As prepared(a). Pre-oxized at 300 °C and annealed at 800 °C (b). Data from ref. [15]

Since the oxygen gas easily diffuses through PS, a full thermal oxidation of the material can be achieved even for a very thick layer (~100μm). However, the chemical reactivity of porous silicon in a liquid phase is completely different from its reactivity in a gaseous phase. In fact, some liquids do not penetrate the pore and only the top surface of the PS layer is involved in the chemical reaction. These phenomena are discussed below.

6.5 Chemical dissolution and wettability phenomena

It is well known that a thick (few tens of μm) PS layer can be detached from its silicon substrate by electropolishing at the interface between PS and the substrate [18]. Using these resulting free-standing PS films some funny and simple experiments can be performed. If a drop of ethanol is put on the top surface of such a film, the liquid infiltrates the pores and crosses the layer. However, when a drop of pure water is put on the top surface of the film, this drop remains on the surface even for a long period of time. These simple experiments seem to show that ethanol infiltrates the pores while water (or any purely aqueous solution) does not. This behavior can be explained by wettability and capillary phenomena which occur in porous silicon, which is known to be a highly hydrophobic and organophilic material.

Figure 18 shows the significant effect of these phenomena on the chemical dissolution of PS in HF solution. This figure shows that the etch rate of a PS layer in a purely aqueous solution of HF is very low compared to that obtained with an ethanoic solution (a solution which contains ethanol). When a purely aqueous HF solution is used, the liquid does not infiltrate the pores and only the top surface is etched, resulting in very low weight loss. However, when some ethanol is added to the solution, the wettability of the material is then noticeably increased. In this case, the ethanoic HF solution completely infiltrates the pores and the whole inner surface is involved, resulting in a higher mass loss. Concerning the chemistry of the dissolution process, more details have been published elsewhere [5]. In summary, ethanol is not directly involved in the chemical dissolution process, its role is only to improve the liquid infiltration. The etch rate of PS in an HF solution is not directly correlated with the HF concentration but with the pH of the solution. The etch rate increases with increasing pH (OH⁻ concentration) of the solution.

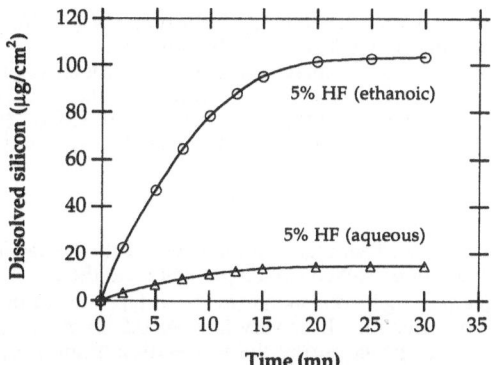

Figure 18: Mass of dissolved silicon from a PS layer (1μm-thick, 65% porosity) as a function of the immersion time in two different HF solutions (aqueous and ethanoic).

We have recently demonstrated [5] that it is possible to determine the specific surface area of a porous silicon layer by comparison of its etch rate in HF with that of bulk silicon. Figs.19a and b show the mass of dissolved silicon from bulk silicon and a PS layer in a ethanoic solution of 5% HF.

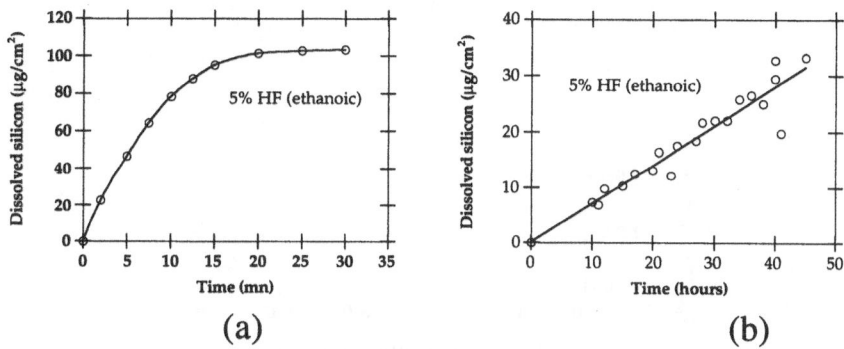

(a)   (b)

Figure 19: Mass of dissolved silicon from a PS layer (a) and bulk silicon (b) in an ethanoic HF solution.

When comparing the chemical dissolution of PS (Fig. 19a) and bulk silicon (Fig.19b) it appears that (i) the dissolution of bulk silicon is very low compared to that of PS; and (ii) the shape of the curve corresponding to bulk Si is linear while that of PS is not and exhibits a saturation regime.

The etch rate of bulk silicon can be defined as: $A=dm/dt$, where m is the mass of dissolved silicon per unit area and t the immersion time. The value of the etch rate can be deduced from Fig. 19b and we find $A= 1.16 \times 10^{-8}$ g/cm^2/mn. In terms of thickness change, a value of 0.5 Å/mn can be deduced from this etch rate ($A/r=0.5$Å/mn, where r is the bulk silicon density). A value of 0.3Å/mn has already been reported [19] for the dissolution of (111)-oriented n-type silicon in a 48 wt.%. aqueous HF solution.

We believe that the mass of dissolved Si from PS is much higher, as the material contains an enormous specific surface area. If we assume that the silicon crystallites which constitute PS are chemically dissolved at the same rate as bulk silicon (no size effect), the mass M of dissolved silicon per unit area for the PS layer can be expressed as:

$$dM = ASWdt \qquad\qquad (3)$$

where S and W are the specific surface area and thickness of the PS layer, respectively. We should note that M is the mass of dissolved silicon per unit area: the experimentally measured mass loss is divided by 58 cm^2 which corresponds to the surface area of the electrode.

Since the etch rate of silicon in HF is very low (A=0.5Å/mn) we can assume that the thickness, W, of the layer is constant during the whole dissolution process (~ 30 mn). To verify this assumption we measured the thickness of a 50%-porosity layer before and after etching up to a porosity of ~80% and it was found that the thickness is almost constant. As a result, the reason for which the curves corresponding to PS dissolution are not linear and saturate is that the specific surface area decreases with increasing porosity, *i.e.* increasing immersion time.

Chemical dissolution was also performed on two PS layers of the same porosity and different thicknesses. We found that when the thickness is increased from 0.5 to 1μm, the mass of dissolved silicon is doubled. This result is in agreement with formula (3) and also means that for both cases (0.5 and 1μm), the HF solutions completely infiltrate the pores and reach the interface between PS and the Si substrate.

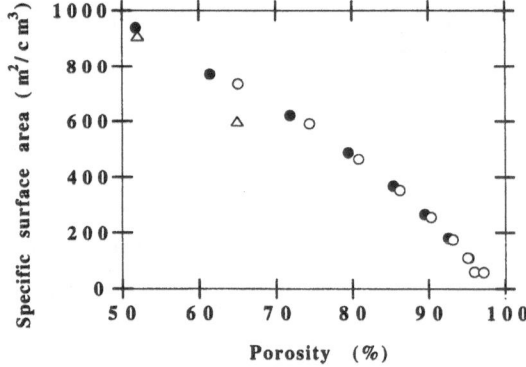

Figure 20: Specific surface area as a function of PS layer porosity. Data from 1μm-thick PS layer and starting porosity of 51% (●) and 65% (o). (Δ): data from Refs [10] and [20].

The change in specific surface area during the chemical dissolution process can be deduced from formula (3) . The values of dM/dt can be obtained from the chemical dissolution curves M(t) (Fig. 19a). To do this, the curve M(t) is first approximated to a polynomial of the 3rd degree using a curve fitting software and the resulting polynomial is then differentiated to extract dM/dt for the desired value of t. The results obtained for PS layers of different starting porosities of 50 and 65% and the same thickness of 1μm are shown in Figure 19. Also presented in this same figure are data from ref. [10] and [20] which are consistent with our measurements. It appears that, the specific surface area decreases with increasing porosity. Furthermore, for a given porosity the two layers used (50 and 65%

starting porosities) led to the same specific surface area: this means that for PS layers obtained from *lightly doped p-type Si,* the relevant parameter is the porosity, independently of the anodization conditions. Recently, Robinson *et al* [21] have predicted, from Fourier transform infrared measurements, that the specific surface area of PS should decrease with increasing porosity.

Many workers [22, 23] have shown that when the porosity is increased up to ~85%, the PL intensity is dramatically increased. Since the surface area decreases with increasing porosity, the luminescence of surface species, as often suggested [24], can be ruled out. It has been suggested recently [25] that many parameters, such as the low refractive index of PS, the carrier confinement and the remarkable passivation of the surface by hydrogen termination, contribute significantly to the enhancement of the PL. Although these parameters are likely to be involved, we believe that the decrease in specific surface area greatly contributes to the PL increase when the porosity of the layers is increased. In fact, when the surface area decreases, the number of nonradiative recombination centers at the surface decreases, thus leading to an increase in luminescence efficiency.

6.6 Optical absorption of free standing porous silicon films

PS layers obtained from lightly doped p-type silicon exhibit photoluminescence spectra which blue-shifted with increasing porosity [22, 23]. TEM studies have shown that when the porosity of the layer is increased, the dimension of the Si crystallites decreases [26, 27]. In the quantum confinement model, the blue shift of the PL spectra could be qualitatively explained by the increase in confinement energy with decreasing the crystallite dimensions.

The size effect, which is assumed to be at the origin of the PS luminescence, is more clearly seen on the optical transmission of free-standing PS films. Fig. 21 shows transmission coefficients (T) versus energy of 40 µm - thick free-standing PS films of different porosities made on p type substrates. T measured on PS layers of intermediate porosities, i.e., 62, 70% are not shown in the figure for the sake of clarity. However, the T versus energy curves also follow the shift towards higher energy with increasing porosity and slip in well between the curves shown.

In these figures, we first observed that the transmission spectra of PS are shifted towards higher energy compared to that of bulk silicon. Absorption coefficients (α) calculated from the transmission data by taking into account the quantity of matter in the films, which is proportional to one minus the porosity, are also shifted to higher energy and this shift increased with increasing porosity (α is given by $T=T_0 \exp[-\alpha w(1-p)]$ where T_0 is the value of the plateau in the low energy region, w and p being the film thickness and porosity, respectively). A possible explanation for the shift in absorption edge is a quantum size effect as already proposed [3, 18]. This explanation is confirmed by comparing the transmission coefficients (Fig. 22) of two PS films of the same thickness and porosity but obtained from lightly (p-) and heavily (p+) doped silicon substrates. As revealed by the HRSEM microphotographs (Fig. 15), the difference between these two layers is the dimension of the silicon crystallites which constitute the PS layer (the average size of the crystallites is about 10 nm for the p+ material and 3 nm for the p- one). Fig. 22 shows that the transmission coefficient of the p- material is shifted towards higher energy as expected by a quantum confinement model.

Although the observed shifts of the absorption edge are qualitatively consistent with a quantum size effect, we have recently demonstrated that the hydrogen coverage of the inner surface contributes significantly to these shifts. When hydrogen is desorbed from the surface by a thermal annealing in an ultra high vacuum chamber (~400 °C), the absorption coefficient increased (red shift of the absorption edge). When the hydrogen coverage is restored by a short HF dip, the absorption coefficient decreases and recovers it initial value. This important result, which requires further investigation, seems to show that both the small size of the crystallites and their hydrogen coverage are involved in the bandgap widening of porous silicon.

A. Halimaoui

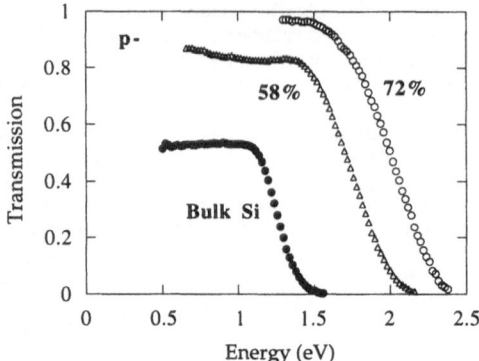

Figure 21: *Transmission coefficient versus photon energy measured at 300K for 40 μm-thick free-standing porous Si films of different porosities obtained from lightly doped p- type substrate (1Ω.cm).*

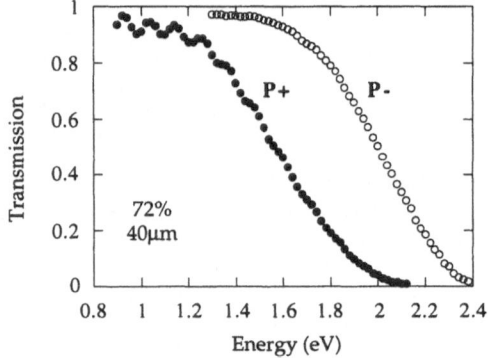

Figure 22: *Transmission coefficient versus photon energy for free-standing porous Si films of the same thickness (40μm) and porosity (72%) obtained from lightly doped (p⁻, 1Ω.cm) and heavily doped (p⁺, 0.01 Ω.cm) substrates.*

7. APPLICATIONS OF POROUS SILICON

Since the discovery of the PS luminescence, many workers have evoked the potential application of porous silicon in optoelectronics. Such an application requires the processing of an efficient electroluminescent device. At present, the quantum efficiency of the PS-based EL devices is still low for any application [28, 29]. Further research work is needed to improve this efficiency.

During the last decade, extensive studies have been devoted to the application of porous silicon in the dielectric isolation of integrated devices, in particular the fabrication of Silicon On Insulator (SOI) substrate [11]. This application is based on the thermal oxidation of porous silicon. The aim is to form crystalline silicon islands on a silicon dioxide layer. For this application, different approaches have been suggested.

The first one is based on the selectivity of the electrochemical reaction between n and n+ doped regions. In this technique, porous silicon is only formed in a thin (1μm) n+ buried layer which is accessible through the upper layer (Fig. 23). A lightly doped (~10Ω.cm) n-type substrate is given a high dose implant of antimony. The implantation dose and energy are chosen so as to produce, after the redistribution annealing, a 1 μm-thick n+ layer with a doping level in the 10^{18} to 10^{19} cm^{-3} range. A standard epitaxial growth of Si is then performed on the implanted substrate. The wafers are patterned with ~50μm wide islands separated by 5μm wide trenches. Porous silicon is formed selectively in the n+ region through the trenches. The as anodized wafers are then subjected to a thermal oxidation, thus leading to silicon islands over a silicon dioxide layer.

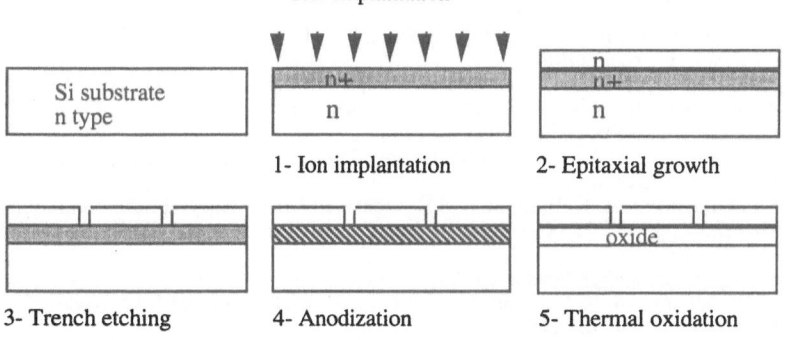

Ion implantation

1- Ion implantation   2- Epitaxial growth

3- Trench etching   4- Anodization   5- Thermal oxidation

Figure 23: Schematic of the different steps involved in the fabrication of an SOI structure using the selective anodization of a buried n+ Si layer.

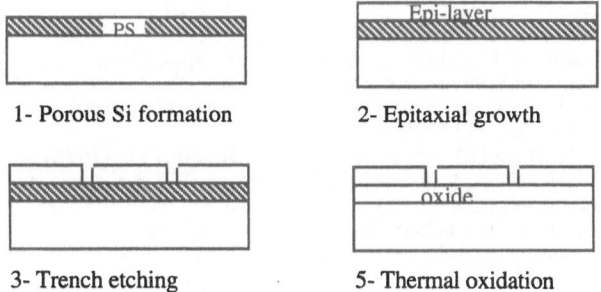

1- Porous Si formation   2- Epitaxial growth

3- Trench etching   5- Thermal oxidation

Figure 24: Schematic of the different steps involved in the fabrication of an SOI structure using the epitaxial growth of Si over porous silicon.

The second approach is based on the fact that porous silicon layers prepared under certain conditions maintain the monocrystalline characteristic of the silicon substrate.

Consequently, it is possible to grow an epitaxial silicon layer over the PS film [30]. The epi-layer is then patterned to define silicon islands and trenches. The buried PS layer is thermally oxidized to form the SOI structures (Fig. 24).

Using this type of SOI structure, circuits with good electrical characteristics have already been fabricated in a CMOS technology [31].

References

1.   D. R. Turner, J. Electrochem. Soc. **105**, 402 (1958).
2.   R. Memming, and G. Schwandt, Surf. Sci. **4**, 109 (1966).
3.   V. Lehman, and U. Gösele, Appl. Phys. Lett. **58**, 856 (1991).
4.   A. Halimaoui, Appl. Phys. Lett. **63**, 1264 (1993).
5.   A. Halimaoui, Surf. Sci. **306**, 550 (1994).
6.   F. Gaspard, A. Bsiesy, M. Ligeon, F. Muller, and R. Herino,
      J. Electrochem. Soc. **136**, 3043 (1989).
7.   M. I. J. Beale, J. D. Benjamini, M. J. Uren, N. G. Chew, and A. G. Cullis
      J. Cryst. Growth **73**, 622 (1985).
8.   C. Pickering, M. I. J. Beale, D. J. Robbins, P. J. Pearson, and R. Greej
      J. Phys. C **17**, 6535 (1984).
9.   F. Ferrieu, A. Halimaoui, and D. Bensahel, Solid State comm. **84**, 293 (1992).
10.   R. Herino, G. Bomchil, K. Barla, C. Bertrand, and J. L. Ginoux
      J. Electrochem. Soc. **134**, 1994 (1987).
11.   G. Bomchil, A. Halimaoui, and R. Herino, Microelectronic Engineering **8**, 293 (1988),
      and references therein.
12.   J. P. Gonchond, A. Halimaoui, and K. Ogura, in Microscopy of Semiconducting
      Materials, pp. 235-238 (1991). Edited by A. G. Cullis and N. J. Long, IOP Publishing
      Ltd., Bristol.
13.   H. Sugiyama, and O. Nittono, J. Cryst. Growth **103**, 156 (1990).
14.   L. T. Canham, M. R. Houlton, W. Y. Leong, C. Pickering, and J. M. Keen,
      J. Appl. Phys. **70**, 422 (1991).
15.   R. Herino, A. Perio, K. Barla, and G. Bomchil, Mater. Lett. **2**, 519 (1984).
16.   V. Lubanov, V. Bondarenko, L. Gurenko, A. Dorofeev, and L. Tabulina
      Thin Solid Films **137**, 123 (1986).
17.   P. Gupta, V. L. Colvin, and S. M. George, Phys. Rev. B**37**, 8234 (1988).
18.   I. Sagnes, A. Halimaoui, G. Vincent, and P. A. Badoz,
      Appl. Phys. Lett. **62** 1155 (1993).
19.   S. M. Hu, and D. R. Kerr, J. Electrochem. Soc. **114**, 414 (1967).
20.   L. T. Canham, and A. J. Groszek, J. Appl. Phys. **72**, 1558 (1992).
21.   M. B. Robinson, A. C. Dillon, and S. M. George, Appl. Phys. Lett. **62**, 1493 (1993).
22.   L. T. Canham, Appl. Phys. Lett. **57**, 1046 (1990).
23.   M. Voss, Ph. Uzan, C. Delalande, G. Bastard, and A. Halimaoui
      Appl. Phys. Lett. **61**, 1213 (1992).
24.   Z. Y. Xu, M. Gal, and M. Gross, Appl. Phys. Lett. **60**, 1375 (1992).
25.   A. J. Read, R. J. Needs, K. J. Nash, L. T. Canham, P. D. J. Calcott, and A. Qteish
      Phys. Rev. Lett. **69**, 1232 (1992).
26.   A. G. Cullis, and L. T. Canham, Nature **353**, 335 (1991).
27.   I. Berbezier, and A. Halimaoui, J. Appl. Phys. **74**, 5421 (1993).
28.   N. Koshida, and H. Koyama, Appl. Phys. Lett. **60**, 347 (1992).
29.   A. Richter, P. Steiner, F. Kozlowski, and W. Lang,
      IEE Electron. Dev. Lett. **12**, 691 (1991).
30.   C. Oules, A. Halimaoui, J. L. Regolini, A. Perio, and G. Bomchil,
      J. Electrochem. Soc. **139**, 3595 (1992).
31.   K. Barla, G. Bomchil, R. Herino, A. Monroy, and Y. Gris, Electron. Lett. **22**, 1291
      (1986).

LECTURE 4

Luminescence of porous silicon after electrochemical oxidation

R. Herino

*Laboratoire de Spectrométrie Physique,
Université Joseph Fourier de Grenoble, BP. 87,
38402 Saint Martin d'Hères cedex, France*

1. INTRODUCTION

One of the most commonly used hypothesis to explain the light emission in the visible range from porous silicon is the confinement of charge carriers in the quantum sized crystallites which are formed in the material [1] [2] [3]. The rather good emission efficiency which is obtained also supposes an efficient passivation of the large specific surface of the material. Such passivation is readily obtained just after formation of the porous layer, and it is provided by the Si-H surface coverage which results from the anodic attack of the crystalline silicon in hydrofluoric acid (HF) solutions [4]. However, this passivation is not permanent, and the result is that photoluminescence degradation is often observed, for example upon annealing [5] or under illumination.

Different attempts have been made in order to stabilize the surface passivation of the material, and it has been shown that partial oxidation of the layer can be an efficient procedure. A well appropriate technique to realize this is the electrochemical oxidation of the porous layer which is performed under anodic polarization of the sample in an aqueous electrolyte [3].

In this chapter, we will describe in some details the anodic oxidation of porous layers, and show how it leads to a large increase of the luminescence efficiency. Another interest of this electrochemical technique has been to demonstrate for the first time that efficient electroluminescence could be obtained from porous silicon. This chapter will also present an analysis of the different characteristics of this light emission obtained under anodic polarization.

2. ANODIC OXIDATION OF POROUS SILICON

2.1 Experimental aspects

The experimental set-up to realize the electrochemical oxidation of a porous layer is the same than the one used to form the porous material. Figure 1 shows a schematic description of a cell which can be used. The electrical contact with the substrate is achieved by evaporation of a metallic film on the back side of the wafer. The counter electrode is generally a platinum grid, and the silicon potential can be monitored by using a reference electrode in the electrolyte. Samples can be anodized either under galvanostatic or potentiostatic conditions by using a potentiostat. The electrolyte can be pure water, but most often, a supporting inert electrolyte (like KNO_3 or HCl) is used in order to make negligible the ohmic drop in the solution.

Anodic oxidation of porous films must be performed in an electrolyte absolutely free of any traces of HF. For this reason, as-formed porous samples are carefully rinsed under a flow of deionized water for 10 minutes. After the samples have been dried and before anodic

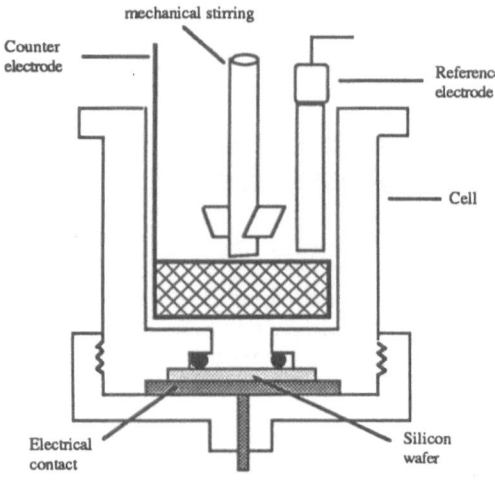

Figure 1 : schematic description of a cell for the electrochemical treatment of silicon

oxidation, it is generally found that about 15% of the porous layer has been oxidized in air. In fact, this amount corresponds to the growth of about one oxide monolayer on the whole internal surface of the material. Similar oxide growth of a few angströms has also been found at the surface of non-porous silicon samples after a ten minute exposition to the atmosphere [6]. The presence of this thin native oxide on the pore walls has a strong influence on the characteristics observed upon anodic oxidation. Its formation can be avoided by keeping some liquid covering the sample during the operations between formation and anodic oxidation, as no native oxide growth is observed on silicon immersed in pure water even for periods of several minutes. This can be explained because at equilibrium, the silicon surface in contact with water is depleted of holes, as shown by impedance measurements [7] and therefore, as silicon oxidation requires holes, the reaction is strongly hindered. Another explanation could be found in a chemical passivation due to the hydrogen coating of the porous silicon surface [4].

2.2 Main characteristics of the electrochemical oxidation

The variation of the silicon potential during porous silicon anodization at constant current density is shown on figure 2. Two different regimes are observed independantly of the layer thickness or of the anodic current density : initially a slow increase in potential is observed, followed by a sharp rise when the exchanged charge exceeds a certain value, Q_0. The value of Q_0, which increases with the layer thickness is related to the end of the oxidation regime of the porous layer. On further oxidation, there is no further increase of the oxygen amount incorporated in the layer, as

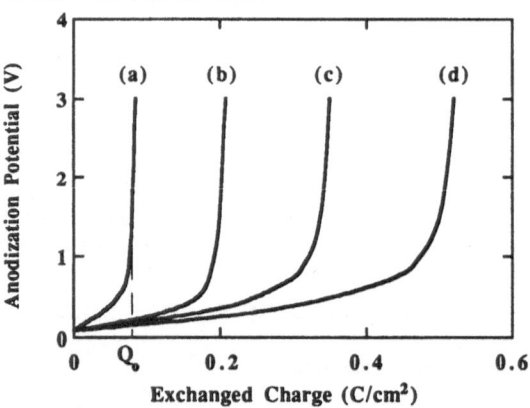

Figure 2 : Variations of the anodic potential as a function dfe exchanged charge. Anodic current density : 0.1 mA/cm2. Layer thickness : (a) 0.3, (b) 1, (c) 3 , and (d) 5 microns.

measured by RBS analysis [8]. The exchanged charge beyond Q_0 no longer contributes to silicon oxidation, but mostly corresponds to solvent oxidation, as the silicon electrode potential rises far beyond one volt. However, the charge Q_0 is always much below that which would be necessary to oxidize all the silicon atoms of the porous layer assuming a 100% current efficiency of SiO_2 formation and a four-electron reaction. This leads to the conclusion that at a transferred charge equal to Q_O, there is a break in the electrical contact between bulk silicon and the partially oxidized porous silicon layer.

2.3 Influence of the anodizing current density

Figure 3 shows the effect of the value of the anodizing current density on the charge Q_O. It appears that Q_O which measures the amount of oxidized silicon in the layer increases with the current density only for values below a certain current density, j_{sat}, which depends on the layer thickness. Beyond j_{sat}, there is no further increase in the oxidation level, although the exchanged charge is still lower than that corresponding to the complete oxidation of the porous film.

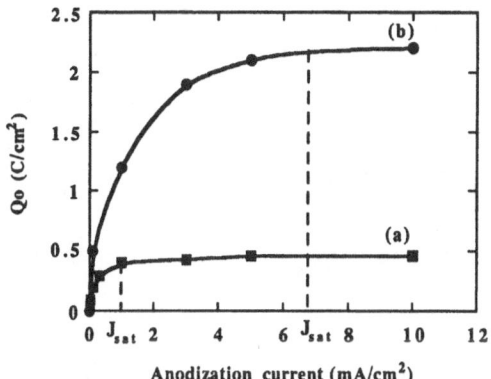

Figure 3 : Variations of the total exchanged charge Qo as a function of the anodic current density. Sample Thickness : (a) 1 and (b) 5 microns

The analysis of the quantity of oxygen incorporated in the porous samples after anodic oxidation allows to clarify this behavior. Figure 4 shows the variations with depth of the percentage of oxidized silicon atoms in the porous layer, deduced from oxygen profiles obtained by RBS analysis of one micron thick layers. These profiles are mostly flat for current densities of 0.3 and 3 mA/cm^2, with oxidized fractions between 40 and 60%. This is in good agreement with the values deduced from coulometry, showing that the current efficiency of SiO_2 formation is equal to 100% for exchanged charge amounts up to Q_O.

The profile obtained for the current density of 0.03 mA/cm^2 is quite different, and can be divided in two different regions. From the surface down to about 0.6 µm, a nearly constant oxidized fraction of about 15% is found ; in the lower part of the film, from 0.6 to 1 µm, an increase up to about 30% at the bottom of the porous layer is observed. It must be noticed that a test sample with the same characteristics, but not submitted to electrochemical oxidation, also shows a uniform oxygen content corresponding to 15% of oxidized silicon atoms, which is caused by air exposure in-between porous silicon formation and the RBS experiment. Then, the first part of the oxidation profile found for the sample anodized at 0.03 mA/cm^2 can be attributed to this native silicon oxide growth. In contrast, the second part is dependent upon the anodization process and shows that there is preferential oxidation at the bottom

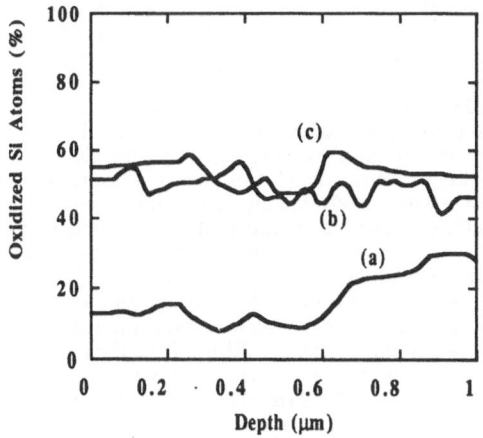

Figure 4 : Depth profiles of the oxidized silicon atom percentage for different anodic current densities : (a) 0.03, (b) 0.3, and (c) 3 mA/cm2. Sample thickness : 1 micron

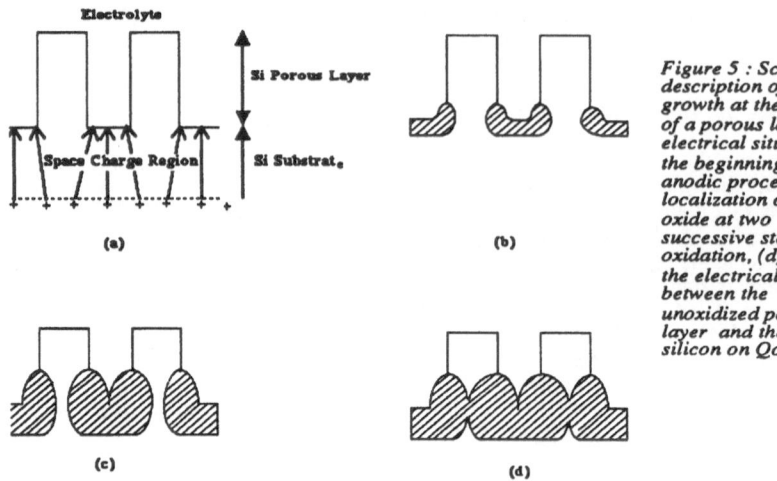

Figure 5 : Schematic description of SiO2 growth at the bottom of a porous layer : (a) electrical situation at the beginning of the anodic process, (b,c) localization of the oxide at two successive steps of oxidation, (d) break of the electrical contact between the unoxidized porous layer and the bulk silicon on Q_0

of the porous layer when the anodic current density is much below j_{sat}.

When the sample has been dried before anodic oxidation, a very thin native oxide covers the whole internal surface of the sample, which significantly affects the anodic oxidation regime. In that case, Q_0 varies linearly with the porous silicon thickness with a slope which is not influenced by the current density value. Furthermore, the profiles of oxygen incorporated in the porous films are flat, showing that the anodic oxidation always proceeds throughout the whole porous layer independently of the current density value [8].

2.4 A phenomenological model for the anodic oxidation of porous films

The limiting process in the oxidation of monocrystalline silicon depends on the formed oxide thickness [7]. As silicon oxidation involves hole exchange with the valence band, for the early stages of oxidation when the oxide thickness is below 4-5 Angströms the limiting parameter is the surface hole concentration. In this regime, the semiconductor surface is depleted of majority carriers and the oxide growth is limited by the hole supply at the silicon surface over the Schottky barrier which is formed at the interface. However, when a continuous oxide film of a few angströms has been formed, the oxidation reaction becomes limited by the oxidizing species migration across the oxide. As the oxidant transport results from a hopping process under high electric fields, this regime corresponds to an anodic polarization of the silicon electrode which is always high enough to accumulate majority carriers at the interface. These two different regimes of oxidation are related to two different situations of the silicon surface, hole depletion or accumulation. This model provides a reasonable explanation of the behavior of porous silicon upon anodic oxidation.

When there is no oxide on the porous sample, at the beginning of the process under low current density, the oxidation reaction is limited by the hole concentration at the silicon surface. The majority charge carriers diffuse from the bulk silicon across the space charge region which exists at the interface with the electrolyte (figure 5-a), that is at the bottom of the pores. Holes are thus preferentially available at the pore tips, and the oxidation first occurs in this region. When the pore bottom is covered by one oxide monolayer, a slight increase of the anodic silicon polarization is necessary to keep the current constant. The oxidation then takes place simultaneously at the pore tips, across the oxide layer, and on the silicon pore walls, near the bottom, where holes are available (figure 5-b and 5-c). The silicon potential will continue to increase slightly because the thickness of the oxide at the pore tips increases and oxidation of the pore walls will progress towards the top of the porous layer. However, the end of the oxidation process occurs before complete oxidation of the layer because the electrical contact between the pores and the bulk semiconductor is broken when a continuous oxide layer has been formed at the bottom of the layer (figure 5-d). This situation corresponds to the sharp rise of the electrode potential observed in galvanostatic conditions.

This model explains the dependence of Q_O upon the anodic current density. When the oxidation current is increased by increasing the silicon potential, the hole supply is enhanced and there will be simultaneous oxidation of the pore tips and of the pore walls on a depth which is greater than that oxidized at a lower current density. Consequently, at the end of the oxidation process, when the electrical contact is broken, the percentage of silicon atoms oxidized in the porous film, and thus Q_O, will be greater when the anodic current density is greater. However, for a given sample thickness, this Q_O increase is necessarily limited to the situation where the whole porous layer has been driven to hole accumulation. This condition, which is obtained on and beyond a certain current density for a given film thickness induces simultaneous oxidation in the whole layer depth leading to a value which is independent of the current density, as experimentally observed (figure 4).

In the case of porous layers covered by a native oxide, the potential which must be applied to the silicon electrode in order to obtain a given current density is always greater than that applied to bare silicon, and is high enough to drive into accumulation the pore walls of the whole porous layer. Consequently, oxide growth occurs mostly with the same kinetics within the depth of the porous layer and this is why nearly flat oxygen profiles are found at any current density for these samples. This also explains why the total exchanged charge Q_O increases linearly with the layer thickness whereas it is almost independent of the current density value for a given film thickness.

3. LUMINESCENCE OF ELECTRO-OXIDIZED POROUS LAYERS FORMED ON P-DOPED SUBSTRATES.

The luminescent properties of porous films appear to be strongly modified by electrochemical oxidation [3]. The oxidation process itself is accompanied with a bright emission of visible light showing that electroluminescence can be obtained from porous films [9]. Large intensity variations and spectral shifts are observed for both photoluminescence and electroluminescence when increasing the exchanged charge, and the aim of this part is to describe these changes and analyse their origin. In a first paragraph, we will first describe the electroluminescence observed under anodic polarization. Then, we will detail the different changes observed in the emission characteristics with the oxidation level and a model allowing to understand these changes will be described. Finally, we will present some characteristics of the electroluminescence observed during anodic oxidation of porous layers formed on heavily p-doped (P+) substrates.

3.1 Main characteristics of the electroluminescence observed under electrochemical oxidation

Figure 6 shows the time evolution of both the anodization potential and the electroluminescent intensity during the anodic oxidation of a lightly p-doped porous film under galvanostatic conditions. A first light emission is observed during the regime where the potential only presents a slight increase corresponding to the progressive in-depth oxidation of the porous layer. The light emission becomes noticeable after a delay which increases for samples of lower porosities. Then, a large increase of the emitted light intensity is obtained up to a maximum, and the emission finally vanishes on Q_O when the anodization potential starts to rise very sharply.

On further anodization, the anodization potential increases quite rapidly and during this second step, a new light emission is observed, which starts for potentials above about 30V, and the emitted intensity increases continuously just as the anodization potential. Further, both the emission and anodization potential become unstable because of the appearance of electrical breakdowns in the oxide layer. In fact, a very similar light emission is obtained during the anodic oxidation of bare silicon in the same conditions [10], which can be related to radiative recombination processes within the oxide during the anodization. This second electroluminescence regime is clearly not specific of the porous material and we will concentrate in the following on the first regime which is undoubtly characteristic of the porous layer.

Figure 7 shows the different spectra recorded as a function of time during the layer anodization. Large intensity variations and a spectral blue shift are observed with the increase of the exchanged charge, which will be analysed in the next section. All along the electrochemical process, the spectral characteristics remain very close to that of the photoluminescence of the same samples, suggesting that the same mechanisms are at the origin of the emission.

Following the confinement model, the observation of a simultaneous emission of light during anodic oxidation requires that

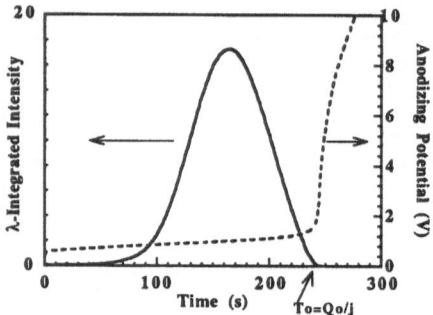

Figure 6 : Time evolution of the anodizing potential and of the light intensity during the anodic oxidation of a 65% porosity layer. Current density : 1 mA/cm2

radiative recombination of carriers takes place within the quantum-sized crystallites in the porous layer. Because holes are supplied by the anodic polarization of the silicon electrode, this implies that electrons are simultaneously injected into the crystallites, from the electrolyte or from the interface. The origin of this electronic injection is not clear, as the involved anodization potentials remains below the oxidation potential of the species present in the electrolyte. This charge injection might result from the oxidation of the Si-H bonds which cover the porous silicon surface after formation. Another possibility might be that this electron injection results from the silicon oxidation itself [11]: in the overall process leading to the oxidation of surface silicon atoms, intermediate species can be formed with energy levels high enough to allow electronic injection into the confined energy levels. In any case, the light emission can be observed only if these charge exchanges take place in the volume of the porous layer and involve quantum-sized crystallites still electrically connected to the substrate.

When approaching the exchanged charge Q_0, the oxidation level in the porous layer is such that the hole supply towards the crystallites becomes severely hindered : finally, upon Q_0, the electrical contact with the bulk is broken and the light emission stops with the charge exchange at the crystallite surface. Of course, this break does not occur at the same time for all the crystallites (which are highly interconnected) and instead of observing a sharp disappearing of the emission, a progressive decrease in the emitted intensity is obtained resulting from the progressive increase in the number of disconnected crystallites.

For a given porosity, the total emitted light as well as the λ-integrated peak intensity are proportionnal to the layer thickness, provided the same "volumic" current density is used, that is to say that the anodic current has to be increased proportionnally with the thickness. This well confirms that the emission is closely related to the local current and that it is actually a volumic phenomenon.

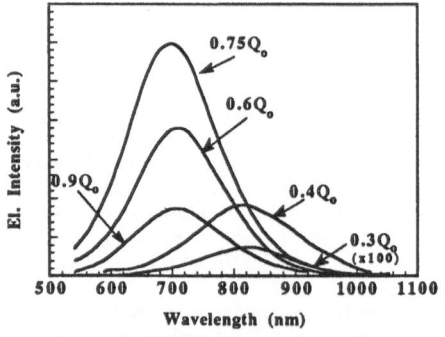

Figure 7 : Evolution of the electroluminescence spectra as a function of the exchanged charge. Sample porosity : 70%, thickness : 2 microns, current density : 1 mA/cm2

Because it is associated with the oxidation of the material, this anodic electroluminescence cannot be permanent, and thus cannot lead to technological applications. However, its interest is not only historical in the way that it was the first demonstration of voltage induced emission from porous silicon ; this emission, when compared to photoluminescence is also useful to study and analyse the material emitting properties.

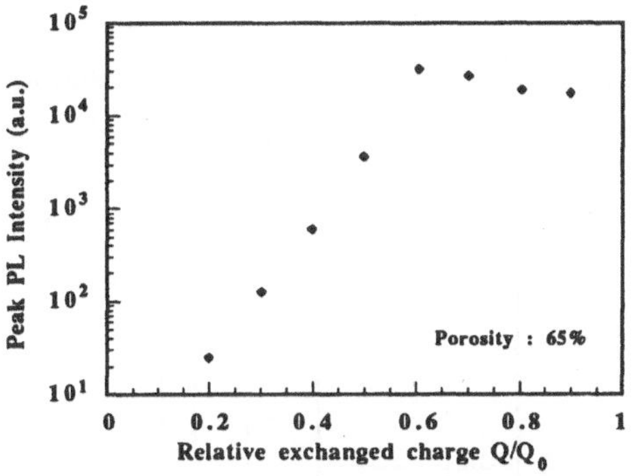

Figure 8 : Variations of the photoluminescence peak intensity as a function of the exchanged charge during the anodic oxidation of a 65% porosity layer. Current density : 1 mA/cm2

3.2 Luminescence modifications under electrochemical oxidation

During the first part of the anodic oxidation, a very large increase in the emitted intensity is observed, as well for the photoluminescence as for the electroluminescence signals [12]. Figure 8 shows that the photoluminescence intensity of 65 % porosity layers is increased by more than three orders of magnitude after anodic oxidation. A very similar behaviour is found for 70% porosity samples. In both cases, this regime corresponds to an exponential variation of the intensity with the exchanged charge Q. If we analyse in the same way the increase of the electroluminescence signal which is observed in the first part of the electrochemical process, the same type of exponential increase is obtained whatever the anodizing current density (figure 9).

This behavior can be explained when looking at the physical mechanisms which govern the emission.

3. 3 Mechanisms governing the emission efficiency

Following the confinement model, the light emission from porous layers results from the radiative recombination of charge carriers in the quantum-sized crystallites which form the material. The relatively low quantum efficiencies of photoluminescence measured at room temperature strongly suggest that the radiative recombination of carriers in porous silicon is

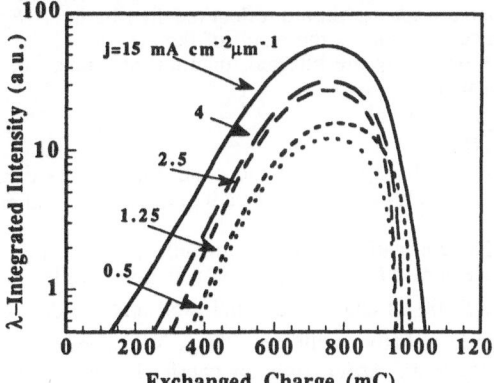

Figure 9 : Logarithmic plot of the λ-integrated light intensity as a function of the exchanged charge upon anodic oxidation at different current densities. Porosity 70%, layer thickness 2μm.

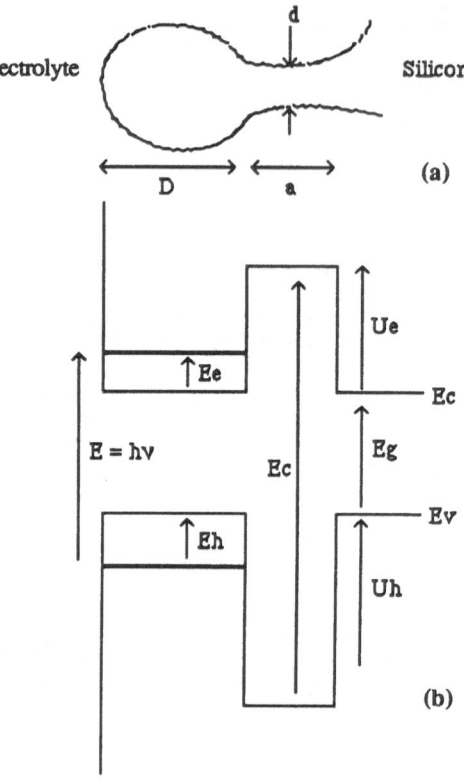

Figure 10: Schematic description of the electrolyte-porous silicon interface.
a) idealized representation of a quantum size crystallite (diameter D) linked to a non-radiative silicon cell by a narrowed region (length a, diameter d).
b) corresponding energetic diagram in the flat band conditions.
Eg : silicon gap energy
Ee and Eh : electron and hole confinement energy in the crystallite
Ue and Uh : electron and hole confinement energy in the narrowed region
E = hv = Eg + Ee + Eh : energy of the emitted light.

countered by non radiative processes, which may involve either volume or surface recombination centres [13]. The high porosity material which shows luminescent properties can be described as an assembly of small size silicon crystallites which are highly interconnected between them by narrowed regions. Because of the very small size involved, it can be postulated that a given crystallite is able to emit light only if it is perfectly passivated, this means zero recombination center, and then the spectra should reflect the distribution $N(E)$ of such crystallites. In fact, the luminescence spectrum is the result of the product of the size distribution of the emitting crystallites $N(E)$ by the internal quantum efficiency of the crystallites. The emitted intensity can be written :

$$I(E) = A_o \, N(E) \, \frac{W_r(E)}{W_r(E) + W_{nr}(E)} \qquad (1)$$

where A_o is a proportionality constant which depends on the conditions of the charge generation in the crystallites, and $W_r(E)$ and $W_{nr}(E)$ are the radiative and non-radiative recombination rates respectively.

Time resolved photoluminescence characterization brings useful information about the internal efficiency [13]. Photoluminescence decays following pulsed excitation have led to the determination of lifetimes in the range 1-100 µs and to the conclusion that the dominant mechanism for the emission relaxation is non radiative, with $\tau = 1/W_{nr}$. We believe that this non-radiative mechanism involve the escape of the carriers from the confined zone where they are generated towards more extended and less passivated neighbouring crystallites where non-radiative recombination can occur [13]. For this escape, the carriers have to experience the energy barrier which corresponds to the short narrowed regions which link the crystallites between them (figure 10). The exponential dependence of the non-radiative recombination rates with the emission energy well supports a tunnelling mechanism for such non-radiative leaks. Then, the lifetime is expected to depend on the barrier transparency $T(E)$ according to :

$$\tau = \frac{1}{W_{nr}} = \frac{1}{W_o \, T(E)}$$

(2)

with :

$$T(E) \propto exp\left[-\frac{2a}{h} \sqrt{2m \, (U_e - E_e)} \right]$$

(3)

where $E = E_e + E_h + E_g$ (E_e is the confinement energy of the electron, E_h is the confinement energy of the hole and E_g the bulk silicon gap), a and U_e are the barrier thickness and the barrier height respectively and m the carrier effective mass. We will here only consider the electron escape for which the transparency is greater ; in fact, as soon as the electron has tunnelled through the barrier, the hole is attracted by coulombic interaction and leaves the crystallite to recombine non-radiatively.

Finally, taking into account this tunnelling model, the emitted intensity can be expressed as :

$$I(E) = A \; N(E) \; \frac{W_r}{W_o} \; exp\left[\frac{2a}{h} \sqrt{2m \, (U_e - E_e)} \right]$$

(4)

This expression shows that exponential intensity variations are expected from any modification of the energy barrier.

It allows to explain the exponential increase of the photoluminescence and electroluminescence intensity upon electrochemical oxidation by the improvement of the barrier efficiency towards the non-radiative leaks. If we remind that the energy barrier is attributed to the silicon narrowed regions which link the crystallites, it corresponds to much smaller dimensions than that of the crystallite. Then, the oxide growth can more significantly affect the sizes of these narrowed regions. The oxidation of the surface of these narrowings will induce an increase in the energy E_c, by rising the confinement energy related to these restricted regions. It is of course difficult to express how this energy E_c is modified as a function of Q, but the experimental results suggest that the barrier modifications are proportional to the exchanged charge Q, which seems quite reasonable.

3.4 Spectral shifts resulting from anodic oxidation

Figure 11 shows the variations of the photoluminescence and electroluminescence peak wavelength as a function of the exchanged charge during the electrochemical oxidation of 65% porosity layers. A quite large peak wavelength decrease is observed, attesting for the blue shift of the spectra with the oxidation level. It is interesting to notice the very similar variations which are obtained for both kinds of emission, which suggest that the same crystallite distribution is involved in both phenomena. This is not necessarily obvious because two different types of emitting crystallites have to be considered. There are in the material crystallites which are "electrically disconnected" from the substrate. This means that it is not possible to inject charge carriers into these crystallites by polarizing the silicon substrate. Consequently, these crystallites will only shine under optical excitation, and they will only contribute to the photoluminescence. On the other hand, there are also crystallites which are still electrically "connected" to the substrate, into which charge carriers can be injected according to the substrate polarization. For this population, the emission excitation can be either optical or electrical, so that it can contribute to both photoluminescence and electroluminescence.

Here again, the spectral blue-shift can be explained by the enhancement of the energy barrier. For a given barrier, the charge escape probability is greater for the more confined crystallites, which emit in the blue part of the spectra. If the barrier height is increased, the reduction of the non-radiative leaks is more important for these crystallites : the increase of the radiative efficiency is then stronger for the more confined crystallites, and this results in the observed blue-shift.

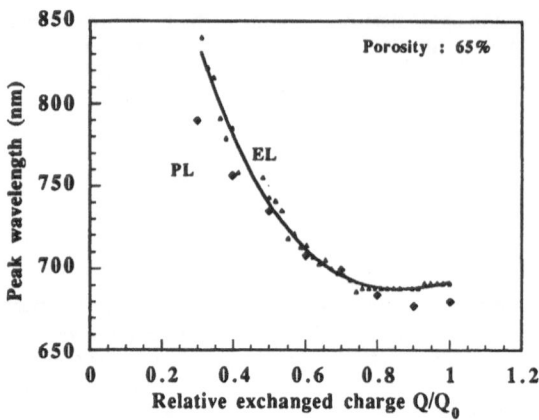

Figure 11 :
Variations of the
peak wavelength of
photoluminescence
and
electroluminescence
during anodic
oxidation of a 65%
porosity layer.
Current density :
1 mA/cmt

Figure 12 shows that similar trends are obtained on 70% porosity layers. However, in that case, in the first part of the anodic treatment, there is a large difference in the peak wavelength of photoluminescence and electroluminescence at a given oxidation level. This feature indicates that different populations of emitting crystallites are involved. All connected and disconnected crystallites are contributing to the photo emission, when it is only the connected crystallites which are involved in the electroluminescence. At the beginning of the electrochemical oxidation, the anodic polarization is quite low, and according to the Lehmann model [2], holes are mostly supplied towards the less confined crystallites (the "red" emitting ones). On further oxidation, the anodic polarization increases, and holes can be also injected into smaller crystallites, leading to a spectral blue shift as well as the barrier enhancement provided by the oxide growth. In the second part of the electrochemical treatment, the photoluminescence and electroluminescence spectra well superpose : all the connected crystallites are similarly excited for both phenomena, and the contribution of disconnected crystallites to the photoluminescence does not provoke major differences in the spectra.

Figure 13 shows the spectral shifts recorded for 80% porosity layers. In that case, it is only at the end of the anodic treatment that a good spectral agreement is observed for both types of emission. These differences account for the evolution of the material structure with the porosity : in the 80% porosity layers, the population of disconnected crystallites is likely more important, and then, the photoluminescence spectra are almost determined by this kind of

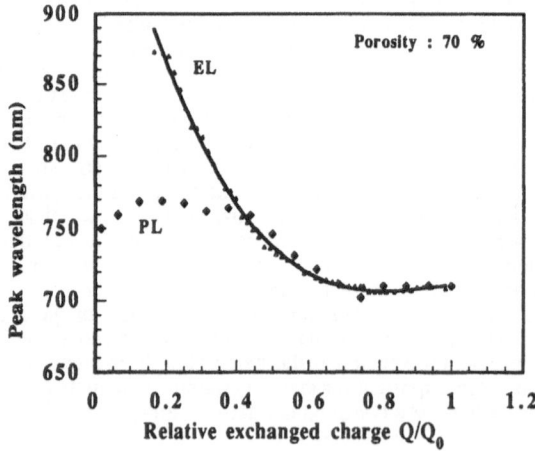

Figure 12 : Variations
of the peak
wavelength of
photoluminescence
and
electroluminescence
during anodization of
a 70% porosity layer.
Current density :
1 mA/cm2

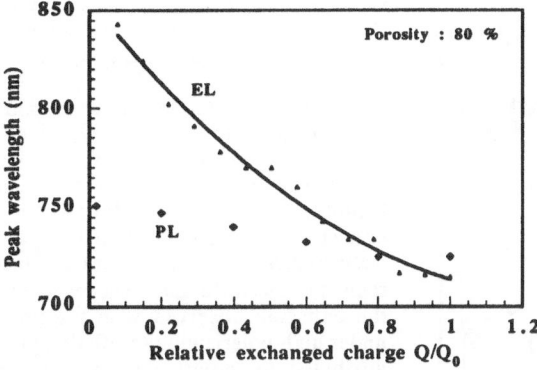

Figure 13 : Variations of the peak wavelength of photoluminescence and electroluminescence during anodic oxidation of a 80% porous layer. Current density : 1 mA/cm2

crystallites. This feature might also explain why the electrochemical treatment of 80% porosity layers only increases the photoluminescence intensity by a factor 5 (after exchange of Q_O). Indeed, the barrier enhancement can only concern the connected crystallites which seem to be in much lower proportion than the disconnected ones in these very high porosity layers.

3.5 Electroluminescence from porous layers formed on heavily p-doped substrates

High porosity layers formed on heavily p-doped silicon (P+) substrates (10^{-2} Ω-cm resistivity) also show an electroluminescence signal during anodic oxidation, which has very similar characteristics to that observed on lightly p-doped porous layers. The general shape for the anodic potential variations during oxidation under galvanostatic conditions is similar to that obtained with lightly p-doped substrates. However, the exchanged charge Q_O measured when the potential starts to rise quite sharply is much greater for the heavily doped samples. Q_O is found to be independant of the current density, and the oxygen profiles show that the whole porous layer is uniformly oxidized. However, the oxidation is still uncomplete, and there is again about 50% of the silicon atoms in the layer which have not been oxidized upon Q_O. Then, the charge Q_O corresponds to the break of the electrical contact between the bulk silicon and the porous layer, just as for lightly p-doped substrates.

As shown on figure 14, the emitted light intensity variations with time also present the

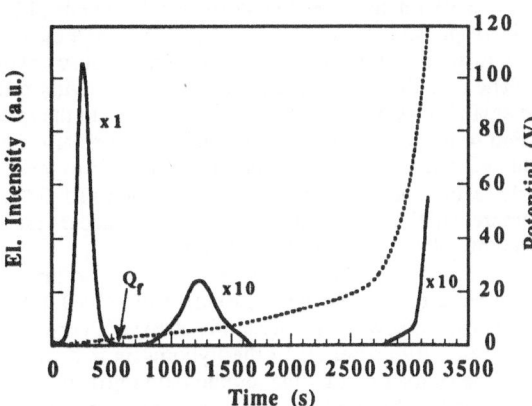

Figure 14: Time evolution of the anodizing potential and of the λ-integrated light intensity during the anodic oxidation of a 70% porosity P+ layer under constant current density (0.5mA.cm⁻².μm⁻¹)

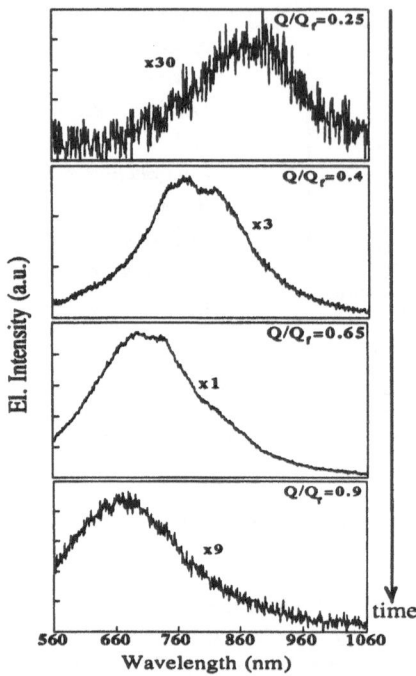

Figure 15: Evolution of the electroluminescence spectra for different times of oxidation (heavily doped P type, 70% porosity sample, current density 0.5mA.cm^{-2}.μm^{-1}).

same general behavior : a large increase is obtained in a first regime, up to a maximum, followed by a decrease upon further oxidation. However, in this case, the end of the emission occurs much earlier and does not correspond to the charge Q_O. For longer times, the potential keeps increasing slightly and a second emission of much lower brightness is obtained. Finally, a third emission is observed when the potential starts to increase sharply after Q_O. This third regime of electroluminescence is identical to the emission observed with p-type layers under high polarization (> 30 V) and can be attributed to radiative phenomena in the oxide layer at the interface between the porous layer and the substrate. The first electroluminescence is of much lower intensity than that observed with ligthly p-doped samples, but it presents very similar spectral characteristics. A blue-shift is also obtained with increasing exchanged charge, and the peak intensity is shifted of about 250 nm during the whole process (figure 15).

The observation of luminescence from highly p-doped samples may seem surprising because it is well known that in this type of porous material, the porous texture is not as thin as in the case of lightly doped substrates. The pore distribution as determined by gas adsorption experiments corresponds to pore radii between 4 and 10 nm, whereas the pore sizes are below 2 nm for lightly doped porous materials [14]. Transmission Electron Microscopy also shows that the crystallite dimensions are greater and may not induce important confinement effects. In fact, a more detailed analysis of the size distribution from Raman experiments [15] suggests that the heavily doped porous layers are formed of two different populations ; one is formed with crystallites with diameters between 4 and 10 nm, and the other corresponds to cells of 2-3 nm diameter. This thinner structure can be responsible for the observed emission, and then lead to spectral characteristics very similar to that of lightly p-doped layers, because involving crystallites of about the same sizes. The much lower intensity recorded from highly p-doped samples might then reflect the much smaller proportion of emitting crystallites in the material. For lightly p-doped substrates, the end of the electroluminescence has been explained by the break of the electrical connexion between the bulk and the porous layer due to the barrier efficiency enhancement by the anodic oxide. The same behavior can take place for P$^+$ porous layers, but it is only the thinner structure which becomes disconnected : this occurs far below the charge Q_O and this does not affect noticeably the anodization potential.

However, this interpretation is severely questionned by the following observation. If after the end of the first electroluminescence regime, the anodic oxide is dissolved in HF, and the electrochemical oxidation is began again, a new light emission is observed, very similar to the previous one, with no noticeable intensity decrease. In the hypothesis that the thinner structure has been disconnected from the coarser one by the anodic oxide growth, the oxide dissolution by the HF treatment should lead to the elimination of the major part of the thinner structure. In that case, the observation of a new emission of about the same intensity after the HF attack is very unlikely. Then, it seems that the end of the emission does not result from the hole injection hindrance due to the oxide growth, as in the case of lightly p-doped layers. This behavior might reveal that the end of the light emission for the P+ porous layers is related to the consumption of the electron injecting species at the crystallite surface. These species can be regenerated by the HF treatment, as the Si-H bonds which cover the porous surface, and allow a new light emission on further anodization. Another possibility might be the de-passivation of the crystallite surface by the anodic treatment, which would lead to the decrease of the number of emitting crystallites. This is supported by the fact that the photoluminescence intensity also decreases with the electroluminescence intensity, and is regenerated by the HF treatment which is known to provide an effective passivation. In that case, it seems difficult to explain why anodic oxidation should alter the passivation of P+ porous layers, and improve that of the lightly p-doped layers.

The second emission regime is also quite difficult to understand. It shows particular characteristics : the emitted intensity can be drastically modified by a change in the composition of the electrolytic solution, and the layer thickness dependence is different from that of the first regime. Moreover, the emitted spectra show a red-shift with increasing exchanged charge. Further work has to be developed in order to get a more detailed description of these different effects and to understand the origin of this second electroluminescence regime.

4. CONCLUSION.

In this paper, we have shown that the electrochemical treatment of porous layers was a process quite easy to achieve to realize a well controlled partial oxidation at room temperature of porous silicon.Its characteristics are reasonably well understood by considering the electrical situation of the interface between the material and the electrolyte. For example, when the current density drives into accumulation the pore walls, a uniform in-depth oxidation level is obtained. Anodic oxidation strongly improves the emission efficiency of porous films, mostly by reducing the non-radiative leaks, and then can be considered as an efficient post-treatment to control the luminescent properties of porous silicon. It also has largely contributed to increase the interest for the material by showing that electroluminescence with a very attractive emission efficiency can be obtained . Unfortunately, the emission obtained under anodic polarization is not permanent and cannot lead to practical applications. However, it has contributed to provide a better understanding of the charge exchange mechanisms which allow to observe the electroluminescence signal, and opened the door to further research in order to obtain a long-term electrically induced light emission.

Acknowledgments.

The author is very grateful to his colleagues from the Laboratoire de Spectrométrie Physique who have largely contributed to the results presented in this paper. Many thanks to S. Billat, A. Bsiesy, F. Gaspard, M. Ligeon, I. Mihalcescu, F. Muller, R. Romestain and J.C. Vial.

References

1. L. T. Canham, Appl. Phys. Lett. **57**, 1046 (1990).
2. V. Lehmann, H. Föll, J. Electrochem. Soc. **137**, 653 (1990).
3. A. Bsiesy, J. C. Vial, F. Gaspard, R. Hérino, M. Ligeon, F. Muller,
   R. Romestain, A. Wasiela, A. Halimaoui, G. Bomchil, Surf. Sci. **254**, 195 (1991) .
4. A. Venkateswara Rao, F. Ozanam, J. N. CHazalviel, J. Electrochem. Soc. **138**, 153 (1991).
5. C. Tsai, K. H. Li, J. Sarathy, S. Shih, J. C. Campbell, B. K. Hance, J. M. White, Appl. Phys. Lett. **59**, 2814 (1991)
6. M. Morita, T. Ohmi, E. Hasegawa, M. Kawakami, K. Suma, Appl. Phys. Lett. **55**, 562 (1989).

7. A. Halimaoui, Thèse de Doctorat d'Etat, Grenoble University, 1991.
8. A. Bsiesy, F. Gaspard, R. Hérino, M. Ligeon, F. Muller, J.C. Oberlin, J. Electrochem. Soc, **138**, 3450 (1991).
9. M. Ligeon, F. Muller, R. Hérino, F. Gaspard, J.C. Vial, R. Romestain, S. Billat, A. Bsiesy, J. Electrochem. Soc. **74**, 1265 (1993)
10. W. Waring and E.A. Benjamini, J. Electrochem. Soc. **111**, 1256 (1964)
11. J.N. Chazalviel, F. Ozanam in Microcrystalline Semiconductors : Material Science & Devices, edited by P.M. Fauchet, C.C. Tsai, L.T. Canham, I. Shimizu, Y. Aoyagi (Mater. Res. Soc. Proc. **283**, Boston, USA, 1992) pp.359-364
12. F. Muller, R. Hérino, M. Ligeon, F. Gaspard, R. Romestain, J.C. Vial, A. Bsiesy, J. Lumin. **57**, 283 (1993)
13. J. C. Vial, A. Bsiesy, F. Gaspard, R. Hérino, M. Ligeon, F. Muller, R. Romestain, Mac Farlane, Phys. Rev. B **45**, 14171 (1992)
14. R. Hérino, G. Bomchil, K. Barla, C. Bertrand, J.L. Ginoux, J. Electrochem. Soc. **134**, 1994 (1987)
15. H. Munder, C. Andrzejak, M. G. Berger, U. Klemradt, H. Luth, R. Hérino, M. Ligeon, Thin Sol. Films **221,** 27 (1992)

Mechanism for light emission from nanoscale silicon

M. S. Hybertsen

AT&T Bell Laboratories, 600 Mountain Avenue,
Murray Hill, New Jersey 07974, U.S.A.

1. INTRODUCTION

Silicon is the basis for the majority of integrated electronic devices. However, due to the indirect band gap in its electronic structure, bulk Si exhibits very weak luminescence. Therefore Si has not been a useful material for the manufacture of active optical devices, e.g. light emitting diodes or laser diodes. Over the past ten years, there has been a rising level of research work focused on the goal of improving the efficiency of light emission from Si based materials through various schemes to produce artificial structures. Examples include special dopants (e.g. Erbium [1]), planar heterostructures (e.g. based on epitaxial growth of Ge containing layers [2]) and most recently etching of Si wafers to produce luminescent porous Si [3]. The common theme is the fundamental alteration of the electronic states by a relatively short range perturbation i.e. the dopant potential, the heterointerface potential or the boundary of a nanoscale crystallite. The bright luminescence from appropriately prepared porous Si has stimulated a burst of activity recently [4]. In particular, it has refocussed attention on fundamental questions concerning light emission from nanoscale Si structures.

As prepared, porous Si is a heterogeneous materials system. Even in its idealized form, a network of crystalline Si filaments with hydride or oxide surface passivation [5], the possibility remains that the luminescence may derive from the large internal surface, rather than from the nanoscale Si structure itself. In addition, other Si based materials such as poly-silanes or siloxenes may be produced which exhibit strong intrinsic luminescence [6]. Thus, chemical identification of the luminescing species in porous Si samples has been quite controversial, with many competing claims [4]. In parallel to the wide interest in porous Si, schemes for direct chemical synthesis of Si nanocrystallites with passivated surfaces have been developed [7]. Studies of these samples show that the optical properties of Si nanocrystallites are very similar to those of luminescent porous Si samples. Both show a strong "red" luminescence band with the energy of the peak luminescence correlated to a measure of the Si feature size. The luminescence is substantially blue shifted (by 0.3 - 1.0 eV) from the band gap of bulk Si (1.1 eV). Thus, it is natural to start from a nanocrystallite model where the electronic states responsible for luminescence are confined by the crystallite boundaries.

A chemically realistic description of the surface will play a crucial role in understanding the details of the optical properties of Si nanocrystallites [8]. However, an understanding of

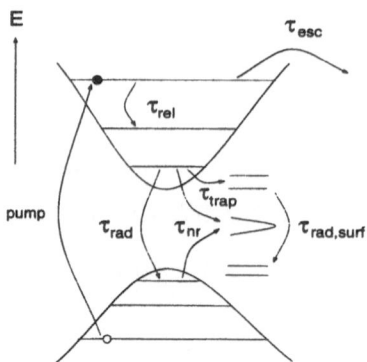

Figure 1. Schematic indication of electrons and holes confined in a crystallite by potential wells giving the discrete levels indicated. Possible states localized in the surface region for electrons and holes are indicated to the right. Dangling bonds can lead to levels deep in the gap. Various possible processes for electrons and holes are sketched.

the electronic states confined to the body of a Si nanocrystallite, and their associated radiative properties, is fundamental to a more realistic picture. The main goal of the present work is to develop the predictions of the nanocrystallite model. In particular, it is important to address experimentally accessible features. After detailed comparison to experimental results, the role of the realistic chemical passivation can be revisited.

For a finite crystallite, the electronic levels will form a discrete set, as in a molecule. The quantization due to the boundaries of the crystallite breaks up the continua of states of the extended crystal lattice. This may be visualized in terms of a confining potential for the electron states and the hole states as sketched in Fig. 1. These potential wells yield discrete levels. Optical transitions will occur between these levels. The lowest corresponding energy is blue shifted with respect to the bulk band gap (represented by the separation of the minima of the wells) as mentioned above. The first theoretical problem is to calculate the electronic levels for Si nanocrystallites. A variety of schemes have been applied. The situation will be briefly reviewed in Sect. 2.

The most widely applied experimental probe of porous Si is photoluminescence (PL). However, from a theoretical perspective, the PL measurement involves a potentially complicated sequence of physical processes. Some of the relevant points are illustrated in Fig. 1. The incident pump beam is absorbed creating an electron and an hole, generally in excited states. Subsequently, these particles can undergo a variety of processes. If the nanocrystallites are connected, or in a medium, an electron can escape, i.e. by tunneling or thermionic emission. The excited particles decay to lower states by scattering from lattice vibrations. The electron-hole pair could radiatively recombine from excited states. If the electron and hole reach their respective lowest states in the body of the nanocrystallite, several further processes are possible. Referring to the surface coating, it is possible that states are available which are further localized in the region of the surface layer, and scattering into these trap states can occur. There may also be deep trap states associated to unpassivated dangling bonds. These in particular can mediate non-radiative recombination of the electron-hole pair. Finally, the electron-hole pair could radiatively recombine to generate the photon which is detected. This radiative process could also involve the surface related states.

It must be stressed that "efficient" luminescence characterizes this competition between radiative and non-radiative processes. The efficiency may be high either because the radiative process is relatively fast or because the non-radiative processes are quenched (slow), i.e. due to effective passivation. Another non-radiative channel is through auger recombination which becomes important at sufficiently high density of electrons and holes for extended systems. For localized centers, an auger process can still be important, for example for neutral donor or acceptor bound excitons where three carriers are confined to the same region. An important feature of nanocrystallites is that electron-hole pairs are generated physically isolated from each other which nominally eliminates the auger channel [6].

Each of the processes relevant to photoluminescence is an interesting subject for theoretical investigation. Some ideas have already been developed. For example, the radiative efficiency of porous Si has been discussed in terms of the escape mechanism [9]. The size dependence of non-radiative recombination at dangling bonds has been studied [10,11]. Although not specifically discussed for Si based nanostructures, the relaxation rate can be significantly reduced for low dimensional structures by a "bottleneck" in the available vibrational states [12]. In the present paper, only the radiative recombination will be studied in detail. It will be assumed that the relaxation processes are sufficiently fast that only radiative recombination from the energetically lowest manifold of states is relevant. Non-radiative processes will be incorporated phenomenologically as an aggregate parallel channel, when required in the discussion.

The main theoretical issue discussed here is the mechanism for radiative recombination. In bulk Si, phonon assisted optical transitions are the primary mechanism for absorption, and this channel will also be present for nanocrystallites. In addition, a zero-phonon channel will open up because the necessary momentum can be supplied by scattering from the boundaries. Evidence for this channel is seen in bulk Si for defect bound excitons [13]. In porous Si, photoluminescence following resonant excitation with photon energies in the luminescence band shows structure clearly related to the crystalline Si phonon energies [14,15]. The key theoretical issue is to understand the competition between the zero phonon and phonon assisted radiative channels, and to assess the overall oscillator strength for radiative transitions in Si nanostructures. The radiative time scale competes with the non-radiative channels illustrated in Fig. 1. This competition determines the luminescence efficiency. It is also a fundamental input for the dynamical behavior of an active optical device. In Sect. 3, a unified model incorporating both radiative channels [16] is reviewed.

In Sect. 4, the present model for the radiative processes in Si nanocrystallites is used to make contact with several experimental results for porous Si. From the theoretical perspective, the optical response is described by the imaginary part of the dielectric function. The absorption of porous Si is studied taking account of the heterogeneous sample through an effective medium approach. The radiative emission time is calculated in detail. The τ_{rad} is found to be strongly size dependent. Furthermore, there is a cross over between the phonon assisted channel, dominant for larger sizes, and the zero-phonon channel, which is only important for the smallest sizes i.e. below 15-20 Å. Comparison to experiments suggests that there is indeed an important active role for the surface layer. The consequences of surface trapping are qualitatively discussed. Finally, recent measurements characterize the Si crystallite size as a function of luminescence energy suggesting that the crystallites are much

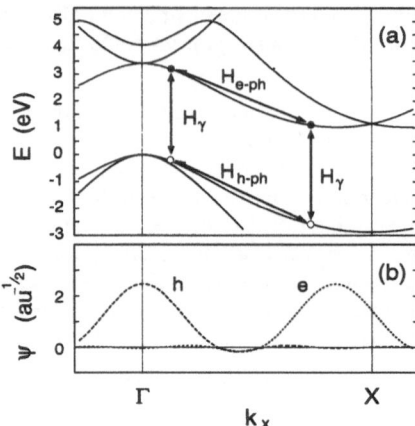

Figure 2. (a) The Si band structure near the gap region along the (100) direction is plotted. The processes for phonon assisted optical transitions are illustrated where H_γ indicates interaction of the light field with the electrons and H_{e-ph} represents electron-phonon scattering. (b) The k_x dependence of the envelope function for an electron and a hole confined to $L_x = 25$ Å is plotted.

smaller than expected from the blue shift [17]. Possible explanations are discussed, but this remains an open problem.

2. ELECTRONIC STATES

The bulk band structure of Si is illustrated in Fig. 2a for states near the fundamental gap, along the k_x direction in momentum space [18]. The highest occupied states occur near $\mathbf{k} = 0$ ("Γ"). This three-fold band complex is in fact split into the light and heavy hole bands, and the split off hole bands by the spin-orbit coupling. However, the spin-orbit splitting is small in Si ($\Delta_0 = 0.046$ eV), and will not be included here. The p-like symmetry of the hole states is an extra degree of freedom in the hole states. Generally this feature complicates detailed calculations of the electronic properties of III-V based heterostructures, e.g. quantum wells. Similar considerations apply here. The lowest empty states occur for large, finite $\mathbf{k}$ near the Brillouin zone edges along the $\pm k_x$, $\pm k_y$, and $\pm k_z$ directions. These six degenerate conduction band valleys have ellipsoidal equal energy surfaces with a relatively heavy mass along the axis and a light mass perpendicular. The valley degeneracy adds an extra degree of complexity to the electron states in nanocrystallites not normally found in III-V or II-VI based crystallites.

The crudest approximation for the electronic states in a nanocrystallite is to treat the surface as an infinite potential barrier and incorporate the modified dispersion through an effective mass for the electron and for the hole. For a spherical crystallite, this predicts that the minimum electron-hole energy is blue shifted according to

$$E_{eh} = E_g + \frac{\hbar^2}{2m_e} \left[\frac{\pi}{R} \right]^2 + \frac{\hbar^2}{2m_h} \left[\frac{\pi}{R} \right]^2 \tag{1}$$

where R is the sphere radius. For a cube, a similar expression holds with R replaced by L, the dimension of the cube side, and a factor of three to account for the confinement in the x, y and z directions. This illustrates that the energy levels are shape dependent for crystallites of similar volume. Detailed calculations for different Si nanocrystallite geometries have been done using this model, including the magnitude of the zero phonon oscillator strength [19,20]. In practice, the strict requirement that the wavefunctions vanish at the edge of the Si region is too severe. The surface passivation of the crystallites involves material with a large, but finite, band gap (i.e. SiO_2) which can be modeled by a finite barrier. This yields a size dependence which is substantially weaker. This type of effective mass approximation is justified starting from crystallites of larger size, and may be problematic for sufficiently small crystallites.

For small crystallites, atomic scale calculations of the electronic levels have been done with varying degrees of approximation [11,21-31]. Different shaped structures have been considered including wires (infinite in one dimension), various finite clusters with up to a thousand Si atoms and two dimensional sheet structures related to silanes and siloxenes. The approximate Hamiltonians used include tight-binding models [11,21,22], empirical pseudopotentials [31] and fully selfconsistent local density functional calculations [23-30]. Each approach has its merits and weakness. One technical point is worth noting: an atomic scale calculation depends specifically on the atomic geometry i.e. bond lengths, bond angles and the choice of surface passivation. This gives chemically specific results. In principle, the atomic geometry should be fixed by minimizing the calculated total energy, which was done in some cases. However, in practice, the cluster geometry in a number of studies was fixed by empirical considerations, which may influence some of the details of the electronic levels. These calculations also test the strength of the zero-phonon radiative channel in a relatively unbiased way.

Figure 3 illustrates an atomic scale calculation for a very small diameter (≈ 8 Å) Si wire with hydrogen passivation of the surface done using the local density functional approach [28]. The geometry of the wire is illustrated in projection in Fig. 3a. The atomic positions were deduced by minimizing the total energy, although the bond lengths and angles do not deviate very much from those in the ideal diamond lattice. The electronic states of the wire, near the region of the fundamental band gap, are shown in Fig. 3b as a function of **k** along the wire. There are several features to observe. First, the band gap is substantially larger than bulk Si, about 3.3 eV for this wire. Other calculations for various wires and clusters show the blue shift to be strongly size dependent. Second, the lowest empty states, and highest occupied states have predominately Si character, having most of their weight in the core of the Si wire, not in the surface region. Third, the six-fold conduction band edge valleys for bulk Si are split. The first four conduction bands derive from the valleys oriented along [100] and [010]. The valleys along [001] are higher in energy. This follows because these valleys present only the light, transverse dispersion to the boundaries. Hence, the kinetic energy of quantization is larger. The valley degeneracy is further split according to the lower (D_{2d}) symmetry of the wire, the interaction fundamentally deriving from the surface region. The valley splitting generally varies with shape and size of the wire or cluster. Fourth, the degenerate hole states are also substantially split. Finally, the lowest states are coupled by a non-zero dipole, as illustrated in Fig. 3c. The magnitude of the coupling is very sensitive to size, as well as the symmetry of the states. These general features are found in all the atomic

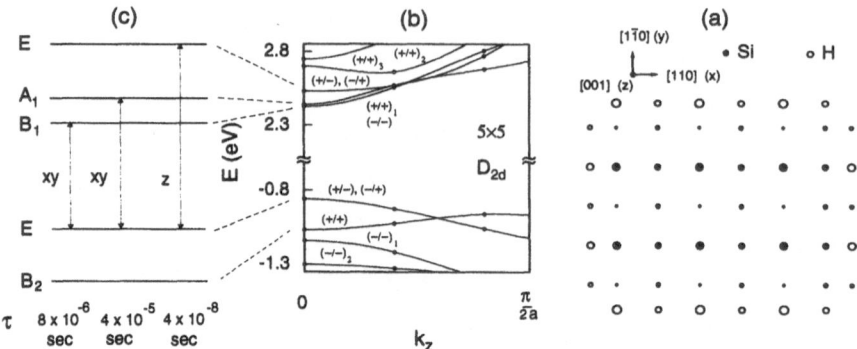

Figure 3.   (a) The cross section of an idealized Si quantum wire with H passivation of the surface, denoted 6×5 according to the arrangement of Si atoms in this projection. Four monolayers are projected along the [001] axis with the sizes of the circles indicative of relative height in the unit cell. (b) The energy bands for a 5×5 wire near the region of the fundamental gap, plotted as a function of **k** along the wire axis. (c) The first few states at $k_z = 0$ abstracted from the band structure as indicated, and labeled according to the square symmetry. Some allowed zero phonon radiative transitions are indicated with their polarization and calculated lifetimes.

scale calculations for wires and clusters. In the limit of Si based chains and sheets with attached ligands, the lowest optical transition can have a strong dipole [29,30].

In order to better understand the observed features in porous Si, a realistic range of nanocrystallite geometries and sizes must be studied. Survey studies of the electronic states, and zero-phonon transitions, have now been done with several methods [11,19,20,26,31]. Since incorporation of the phonon assisted transitions is important, a simplified envelope function approach has been developed. The electron-phonon interaction can be treated with atomic scale models, but this has not yet been applied to Si crystallites.

The electronic states are calculated separately for electrons and holes in effective one-band models. The conduction band valleys are described by a longitudinal mass $(0.92m_e)$ and a transverse mass $(0.19m_e)$ [32]. The hole manifold is replaced by a three-fold degenerate effective band with mass $0.54m_e$ [32]. In addition, analysis of the bulk Si density of states shows that non-parabolicity is significant over the energy relevant for porous Si [33]. This is incorporated via an energy dependent mass. The surface is treated as a finite potential barrier with offset 3 eV, roughly as found for SiO_2. Finally, to further simplify some of the numerical calculations, the crystallites are assumed to have (100) symmetry faces. This is crude, but should capture the main features of the problem. Then the wavefunctions all factor and can be indexed by a set of three integers (n_x, n_y, n_z). In addition, there is a valley index for the electrons. In Fig. 4, the calculated blue shift in this model is shown as a function of crystallite size for a sequence of cubic crystallites. For comparison, the fitted size dependence from a set of tightbinding calculations for passivated clusters is also shown [11]. The blue shift is underestimated in the present calculations. This probably traces to the crude treatment of the hole manifold. More detailed calculations give a larger kinetic energy of confinement for holes [34]. However, the scaling of the blue shift with crystallite size is very similar to the tight-binding result, going roughly as $1/L^\alpha$ with α in the range 1.2-1.8, although the log-

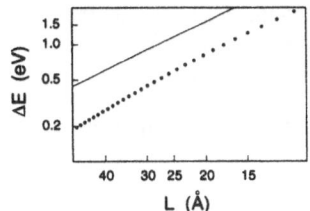

Figure 4. The blue shift for a series of cubic Si nanocrystallites versus size based on the present envelope function approach (dots). For comparison, the fit to tight binding results for Si clusters [11] is also shown (line).

log plot in Fig. 4 shows that the blue shift is not strictly a power law. Assertions in the literature that the "effective mass model" gives a scaling of $1/L^2$ apply only to the simplistic model of an infinite barrier potential, as is well known from studies of quantum wells in III-V materials.

For the electron-hole pairs which recombine, there is additionally a Coulomb interaction. In a crystallite with all dimensions comparable to, or smaller than, the bulk exciton diameter (estimated to be ≈60 Å), this predominately leads to an energetic shift. The enhancement of oscillator strength which is familiar from two and one dimensional structures is not too important for crystallites. That is because the exciton wavefunctions in the nanocrystallite are dominated by the kinetic energy. Relatively little additional electron-hole overlap is induced by the Coulomb attraction [35]. In the present work, the shift has been estimated using a simple screened Coulomb interaction to be $E_{eh} \approx -3e^2/\varepsilon_0 L$ for a cube of size L. Reduction of screening due to the finite size of the crystallite may also play a role. For molecular systems, such as polysilanes, there is evidence for large (of order 1 eV) exciton shifts [36,37]. The case of crystallites is not yet fully studied. Some estimates are of order a few tenths of eV [38], although larger shifts may be possible.

Another interesting aspect of the exciton manifold concerns the low energy splitting. Take Fig. 3 for example: the lowest exciton manifold is made up of the E symmetry hole states and the E, A_1 and B_1 electron states. Including spin degrees of freedom, this results in 32 exciton states. These are split by valley orbit interactions (included in Fig. 3) and the spin-orbit splitting (not included). In addition, there is an electron-hole exchange interaction, which is usually small in the bulk, but can be enhanced in nanocrystallites [14]. These interactions must be studied together to fully understand the lowest manifold of excitons. In practice, for asymmetric crystallites, the orbital momenta (valley for electrons, p-like for holes) are quenched and a four-fold exciton manifold results. However, quantitative results still require a detailed study, just now being considered on the basis of a tight-binding approach [38].

3. MECHANISM FOR RADIATIVE TRANSITIONS

In bulk Si, absorption of light with energy near the fundamental gap proceeds with the assistance of phonons. The process is illustrated in Fig. 2a. For the perfect lattice, the first non-zero coupling to photons is found in second order, with the needed momentum supplied through the electron-phonon interaction. For example, the photon induces a virtual, vertical transition near $\mathbf{k}=0$ which is followed by an electron-phonon scattering process to yield the final state consisting of a hole near the zone center, an electron in the conduction band valley at large k_x, and a change in the phonon population. The similar process involving hole-phonon scattering must be included coherently with the resultant second order matrix element being

$$p_{cv} R_\lambda / N_a^{1/2} = \sum_i \frac{\langle vk|\hat{e}\cdot\mathbf{p}|ik\rangle\langle ik|H_{e-ph}^\lambda|ck+q\rangle}{E_{ck+q} - E_{ik} - \hbar\omega_{\lambda q}}$$

$$+ \sum_j \frac{\langle vk|H_{h-ph}^\lambda|jk+q\rangle\langle jk+q|\hat{e}\cdot\mathbf{p}|ck+q\rangle}{E_{vk} - E_{jk+q} - \hbar\omega_{\lambda q}}. \tag{2}$$

Here the interaction with the light field is proportional to $\hat{e}\cdot\mathbf{p}$ and the valence state $|vk\rangle$ is near the zone center and the conduction state $|ck+q\rangle$ is near one of the conduction band minima at $\mathbf{k}_0$. The matrix element is written for the case of phonon emission. The electron-phonon interaction includes an explicit dependence on the sample size, indicated here by the number of atoms, N_a. The complementary phonon absorption processes have the same form, with reversed sign for phonon momentum and energy. The intermediate states must be summed. Then the square matrix element is averaged over the valence band degeneracy and the conduction band valley degeneracy. Each phonon polarization gives a distinct channel for absorption or emission with the transverse optic (TO) modes dominant and smaller contributions from the longitudinal optic (LO) and transverse acoustic (TA) modes. The relevant mode energies and corresponding relative intensities are well known in bulk Si [39]: 57.5 meV (1.00), 55.3 meV (0.15), and 18.2 meV (0.03).

The fact that the absorption process is second order in bulk Si leads to the well known result that, near threshold, the absorption rises proportional to $(E - E_g - \hbar\omega_{TO})^2$. Conservation of momentum is taken up through the phonon degree of freedom, leaving the convolution of the band edge densities of states. This is to be contrasted with the direct gap case, where absorption is proportional to $(E - E_g)^{1/2}$. There is a parallel difference with regard to radiative recombination in a sample with n_{min} minority and n_{maj} majority carriers introduced through optical pumping or background doping. In the familiar direct gap case, at low temperature, the rate of generation of photons per unit volume per unit time is given by

$$W = \frac{n_{min}}{\tau^{(1)}} \qquad \text{(direct)} \tag{3a}$$

where the radiative recombination time is denoted here by $\tau^{(1)}$ since it derives from a dipole allowed transition, a first order matrix element. The minority carrier population controls the rate due to momentum conservation in vertical transitions. The phonon assisted case can be put in a similar form

$$W = \Omega_c\, n_{maj}\, \frac{n_{min}}{\tau^{(2)}} \qquad \text{(phonon assisted)} \tag{3b}$$

where Ω_c is the unit cell volume in the lattice. The radiative time $\tau^{(2)}$ which enters comes directly from $p_{cv} R_{TO}$ in Eq. (2). The product of majority and minority populations enters, exactly in parallel with the convolution of the densities of states in the absorption cross section. The difference from the usual direct gap case comes in part from a somewhat slower time constant (by about an order of magnitude). But more importantly, the prefactor in Eq. (3b) is the average majority carrier density per unit cell, a number typically less than 10^{-4}.

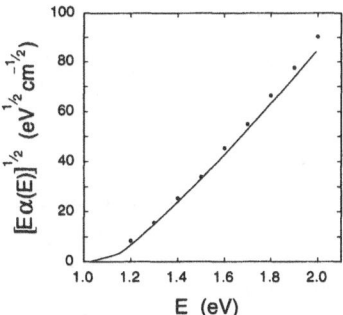

Figure 5. The absorption cross section of bulk Si as a function of photon energy, calculated (solid line) and measured (dots [40]) at 300 K. The calculation includes the LO and TO phonon assisted transitions.

The measured absorption for bulk Si [40] gives a value for the needed matrix elements. Using the effective masses and non-parabolicities described in Sect. 2, the dominant TO matrix element is determined to be $R_{TO} = 0.24$, with the other channels having the relative strengths noted above [41]. The agreement with experiment over a wide energy range is illustrated in Fig. 5. Note that beyond about 2 eV, new transitions are possible in bulk Si involving the conduction band minima near the L point in the Brillouin zone.

The radiative time for electron-hole pairs in bulk Si, $\tau^{(2)}$, is estimated to be about 40 nsec. As noted in Eq. (3b), the radiative rate depends on the density of carriers in the crystal. For the special case of excitonic recombination at low temperature, a radiative lifetime can be estimated from the absorption measurements and detailed balance arguments [42]. The result, 60 μsec, corresponds to the lifetime of an isolate exciton radiatively decaying with the assistance of a TO phonon. This is based on the assumption that the exciton density of states is the same as for the electrons. A more realistic estimate yields 120 μsec, quite similar. In terms of Eq. (3b), the binding of the electron and hole define a region of overlap corresponding to the size of the exciton. Then the effective first order decay kinetics are governed by $\Omega_c n_{ex}/\Omega_{ex} \tau^{(2)}$. Using the estimates for the exciton radius (30 Å) and $\tau^{(2)}$, an effective lifetime of 150 μsec results, in rough agreement with the analysis in Ref. 42.

In a crystallite (or near a lattice defect), the situation changes. The electron and hole are confined to a localized region. This implies that they are delocalized correspondingly in momentum space. The simplist model of localization in a box of dimension $L_x = 25$ Å along the [100] direction is illustrated in Fig. 2b. This has two consequences for the radiative transitions. First, the electron and hole now have non-zero amplitude with equal momenta, where vertical transitions are allowed. Therefore, zero-phonon transitions become allowed, with a matrix element given be the overlap of the envelope functions in momentum space. The simplest physical picture is that scattering from the boundaries of the crystallite supplies the needed momentum. As one might expect, this is strongly size dependent, with zero phonon transitions becoming stronger with decreasing crystallite size. Second, the phonon assisted transitions are modified. They are also size dependent. But, in addition, the phonon momentum is not strictly determined: there can be a combination of momentum transfer from the boundaries and from the phonon. All possible combinations must be summed. To the extent that the phonons are dispersive, this will intrinsically broaden the transition.

This qualitative picture can be analyzed more quantitatively using an envelope function approach [16,33]. The electronic states in the crystallite are expanded in crystal basis, e.g. for the electron:

$$\phi_e(\mathbf{r}) = \int d\mathbf{k}\,\psi_e(\mathbf{k}-\mathbf{k}_0)\,\phi'_{c\mathbf{k}}(\mathbf{r}) \tag{4}$$

where the Fourier transform of the envelope function ψ_e is developed around the band edge position $\mathbf{k}_0$. In order to include the electron-phonon interaction, without introducing unphysical large momentum transfer scattering between envelope function states, the influence of the phonons is incorporated in the crystal basis set. The crystal basis function $\phi'_{c\mathbf{k}}$ is given by the usual Bloch function plus those coupled in through the electron-phonon interaction

$$\phi'_{c\mathbf{k}} = \phi_{c\mathbf{k}} + \sum_{\lambda\mathbf{q}} \frac{\phi_{c\mathbf{k}-\mathbf{q}}\langle c\mathbf{k}-\mathbf{q}|H^\lambda_{e-ph}|c\mathbf{k}\rangle(1+n_{\lambda\mathbf{q}})^{\frac{1}{2}}}{E_{c\mathbf{k}} - E_{c\mathbf{k}-\mathbf{q}} - \hbar\omega_{\lambda\mathbf{q}}} + [\text{phonon absorption}]. \tag{5}$$

The phonon degrees of freedom are described by polarization λ, wavevector $\mathbf{q}$, occupation number $n_{\lambda\mathbf{q}}$ and energy $\hbar\omega_{\lambda\mathbf{q}}$. The influence of confinement on the phonon modes has not been explicitly considered. In order to do so, one would need to additionally incorporate an envelope function for the phonons in the crystallite. This complication will not be included because at the end, the phonon degrees of freedom are going to be summed out explicitly. A similar form for the hole wavefunction is used.

The matrix element for interaction with the light field contains several parts. Consider first the zero phonon contribution:

$$\langle h|\hat{e}\cdot\mathbf{p}|e\rangle = \int d\mathbf{k}d\mathbf{k}'\,\psi_h^*(\mathbf{k})\,\psi_e(\mathbf{k}'-\mathbf{k}_0)\langle v\mathbf{k}|\hat{e}\cdot\mathbf{p}|c\mathbf{k}'\rangle \tag{6}$$

This come from the first term in Eq. (5) and the corresponding term for the hole. The information on the dipole coupling is carried by the full bloch states of the crystal lattice. The usual selection rule of vertical transitions requires $\mathbf{k}' = \mathbf{k}$, giving for each term a factor $p_{cv}(\mathbf{k})$ which is fully allowed. The result is an amplitude for zero phonon transitions proportional to the convolution of the envelope functions, corresponding to the illustration in Fig. 2. The terms involving the electron-phonon interaction have a very similar form. They can be approximately written in terms of the bulk matrix element of Eq. (2) and a convolution of envelope functions evaluated at $\mathbf{q}-\mathbf{k}_0$ where $\mathbf{q}$ is the phonon momentum. This explicitly shows the trade-off between phonon momentum and scattering from the boundary.

Once the matrix elements have been evaluated, they are organized according to the final states and inserted into a conventional expression for the optical response of the crystallites. The result for the imaginary part of the dielectric function of an array of N_c crystallites per unit volume is

$$\epsilon''(\omega) = \frac{8\pi^2 e^2}{\omega^2 m_0^2} N_c \left[\sum_{eh} |p^{(1)}_{eh}|^2 \delta(E_g + E_e + E_h - \hbar\omega) \right.$$
$$+ \sum_{eh,\lambda\mathbf{q}} |p^{(2a)}_{eh,\lambda\mathbf{q}}|^2 n_{\lambda\mathbf{q}}\,\delta(E_g + E_e + E_h - \hbar\omega_{\lambda\mathbf{q}} - \hbar\omega)$$
$$\left. + \sum_{eh,\lambda\mathbf{q}} |p^{(2e)}_{eh,\lambda\mathbf{q}}|^2 (1+n_{\lambda\mathbf{q}})\,\delta(E_g + E_e + E_h + \hbar\omega_{\lambda\mathbf{q}} - \hbar\omega) \right] \tag{7}$$

where the quantized electron and hole energies $E_{e,h}$ enter together with the bulk band gap E_g.

The labels e (h) indicate the electron (hole) levels in the crystallite. The terms correspond to the zero phonon, phonon absorption and phonon emission assisted transitions respectively. Photon polarization is not indicated explicitly. An average matrix element appropriate to unpolarized incident light is implicit.

The matrix elements can be further simplified under some standard assumptions. The dipole matrix elements $p_{cv}(\mathbf{k})$ are assumed to be independent of $\mathbf{k}$. The electron-phonon matrix elements and the energy denominators are assumed to be independent of both $\mathbf{k}$ and $\mathbf{q}$. Neither approximation is strictly correct. The quantitative impact depends on the spread in momentum induced by the finite size. These approximations will tend to break down as the length scale drops below a few unit cells, at which point the whole envelope function approximation is also questionable. Then the zero phonon matrix element is given by

$$p_{eh}^{(1)} \approx p_{cv} M_{eh}(\mathbf{k}_0), \tag{8}$$

where M_{eh} is the convolution of envelope functions introduced in Eq. (6). The phonon assisted processes have amplitude

$$p_{eh,\lambda \mathbf{q}}^{(2e)} \approx p_{eh,\lambda \mathbf{q}}^{(2a)} \approx p_{cv} R_\lambda M_{eh}(\mathbf{q}-\mathbf{k}_0)/N_a^{1/2}. \tag{9}$$

The number of atoms per crystallite N_a enters explicitly from the electron-phonon interaction. The effective bulk electron-phonon coupling from Eq. (2), R_λ, contributes for each phonon branch. The convolution of envelope functions M_{eh} is equivalent to

$$M_{eh}(\mathbf{q}) = \int d\mathbf{r}\, \psi_e^*(\mathbf{r})\, \psi_h(\mathbf{r})\, e^{-i\mathbf{q}\cdot\mathbf{r}} \tag{10}$$

Since the important $\mathbf{q}$ are from a restricted region, the phonon frequencies can be taken to be approximately constant. Then the explicit phonon wavevector sum in Eq. (7) can be done analytically

$$\sum_{\mathbf{q}} |M_{eh}(\mathbf{q})|^2 = \frac{\Omega}{(2\pi)^3} \int d\mathbf{r}\, |\psi_e(\mathbf{r})|^2\, |\psi_h(\mathbf{r})|^2 \tag{11}$$

with Ω the crystallite volume.

At this point, a theory incorporating both the zero phonon and phonon assisted transitions in Si nanocrystallites has been given which is general within the envelope function approach. The weakest point in the theory is precisely the zero phonon contribution. The amplitude derives from the large $\mathbf{k}$ portion of the envelope function. This may be less reliable since the overlap of electron and hole occurs for values of $\mathbf{k}$ approaching a large fraction of the Brillouin size. As is often the case for the envelope function approach, it is best to justify the results empirically.

The matrix element in Eq. (8) is sensitive to geometry, both overall length scale as well as shape, through the overlap factor $M_{eh}(\mathbf{k}_0)$. It can be easily evaluated for the simplified geometry of (100) symmetry faces and under the assumption of infinite barriers. In practice, finite barriers give a minor quantitative correction. The matrix element for the [100] valleys is determined solely by the crystallite dimension in that direction. Similarly for the other valleys. Then for the [100] valleys

$$|M_{eh}|^2 = \frac{1}{\pi^2 k_0^2} \left[\frac{2\pi}{L_x}\right]^6 \left[k_0^2 - \left[\frac{2\pi}{L_x}\right]^2\right]^{-2} \sin^2(k_0 L_x/2) \tag{12}$$

This is ratio of the zero phonon oscillator strength to that which would be found in a

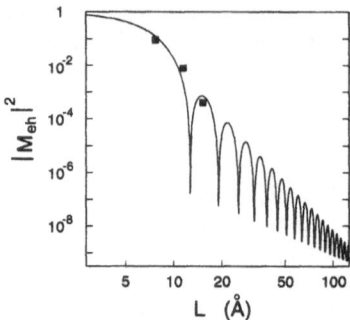

Figure 6. Ratio of the zero-phonon transition intensity to that of a standard dipole allowed transition, calculated from Eq. (12). The solid squares are based on the zero-phonon matrix elements calculated with the local density approach for passivated Si wires of the indicated cross sectional size [25].

conventional bulk semiconductor with an allowed direct transition (i.e. governed by p_{cv}). The result is plotted in Fig. 6 as a function of crystallite dimension. Aside from the oscillatory factor, the overall size dependence is strong: $1/L_x^6$. This specific power law depends on the present choice of shape. For comparison, the results from a first principles calculations of three different size wires are also shown [25]. This is not an entirely representative comparison, since the wire geometry involves (110) surfaces. Nonetheless, the order of magnitude and size sensativity are reproduced. Similarly, strong size and shape dependence for the zero phonon transitions have been calculated for clusters using the tight binding [11,22], the local density [26] and the empirical pseudopotential [31] approaches.

The phonon assisted transitions have a much weaker dependence on crystallite geometry, coming from the integral in Eq. (11). From the explicit dependence on the size of the crystallite in Eq. (9), the intensity for the phonon assisted transitions scale inversely with the crystallite volume, getting stronger for smaller sizes. A simple physical argument for this traces back to the radiative recombination rate in the bulk, Eq. (3b). Consider an intrinsic crystallite which is optically excited with a single electron-hole pair. Then the minority and majority carrier densities in Eq. (3b) become just one per crystallite and one can see that the radiative time effectively scales with crystallite volume due to the prefactor. The general result that the effective electron-phonon coupling in crystallites scales inversely with volume extends to the III-V direct gap case also, where it enters the broadening of the optical response [35].

The present theory for the competition between zero phonon and phonon assisted transitions in Si crystallites is general. In particular, it is clear that the two processes present different size dependencies. Since the phonon assisted channel dominates for large size, this implies that there should be a crossover for some crystallite size. Size is roughly equivalent to blue shift. Beyond some value of the blue shift, the luminescence mechanism crosses over from phonon assisted transitions to be dominated by zero phonon transitions. Calculations to estimate the cross over size will be given below. But in general, at a given luminescence energy, there will be a ratio between the intensities of zero phonon and phonon assisted luminescence lines which is dependent on crystallite size. Furthermore, the radiative recombination time will depend on crystallite size, or blue shift, getting faster for bluer luminescence. This is referred to in luminescence studies as spectral diffusion. The nanocrystallite model predicts this behavior to be intrinsic, as opposed to being related to electron-hole diffusion and relaxation.

Anticipating the discussion below, the present theory also covers the case of traps in the crystallite. The only condition is that the electron or hole wavefunction of the trap still be predominately in regions of tetrahedral Si. Then the envelope function approach still applies. An obvious exception would be a case such as the deep levels associated with a dangling bond defect. The theory also should extend to the case of defect bound excitons in the bulk. Electron-hole correlations need to be taken into account, but the basic approach still applies, giving the ratio of the zero phonon to phonon assisted radiative intensities.

Estimates for the oscillator strength of the various radiative channels will be made using the present simplified effective mass model. For the crystallites with (001) symmetry surfaces, the selection rules are quite simple. If the conduction band valley is along [100] then the transverse quantum numbers must be conserved in the zero phonon transitions: $n_{y,e} = n_{y,h}$ and $n_{z,e} = n_{z,h}$. However, all combinations of $n_{x,e}$ and $n_{x,h}$ are allowed. This contrasts with the usual case for a direct band gap semiconductor, where $n_{x,e} = n_{x,h}$ would be required. The phonon assisted transitions are generally allowed. The integral in Eq. (11) is simple for the case of infinite barriers: each direction contributes a factor $(2 + \delta_{n_{x,e} n_{x,h}})/2$. Finite barriers introduce a minor quantitative correction.

4. DISCUSSION OF THE OPTICAL PROPERTIES OF POROUS SILICON

The intrinsic optical response of Si nanocrystallites is given by the imaginary part of the dielectric function in Eq. (7). This is evaluated for the case of a dense lattice of crystallites by setting $N_c = 1/\Omega$. The result is shown in Fig. 7 for crystallites of dimension near 20 Å. The inset shows the lowest transitions for a non-cubic crystallite. The low symmetry breaks the six fold conduction band valley degeneracy into three doublets. Then each possible radiative channel gives a contribution, as labeled. The phonon assisted lines are shifted up from the zero phonon lines by the respective phonon energies. The main curves show the average over an ensemble where all dimensions vary by about ±1 Å. For this size range, the phonon assisted transitions are still stronger by a factor of three or more. Despite the substantial spread in transition energies, a step with characteristic energy of the TO phonon survives, as indicated by the arrow in Fig. 7. The ratio of phonon assisted to zero phonon contributions is similar to the relative feature sizes observed in resonant photoluminescence near this energy [14].

In order to discuss the optical properties of either porous Si or realistic samples of nanocrystallites, more analysis is required [33]. First, it is assumed that the sample consists of an ensemble of crystallites described by a size distribution. For simplicity, this is taken to have the form

$$P(L) \propto L^2 e^{-(L-L_0)^2/\delta L^2}. \tag{13}$$

The dielectric response will be averaged over the ensemble.

Next, the microscopic dielectric response ε from Eq. (7) must be converted to a macroscopic response ε_m for the sample as a whole. This in general depends on the sample morphology. With the use of a simple effective medium theory [43], the ε_m only depends on

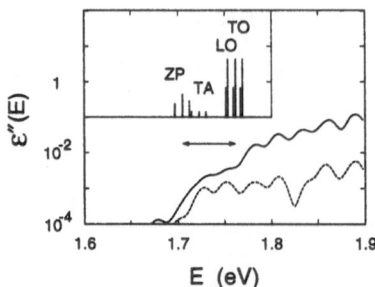

Figure 7. The imaginary part of the dielectric constant (log scale) calculated for ensembles of Si crystallites. The zero phonon contribution is shown separately (dashed line) from the total (solid line). An ensemble of crystallites is averaged with size near 20 Å. A gaussian broadening of 0.015 eV, FWHM, is included. Inset: Near threshold transitions for a 19 Å × 20 Å × 21 Å crystallite labeled by the phonon channel.

the porosity, or relative fraction f of the sample occupied by crystallites. Assuming the sample consists of crystallites plus vacuum, the "self consistent" version of the theory just gives

$$\frac{\varepsilon_m - 1}{3\varepsilon_m} = f\frac{\varepsilon - 1}{2\varepsilon_m + \varepsilon}. \tag{14}$$

This approach is basically valid provided that it is applied in a frequency range where dynamical coupling of the crystallites is unimportant. The local field inside a crystallite differs substantially from the average, macroscopic field. Near threshold, $\varepsilon'' \ll \varepsilon'$ so it is straightforward to correct the absorption and emission. This introduces a local field correction factor for the calculated absorption:

$$\alpha_m(\omega) = \frac{\omega}{n_m c} F(\varepsilon, \varepsilon_m) f \varepsilon''(\omega). \tag{15}$$

The absorption is trivially corrected for the volume fraction f of absorbing material and additionally by the factor $F(\varepsilon, \varepsilon_m)$. In the dilute limit where $f \to 0$, the usual result for a dielectric sphere in vacuum is applicable and

$$F(\varepsilon, \varepsilon_m) = \left[\frac{\varepsilon'_m + 2}{\varepsilon' + 2}\right]^2. \tag{16}$$

Then Eq. (15) reduces to the standard result for small absorbing particles in Mie theory [43]. This limit is applicable to dilute suspensions of Si nanocrystallites generated through direct chemical synthesis. Although porosities vary, luminescent porous Si samples typically have f in the range 0.1 - 0.4. In this case, Eq. (14) must be solved directly for real and imaginary parts to find the function F. The local field inside the crystallites also controls the radiative recombination rate

$$W_{sp}(\omega) = n_m F(\varepsilon, \varepsilon_m) \frac{2}{3} \omega \frac{e^2}{\hbar c} \frac{|p_{eh}^{(j)}|^2}{m_0^2 c^2}. \tag{17}$$

The spontaneous emission rate is indicated for a particular recombination channel, which may or may not involve phonons. This expression includes the average over electron and hole spin, as well as photon polarization. The photon density of states brings in the factor of the macroscopic index of refraction n_m.

The calculated absorption coefficients and radiative recombination times presented here include the local field factor based on $\varepsilon'(\omega)$ for bulk Si and the measured fraction, f. The factor F takes values in the range 0.2-0.3 for f=0.25-0.30 the range relevant for the samples studied in Ref. 33 as well as many other samples. The effective medium approach, used in this way, gives an index of refraction in agreement with the measured value [33].

Direct information on the size distribution is notably lacking, although various techniques can supply some information. A common procedure is to deduce a size distribution from the photoluminescence spectrum. However, it is crucial to incorporate information on the radiative efficiency into the analysis, a step often overlooked. For the present purposes, a simple model is adopted to get the radiative fraction of the ensemble. If there is on average one fast trap per unit volume L_c^3, then the probability that a given crystallite of volume L^3 will have no traps, and hence radiate, is simply given by Poisson statistics:

$$P_{rad}(L) = e^{-L^3/L_c^3}. \tag{18}$$

Then the photoluminescence spectrum is modeled by using the lowest transition energy for a given crystallite and averaging over the ensemble with this radiative probability.

An example of the luminescence and absorption spectrum calculated with this model is shown in Fig. 8. The size distribution and radiative fraction are shown in the inset. The parameters were chosen to represent the measured efficiency and peak luminescence energy for a sample measured in Ref. 33. The average size for this case is about 40 Å, in good agreement with the transmission electron microscopy images for the sample. There are a number of interesting features in Fig. 8. The absorption rises smoothly with energy, showing no sharp onset. This is similar to the behavior in bulk indirect gap semiconductors. It is fully consistent with an individual crystallite having a well defined absorption threshold. The ensemble average introduces a tail at the low energy side corresponding to contributions from a few larger crystallites. There is no shift between the luminescence line and the absorption. At the luminescence energy there is weak absorption. Although implicit in the model, the combined luminescence and absorption measurements in Ref. 33 show the same behavior.

The smooth, slow rise of the absorption has been observed in a number of other studies of porous Si samples [44-47]. In particular, the shift of the absorption spectrum to higher energy correlates to smaller feature sizes in the samples. This is fully consistent with the present model. A common interpretation of the data is to assign the "absorption edge" to the energy where the absorption appears to rise up to significant levels relative to the noise. As illustrated in Fig. 8, this suggests a substantial shift between this apparent absorption edge and the photoluminescence. However, the interpretation is erroneous. In fact there is non-zero absorption at the energy of the luminescence, as demonstrated in the experiments of Ref. 33. The fundamental reason for the slow rise of the absorption is the phonon assisted process for absorbing a photon, which is dominant in this energy range. A similar shape for the absorption spectrum was observed in photoluminescence excitation measurements on chemically prepared Si nanocrystallites [7]. In Ref. 7, this was compared to the absorption of

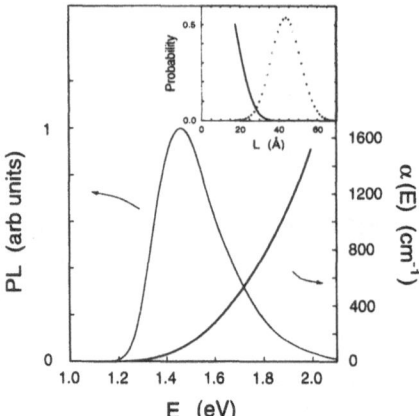

Figure 8.   The calculated absorption cross section for an ensemble of crystallites described by the size distribution shown in the inset. The porosity is assumed to be 70% (f=0.30). The photoluminescence curve is calculated by further assuming a probability of radiative recombination which varies with size as shown by the curve in the inset.

bulk Si with the conclusion that the nanocrystallites behaved like an indirect gap semiconductor. The present theory supports this conclusion, interpreting the term "indirect" more precisely to mean that the zero-phonon optical transitions are unimportant in this size and energy range.

It is interesting to directly compare the model results with experimental data. This is done on a semi-log plot in Fig. 9. Note that the effective absorption presented experimentally corresponds to dividing out f in Eq. (15), which is also done for the calculated absorption. Except for the interference oscillations, an artifact of the thin film sample, the experimental absorption fits very closely to an exponential form $\exp(E/E_0)$ over a wide range of energy with $E_0 \approx 0.3$ eV. The calculated absorption agrees very well in magnitude with the experiment around 2 eV. Since the overall size distribution in Fig. 8 is relatively realistic for this sample, the comparison in Fig. 9 suggests that the calculated oscillator strength is approximately correct. The overall scale of the absorption is set by the average size of the crystallites. However, the low energy tail of the absorption spectrum is strongly influenced by the distribution of crystallite sizes. If the sample contained more large crystallites than the gaussian distribution used in the model, the model would give more low energy absorption.

The exponential shape of the experimental absorption has been observed in other samples as well [8]. The interpretation is a matter of some debate. It is clear that one could reproduce this behavior using the ideal nanocrystallite model with an appropriate distribution. In general, the porous Si sample morphology is not well characterized. Probes which average over a region of the sample include small angle X-ray scattering [48] and Raman scattering [49,50]. Some give evidence for more than one peak in the size distribution below 100 Å [50]. One could argue that this leaves room for a distribution which yields an exponential absorption edge. Alternatively, it is proposed in Ref. 8 that this characteristic feature is more universal, deriving from a wide distribution of localized states at the surfaces of the

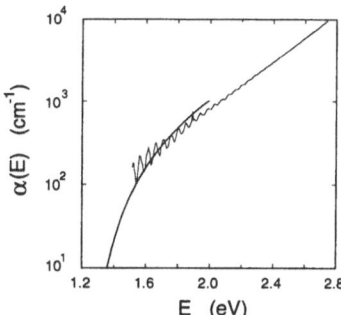

Figure 9.   The calculated absorption cross section from Fig. 8 is replotted (heavy line) in comparison to experimentally deduced absorption data from Ref. 33 (oscillatory, light line). Note the log vertical scale.

crystalline Si skeleton. Experimental evidence for this picture includes the observation of the exponential absorption edge in mesoporous Si, where the characteristic length scale for crystalline Si is closer to a micron.

The radiative emission time is a very interesting signature of the luminescence. This has been calculated for a sample of 70% porosity. The ensemble of crystallites allowed for 20% fluctuations in shape around two geometries: a cube and a "wire" which is represented by an elongated crystallite with aspect ratio near four to one. Then the radiative time is examined as a function of luminescence energy. Since a strong size dependence is anticipated, it is convenient to display this as a function of blue shift of the luminescence. The results are shown in Fig. 10. The zero phonon channel is plotted as a data point for each element of the ensemble. Since the size and shape dependence for the TO phonon assisted channel is much smoother, this is shown by the lines in Fig. 10. The general features discussed at the end of Sect. III are evident: the strong size dependence and the cross over between phonon assisted and zero phonon radiative channels. In the present model, the crossover is estimated to be for a luminescence energy of about 2 eV (blue shift near 1 eV). This corresponds to a size of 15 - 20 Å for the case of a cube. Since Fig. 4 shows that the present effective mass approach yields a blue shift which is too small, one might expect the crossover to occur even further into the blue emission region. From Fig. 10, the crossover energy depends on geometry, so that for a wire-like geometry, it occurs at lower energy and somewhat larger size. The zero phonon channel is basically dominated by the shortest length in the crystallite. Therefore the width of the "wire" determines the radiative time, while the near elimination of the confinement energy along the wire direction reduces the blue shift. The phonon assisted channel, scaling essentially with crystallite volume, is less effected. The net result yields a crossover at smaller blue shift.

The low energy exciton manifold includes states with extremely long decay times. One mechanism for this is the exciton exchange splitting [14,37] discussed briefly in Sect. 2. The present results do not include terms in the Hamiltonian which will split up the low energy manifold in such a way as to lead to "dark" states lower in energy than emitting states, which might then dominate the decay. Put slightly differently, the present results are for the average over the low energy manifold of states. As a practical matter, they apply to the system at a

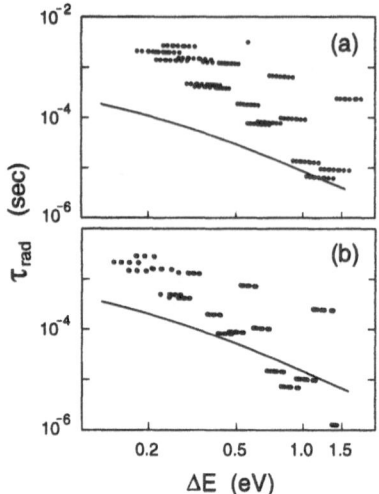

Figure 10. Radiative recombination time as a function of the blue shift of the photon energy from the bulk
Si band edge: zero phonon transitions (dots); TO phonon assisted transitions (line). These scatter plots show
the radiative time for each member of an ensemble uniformly distributed around (a) a cubic geometry and (b)
a wire-like geometry (aspect ratio of four to one along the "wire" axis).

modest temperature. However the ratio of the phonon assisted and zero phonon channels
should be largely unchanged by these perturbations.

It would be very interesting to observe the crossover experimentally. Direct
measurements of the radiative decay time, discussed below, are difficult. However, the
photoluminescence measurements with resonant excitation [14] give results closely related to
the relative oscillator strength of the zero phonon and phonon assisted emission channels.
The sample is pumped with photons tuned to an energy in the luminescence band. Then the
luminescence is observed for energies just below the pump energy. The basic idea is that
only those crystallites in the sample which absorb near their fundamental absorption line will
emit so close in energy. A series of three steps are observed in the luminescence with
characteristic energy separation closely related to the TO phonon mode energy. The situation
is discussed in detail in Ref. 14. A complete analysis using the present model is complicated
by the importance of absorption into low lying excited states of the relevant crystallite. These
derive, for example, from the splitting of the valley degeneracy for low symmetry crystallites.
A reasonable ensemble must be averaged to show the step-like structure. Nonetheless, the
intensity in the first step is related to the zero phonon oscillator strength while the second step
involves the phonon assisted transitions. The experiment has been performed for pump
energies spanning the entire luminescence band. The ratio of the first to second steps various
smoothly across the luminescence band showing that zero phonon transitions become
stronger relative to the phonon assisted transitions as the energy increases. This is fully
consistent with the results in Fig. 10. With the pump energy near 2 eV, the onset of the
second step is hardly resolved, suggesting that this is near the crossover regime.

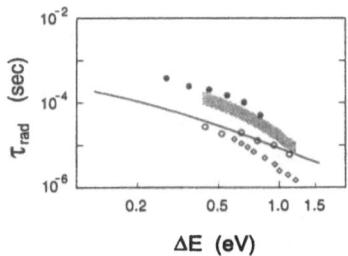

Figure 11. The TO phonon assisted radiative rate from Fig. 10a is replotted in comparison to photoluminescence decay times from various experiments. The data from Ref. 33 taken at room temperature (open circles) and low temperature (filled circles) is for a film of 70% porosity. The range of decay times observed at room temperature for samples of 65% porosity in Ref. 9 is indicated by the gray region. The measured $1/e$ times from Ref. 44 are shown by the open diamonds.

The predicted radiative time scale is in the range of a few to 100 μs over the region of the main luminescence band in porous Si. This is faster than the radiative time estimated for bound exciton decay in bulk Si, as it should be since these crystallites are smaller than the nominal exciton size in bulk Si. The calculated time is about the right order of magnitude in comparison to the measured photoluminescence decay time, as shown in Fig. 11. Several different experimental results are summarized there. In each case, the photoluminescence decay curve is measured for a narrow window of energy to give an energy resolved decay time. The results of Refs. 9 and 33 give an effective time which is an average over the luminescence decay curve. The results shown from Ref. 44 are the $1/e$ time in the decay curve, and are quite similar to those of Ref. 14. In general, the observed luminescence decay is not a simple exponential. However, the chemically prepared nanocrystallites have a more nearly exponential decay [7]. The actual radiative decay time does depend on sample porosity, although the cases represented in Fig. 11 are relatively similar.

The measured times include all the parallel non-radiative decay channels, so care must be taken in the comparison between theory and experiment. This is related to the quantum efficiency of the luminescence. In the standard picture, the total decay rate is the sum of the radiative and non-radiative channels, $W_{rad} + W_{nr}$, while the efficiency is the ratio $W_{rad}/(W_{rad} + W_{nr})$. The non-radiative processes usually depend strongly on temperature, and often are quenched at low temperature. Then the measured decay rate would correspond to the radiative rate at low temperature. This would require the efficiency to approach unity, which is generally not observed in porous Si samples. The chemically prepared nanocrystallites do approach unit efficiency at low temperature [7]. A further complication arises from observations of anomalously slow luminescence decay times at low temperature: the decay time slows down by far more than can be accounted for by changes in efficiency [14,51]. This is also observed for the chemically prepared nanocrystallites. Various explanations for this are possible e.g. exchange splitting of the exciton or trapping of the electron and/or the hole. Taken together, these factors imply that the radiative emission time is hard to accurately characterize experimentally. However, the fact that quantum efficiency for porous Si samples ranges from one to ten percent does imply that the experimental radiative decay times must be one to two orders of magnitude slower than the measured decay rate. The sample measured in Ref. 33 was unusual in that the change in decay rate and efficiency were exactly in parallel as a function of temperature. Therefore, the low temperature decay data plotted in Fig. 11 may be representative of the radiative rate in that sample.

The comparison in Fig. 11 shows that the present theory roughly accounts for the observed decay rates. In particular, the energy dependence of the observed decay rate can be accounted for as an intrinsic property of the nanocrystallite model. However, taking into account the discussion of non-radiative processes and the quantum efficiency, the comparison with experiment suggests that the calculated radiative rate is too fast. If this is correct, it requires explanation in terms of the states responsible for the luminescence. This situation can not be explained by unknown non-radiative processes.

One natural explanation could relate to states localized near the surface of the crystallite by some feature of the chemical passivation layer. The basic point is that such localization of one or both of the electron and hole would force the recombining particles to be separated to some degree in real space. This mechanism was important to understand the radiative rates in CdSe nanocrystallites where the holes were shown to localize in the anion lone pair band at the surface [52]. With reference to Eqs. (10) and (11), it immediately follows that the oscillator strength would be reduced by such separation. For the simplest model of exponential localization on opposite, planar surfaces of a crystallite, the reduction of the zero phonon and phonon assisted intensities would be approximately the same. Then the general picture of a crossover between these channels as a function of blue shift would remain, but the overall radiative time scale would be slower. Comparison between the model calculation and the low temperature decay times measured in Ref. 33 show roughly a factor of five discrepancy. This could be accounted for by relatively shallow traps which would be well within the envelope function approach.

The picture developed in Ref. 8 involves a broader (deeper) distribution of surface related states. A realistic model of passivated crystallites would include not only asymmetric crystallite surfaces, but non-uniform potentials in the interior. For shallow levels, and for small crystallites, the lowest electron-hole pairs could easily be localized to varying extents. Only higher energy states would extend over the entire crystallite. This still fits within the present framework. The exponential absorption and non-exponential luminescence decay would derive from the distribution of this "disorder" with respect to the ideally terminated crystallite. If the important states are localized to the extent of just one to two bond lengths, then an atomic scale model is required fully understand the radiative rates, particularly the lattice vibrations which will be coupled in the radiative process.

A new measurement of the size of the Si nanocrystallites in various porous Si and chemically prepared samples has been made on the basis of near-edge and extended x-ray absorption fine structure [17]. For samples with relatively large crystallites (40 - 60 Å), the correlation to the peak luminescence energy (1.4 - 1.6 eV) fits very well with expectations from the nanocrystallite model. However, for luminescence peak energies approaching 2 eV, the characteristic size found in this measurement is substantially smaller than expected, well below 20 Å. The predicted blue shift in this size range, for example from tight binding calculations [11], is one eV or more too large. Care must be taken when interpreting these results for porous Si samples with a small radiative efficiency. However the results for the porous Si and the chemically prepared Si nanocrystallites (which do exhibit high radiative efficiency at low temperature) are fully consistent. Such a large discrepancy between the luminescence energy for this measured crystallite size and the calculated blue shift can only be resolved by further atomic scale calculations for passivated clusters. There are several possibilities which must be considered, beyond the simple band, or molecular orbital, picture of the excited states of the crystallites. There could be a substantial excitonic shifts for small crystallites or a large change in the coupling to the lattice vibrations in the excited states. Both of these effects can be large in molecules [7]. Alternatively, the emission could be due to relatively deep states related to the surface region, following the line of thought in Ref. 8. Each possibility requires further study.

5. CONCLUDING REMARKS

The optical properties predicted from the intrinsic electronic states of Si nanocrystallites explain a number of the features observed in chemically prepared samples of nanocrystallites, as well as luminescent porous Si. These include the presence of a substantial blue shift relative to bulk Si, the magnitude of absorption and the overall radiative emission time. The model predicts intrinsic spectral diffusion, that is the smooth increase in the radiative rate with increasing luminescence energy. The radiative process is found to be a competition between zero phonon recombination where scattering from the crystallite surface is important and phonon assisted recombination which evolves smoothly from the bulk limit. There is a crossover as a function of crystallite size, or blue shift of the emission energy, with the zero phonon process only becoming important below 15-20 Å. More detailed studies are still required to understand other processes such as relaxation times and non-radiative recombination. A quantitative understanding of the luminescence requires a more detailed treatment of the passivating surface coating. The low energy exciton states must be better characterized in order to understand the temperature dependence of the luminescence, and in particular, to explain the luminescence energy for the smallest crystallites.

Acknowledgements

It is a pleasure to acknowledge extensive discussions with Drs. L.E. Brus, W.L. Wilson and Y.-H. Xie during the course of this work. I have also benefited from discussions with many of the participants in the International School at Les Houches, *Luminescence of Porous Silicon and Silicon Nanostructures*, February, 1994.

References

[1]   Xie Y.H., Fitzgerald E.A. and Mii Y.J., *J. Appl. Phys.* **70** (1991) 3223.

[2]   Bean J.C., *Proc. of IEEE* **80** (1992) 571.

[3]   Canham L.T., *Appl. Phys. Lett.* **57** (1990) 1046.

[4]   Recent symposia proceedings include *Microcrystalline Semiconductors: Materials Sciences and Devices*, Symp. Proc. Vol. **283**, edited by P.M. Fauchet, C.C. Tsai, L.T. Canham, I. Shimizu and Y. Aoyagi (Materials Research Society, Pittsburgh, 1993); *Light Emission from Silicon*, edited by J.C. Vial, L.T. Canham and W. Lang, J. of Luminesc. **57**, 1-6 (1993).

[5]   Cullis A.G. and Canham L.T., *Nature* **353** (1991) 335.

[6] Brus L., *J. Phys. Chem.* **98** (1994) 3575.

[7] Littau K.A., Szajowski P.J., Muller A.J., Kortan A.R. and Brus L.E., *J. Phys. Chem.* **97** (1993) 1224; Wilson W.L., Szajowski P.F. and Brus L.E., *Science* **262** (1993) 1242.

[8] Koch F., Petrova-Koch V., Muschik T., Nikolov A., and Gavrilenko V. in *Microcrystalline Semiconductors: Materials Sciences and Devices*, Symp. Proc. Vol. **283**, edited by P.M. Fauchet, C.C. Tsai, L.T. Canham, I. Shimizu and Y. Aoyagi (Materials Research Society, Pittsburgh, 1993) p. 197; Koch F., Petrova-Koch V. and Muschik T., *J. of Luminesc.* **51** (1993) 271.

[9] Vial J.C., Bsiesy A., Gaspard F., Herino R., Ligeon M., Muller F., Romestain R., and Macfarlane R.M., *Phys. Rev B* **45** (1992) 14171.

[10] Lannoo M., Delerue C. and Allan G., *J. of Luminesc.* **57** (1993) 243.

[11] Delerue C., Allan G. and Lannoo M., *Phys. Rev. B* **48** (1993) 11024; *J. of Luminesc.* **57** (1993) 249.

[12] Bockelmann U. and Bastard G., *Phys. Rev. B* **42** (1990) 8947.

[13] Dean P.J., Haynes J.R. and Flood W.F., *Phys. Rev.* **161** (1967) 711.

[14] Calcott P.D.J., Nash K.J., Canham L.T., Kane M.J., and Brumhead M.D., *J. Phys.: Condens. Matter* **5** (1993) L91; *J. of Luminesc.* **57** (1993) 257.

[15] Suemoto T., Tanaka K., Nakajima A. and Itakura T., *Phys. Rev. Lett.* **70** (1993) 3659; R.T. Collins and M.A. Tischler, unpublished.

[16] Hybertsen M.S., *Phys. Rev. Lett.* **72** (1994) 1514.

[17] Schuppler S., Friedman S.L., Marcus M.A., Adler D.L., Xie Y.-H., Ross F.M., Harris T.D., Brown W.L., Chabal Y.J., Brus L.E. and Citrin P.H., *Phys. Rev. Lett.* **72** (1994) 2648.

[18] Cohen M.L. and Chelikowsky J.R., *Electronic Structure and Optical Properties of Semiconductors*, (Springer-Verlag, New York, 1988).

[19] Hybertsen M.S. in *Light Emission from Silicon*, edited by S.S. Iyer, L.T. Canham and R.T. Collins (Materials Research Society, Pittsburgh, 1992) p. 179.

[20] Takagahara T. and Takeda K., *Phys. Rev. B* **46** (1992) 15578.

[21] Sanders G.D. and Chang Y.-C., *Phys. Rev. B* **45** (1992) 9202.

[22] Proot J.R., Delerue C., and Allan G., *Appl. Phys. Lett.* **61** (1992) 1948.

[23] Read A.J., Needs R.J., Nash K.J., Canham L.T., Calcott P.D.J., and Qtiesh A., *Phys. Rev. Lett.* **69** (1992) 1232.

[24] Buda F., Kohanoff J., and Parrinello M., *Phys. Rev. Lett.* **69** (1992) 1272.

[25] Ohno T., Shiraishi K., and Ogawa T., *Phys. Rev. Lett.* **69** (1992) 2400.

[26] Delley B. and Steigmeier E.F., *Phys. Rev. B* **47** (1993) 1397.

[27] Hirao M., Uda T. and Murayama Y. in *Microcrystalline Semiconductors: Materials Sciences and Devices*, Symp. Proc. Vol. **283**, edited by P.M. Fauchet, C.C. Tsai, L.T. Canham, I. Shimizu and Y. Aoyagi (Materials Research Society, Pittsburgh, 1993), p

425.

[28]   Hybertsen M.S. and Needels M., *Phys. Rev. B* **48** (1993) 4608. Empirical self energy corrections were included to give the physical band gap energy.

[29]   Takeda K. and Shiraishi K., *Phys. Rev. B* **39** (1989) 11028.

[30]   van de Walle C.G. and Northrup J.E., *Phys. Rev. Lett.* **70** (1993) 1116. A full self energy calculation was included to give the physical band gap energy.

[31]   Yeh C.-Y., Zhang S.B. and Zunger A., *Appl. Phys. Lett.* **63**, (1993) 3455; *Phys. Rev. B* (submitted); Wang L.-W. and Zunger A., *J. Phys. Chem.* **98** (1994) 2158.

[32]   *Zahlenwerte und Funktionen aus Naturwissenshaften und Technik*, in Vol. III of Landolt-Bornstein (Springer, New York, 1982), pt. 17a.

[33]   Xie Y.H., Hybertsen M.S., Wilson W.L., Ipri S.A., Carver G.E., Brown W.L., Dons E., Weir B.E., Kortan A.R., Watson G.P. and Liddle A.J., *Phys. Rev. B* **49** (1994) 5386.

[34]   Voos M., Uzan Ph., Delalande C., Bastard G., and Halimaoui A., *Appl. Phys. Lett.* **61** (1992) 1213.

[35]   Schmitt-Rink S., Miller D.A.B., and Chemla D.S., *Phys. Rev. B* **35** (1987) 8113.

[36]   Moritomo Y., Tokura Y., Tachibana H., Kawabata Y. and Miller R.D. *Phys. Rev. B* **43** (1991) 14746.

[37]   Allan G., Delerue C. and Lannoo M., *J. of Luminesc.* **57** (1993) 239.

[38]   Martin E., Delerue C., Allan G. and Lannoo M., unpublished.

[39]   Glembocki O.J. and Pollak F.H., *Phys. Rev. Lett.* **48** (1982) 413; *Phys. Rev. B* **25** (1982) 1193; and references therein.

[40]   Dash W.C. and Newman R., *Phys. Rev* **99** (1955) 1151; Aspnes D.E. and Studna A.A., *Phys. Rev. B* **27** (1983) 985.

[41]   The values given in Ref. 33 were incorrect by a factor of three, although the results reported were calculated with the correct values of R_{TO}.

[42]   Haynes J.R., Lax M. and Flood W.F. in *Proc. Int. Conf. on Semicond. Phys., Prague, 1960* (Academic Press, New York, 1961), p. 423; Cuthbert J.D., *Phys. Rev. B* **1** (1970) 1552.

[43]   Bottcher C.J.F. and. Bordewijk P, *Theory of Electric Polarization*, 2nd Ed., Vol. II (Elsevier, New York, 1978) pp. 476-487.

[44]   Xie Y.H., Wilson W.L., Ross F.M., Mucha J.A., Fitzgerald E.A., Macaulay J.M., and Harris T.D., *J. Appl. Phys.* **71** (1992) 2403.

[45]   Sagnes I., Halimaoui A., Vincent G. and Badoz P.A., *Appl. Phys. Lett.* **62** (1993) 1155.

[46]   Kanemitsu Y., Uto H., Masumoto Y., Matsumoto T., Futagi T. and Mumura H., *Phys. Rev. B* **48** (1993) 2827.

[47]   Lockwood D.J., Wang A., and Bryskiewicz B., *Solid State Commun.* **89** (1994) 587.

[48]   Vezin V., Gaudeau P., Naudon A., Halimaoui A. and Bomchil G., *Appl. Phys. Lett.* **60** (1992) 2625.

[49]   Sui Z.,. Leong P.P, Herman I.P., Higashi G.S., and Temkin H., *Appl. Phys. Lett.* **60** (1992) 2086; Gregora I., Champagnon B. and Halimaoui A., *J. of Luminesc.* **57** (1993) 73.

[50]   Münder H., Berger M.G., Frohnhoff S., Thönissen M. and Lüth H., *J. of Luminesc.* **57** (1993) 5.

[51]   Chen X., Henderson B. and O'Donnell K.P., *Appl. Phys. Lett.* **60** (1992) 2672.

[52]   Bawnedi M.G., Wilson W.L., Rothberg L., Carroll P.J., Jedju T.M., Stegerwald M.L. and Brus L.E., *Phys. Rev. Lett.* **65** (1990) 1623.

LECTURE 6

Theory of silicon crystallites. Part II

C. Delerue, G. Allan, E. Martin and M. Lannoo

*IEMN, Département ISEN,
41 Bd Vauban, 59046 Lille cedex, France*

1. INTRODUCTION

Although the photoluminescence of porous silicon was discovered some time ago [1] its origin is not yet firmly established. The most natural explanation is the quantum confinement in silicon crystallites [2 to 5] but states localized at the surface of the crystallites, e.g. due to dangling bonds should play an important role [2, 6]. Our aim here is thus to develop some aspects of the theory of silicon crystallites which could be helpful for the interpretation of experimental data. The contents is complementary to the Chapter written by M. Hybertsen in the same volume.

We start the first section with a review of the empirical tight binding method as applied to crystallites. We compare our results for the gap to those of other calculations and the predicted optical properties to those observed not only for porous silicon but also for isolated Si crystallites. We treat the case of polysilanes, calculate their Stokes shift, and show that their strong luminescence properties are due to the fact that the infinite chains have a direct gap. We then consider two cases of imperfections : dangling bonds and donor impurities, respectively at the surface and inside the silicon crystallites. We show how one can calculate their main physical characteristics (radiative and non radiative recombination rate, binding energy). The final section is devoted to the importance of exchange and electron-phonon interactions in the optical properties of silicon crystallites.

II. TIGHT BINDING TREATMENT (T. B.)

We first discuss the basic principles of T. B. and its application to the calculation of optical transitions. We then compare the results to those of other calculations and to experimental data. Our calculations do not include phonon assisted transitions which are treated by M. Hybertsen.

As expected the highest confinement energy corresponds to the QD case. The observed dominant luminescence of porous silicon is in the energy range 1.4 to 2.2 eV [3, 8]. From fig. 1 we see that this could be compatible with crystallite sizes between 2.5 and 4.5 nm. Similar results have been obtained by different groups [2, 9, 10] and our results agree with other published band gap energies [9 - 17].

A systematic comparison is performed in ref [18] where it is shown that an accurate pseudopotential approach exactly reproduces our results for Si crystallites. This is to be expected on quite general grounds. Indeed, the solution ψ (E) with energy E of Schrödinger's equation within a crystallite, can be written as a combination of the bulk solution $\psi_{n,b}$ (E) at the same energy :

$$\psi (E) = \sum_{n} C_n \; \psi_{n,b} (E) \tag{5}$$

Such an expansion is valid except in the near vicinity of the surface, the unknown coefficients being determined from the boundary conditions. The conclusion which can be drawn from this is that different methods giving accurate representations of the bulk hamiltonian must lead to identical results for crystallites provided the boundary conditions are correctly taken into account. This justifies the perfect agreement between our results and ref [18] with such different methods as the empirical pseudopotential + plane waves and T. B. Another interesting point to notice is that both methods predict a dependence $d^{-1.39}$ for the energy gap versus crystallite diameter d. This differs appreciably from the d^{-2} law which would be deduced from effective mass theory.

On figure 1 are also reported experimental results [19] on optical bandgaps measured on small hydrogenated silicon crystallites which fall exactly on the predicted curve.

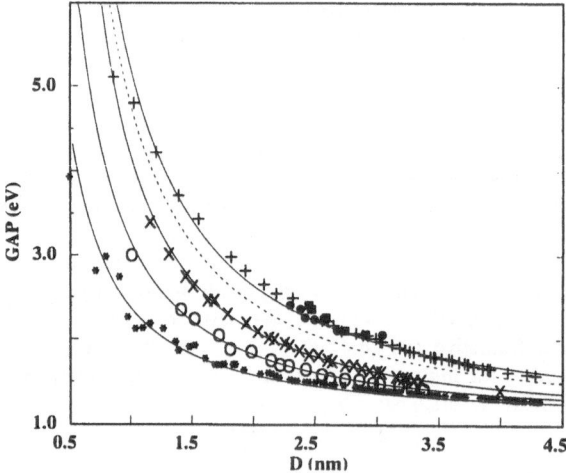

Fig 1 :   Calculated optical band-gap energies for various silicon crystallites (+) or wires
                   (100: x; 110: *; 111: 0) with respect to their diameter d. The continuous lines
                   are an interpolation and an extrapolation of these results by a d^{-n} law.
                   The black dots and squares are the experimental results of Ref. 19.
                   The dashed line is the band-gap energy for the crystallites including the
                   Coulomb interaction between the electron and the hole.

However they differ considerably from more recent experimental data [20, 21] which are almost half the calculated values. The origin of this discrepancy is not yet clear but we think it could be attributed to the fact that silicon crystallites embeddied in a SiO2 matrix should have smaller gaps than hydrogen terminated clusters. Calculations to check this point are in progress.

Optical properties of crystallites

The key quantity is the radiative recombination time which is given by the Fermi golden rule [22] :

$$\frac{1}{\tau} = \frac{16\pi^2}{3} \, n \, \frac{e^2}{h^2 m^2 c^3} \, E_0 \, \left| \langle i_{CB} |p| f_{VB} \rangle \right|^2 \tag{6}$$

where $|i_{CB}>$ and $|f_{VB}>$ are the initial and final states respectively belonging to the conduction and valence bands, E_0 is the transition energy, n the refractive index. A thermal average $<1/\tau>$ over the initial states in (6) is performed so that we can get the dependence upon temperature. Details of the calculation are given in the last reference [2].

Figure 2 shows the result obtained at T = 300 K as a function of the crystallite energy gap. Apart from the oscillations in the predicted values (whose origin is well understood [2]) the recombination rate is consistently lower than experiment, especially for large clusters (small gaps) close to the bulk situation. This is coherent with the fact that, in this limit, one is dealing with forbidden indirect transitions. Recombination could then only occur either by phonon assisted radiative transitions [23] or non radiative processes.

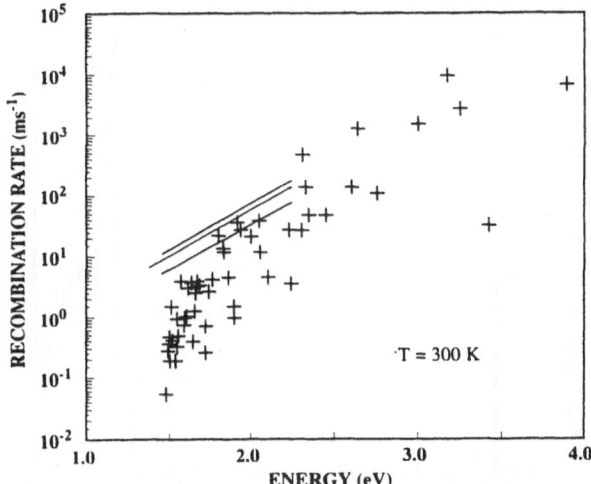

Fig 2 .   Calculated recombination rate (ms-1) of an excited electron-hole pair in silicon crystallites (crosses) with respect to the photon energy at 300 K. The spin degeneracy is not included: its inclusion would divide the calculated recombination rates by factor of 2. Continuous lines plot the experimental dependence of decay rates on photon energy for 65% porosity layers that differ by oxidation level.

It is also instructive to study the optical absorption of silicon crystallites which also exhibits a blueshift compared to the bulk silicon [5,24]. An interesting point is that the observed sharp absorption edge for crystallites of 3nm size occurs at ~ 3.2 eV, much higher than the predicted value for the gap ~ 1.5 eV. To elucidate this point, we have performed a calculation of the optical absorption coefficient at frequency ν, given by:

$$\alpha \left(h\nu \right) \sim \frac{1}{h\nu} \sum_{n,n'} \; \left| <n|\, p\, |n'> \right|^2 \; \delta \left(E_{nn'} - h\nu \right) \qquad (7)$$

This one is plotted in fig 3 for a small crystallite characterized by a band gap of 3.45 eV. We see that the maximum of absorption occurs near 5eV. Anyway there is a visible peak at 3.45 eV indicating that the absorption near the band edge is already rather efficient. The situation of fig 4, corresponding to a larger crystallite of gap 1.67 eV, is completely different. Here the absorption edge starts at ~ 3 eV and no structure is visible near the gap. The reason is that the oscillator strength for transitions between 1.67 eV and 3 eV is several orders of magnitude smaller than above 3eV. This is made clear on fig 5 where α (hν) in this energy region is plotted on a magnified scale. This means that the excitation spectrum of porous silicon reported in ref [25] with an excitation edge above 3eV does not contradict the quantum confinement model. Furthermore the fairly small value of α (hν) near the gap is consistent with the large value of the radiative recombination time discussed above.

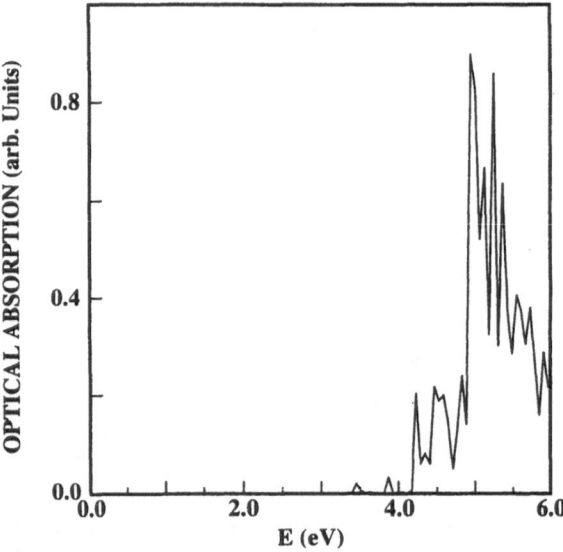

Fig 3 :   Optical -absorption coefficient α with respect to the photon energy hν calculated for
           a silicon crystallite with diameter of 1.56 nm. The one-electron band gap
           is calculated at 3.45 eV.

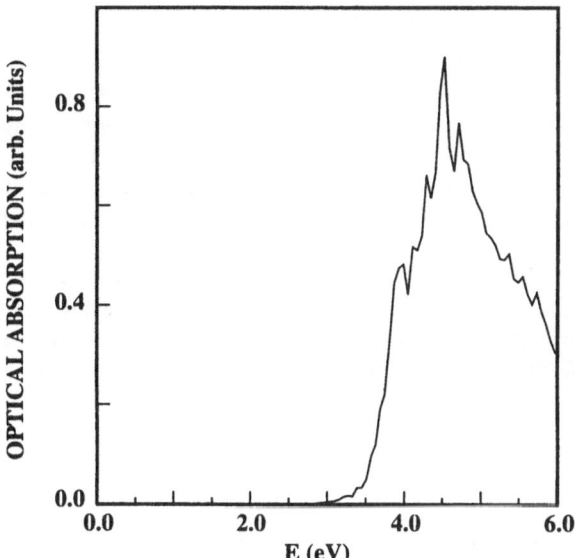

Fig 4 :   Optical-absorption coefficient α with respect to the photon energy hν calculated for
            a silicon crystallite with diameter of 3.86 nm. The one-electron band gap
            is calculated at 1.67 eV.

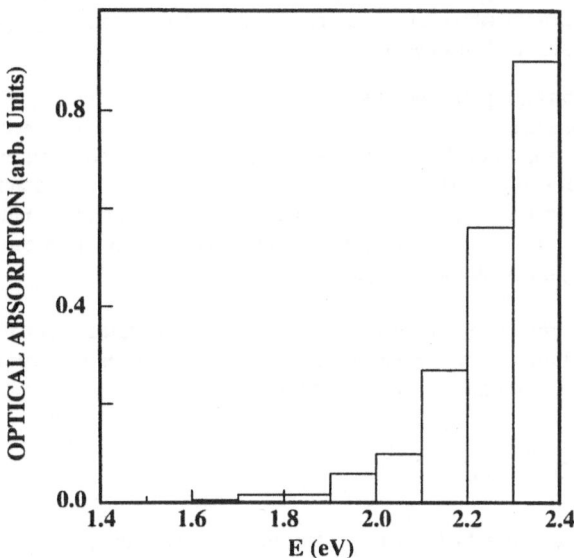

Fig 5 :   Optical-absorption coefficient α with respect to the photon energy hν calculated for
            a silicon crystallite with diameter of 3.86 nm. The one-electron band gap is
            calculated at 1.67 eV. Same as Fig 4 but only the energy region near the
            gap is plotted. Amplitudes of the bars represent the integrated absorption coefficient
            over the width of the bar.

III. OPTICAL PROPERTIES OF POLYSILANES

Recently Kanemitsu et al [26] have studied experimentally the optical properties of chains of silicon atoms, their number (between 5 and 110) being controlled quite accurately by organic synthesis. Such polysilanes have been proposed to be at the origin of the luminescence of porous silicon [27,28], recent molecular calculations showing that polysilanes connecting silicon crystallites could emit light in the visible range [28]. It is thus interesting to discuss their properties, summarising here the detailed calculations performed in reference [29].

The chains of silicon atoms produced and studied in reference [26] have a molecular structure Et 0 - (ϕ - Si - CH_3)$_N$ - O Et where ϕ and Et represent phenyl and ethyl respectively. We will focus our interest here on the excitons in the σ bonded chains represented on figure 6 where the two silicon dangling bonds are saturated by hydrogen atoms.

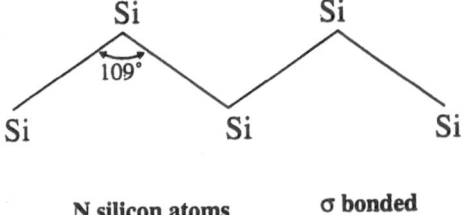

N silicon atoms   **σ bonded**

Fig 6 :   Atomic configuration of the polysilanes studied in the paper. Only the bonds
           between the silicon atoms are represented. The silicon dangling bonds are
           saturated by hydrogen atoms.

Again our calculations are performed using a semi empirical tight binding method, but slightly different from the one used in section II. We use the parameters of reference [30] which give a slightly inferior description of the bulk bands but allow the fairly complex exciton calculations described in the following. The parameters describing the Si - H interactions are those of reference [31]. To calculate the Stokes shift between absorption and luminescence, we make use of a total energy T.B. method, described in reference [29], in which the interatomic terms are taken to vary exponentially with distance.

Let us first describe the method used to calculate the exciton states. For this, we expand the total exciton wave function as a linear combination of basis states corresponding each to one different electron-hole excitation. We write thus

$$\Psi_{exc} = \sum_{v,c} \alpha_{v,c} \, \psi_{v \to c} \tag{8}$$

where $\Psi_{v \to c}$ represents a Slater determinant, deduced from the ground state (filled valence band, empty conduction band) by one electron - hole excitation from the valence band state v to the conduction band state c. As detailed in ref [29], the full calculation reduces to the diagonalization of a matrix which is the sum of two contributions:

i- a diagonal one equal to the difference ε_c-ε_v between the energies of the one particle states obtained directly from the T.B. calculation;

i- a diagonal one equal to the difference ε_c-ε_v between the energies of the one particle states obtained directly from the T.B. calculation;

ii- a second one with diagonal and non diagonal matrix elements due to the Coulomb interactions between electrons.   This one is equivalent [29] to a screened electron - hole attraction - $e^2/\varepsilon\ r_{eh}$ where ε is the dielectric constant and r_{eh} the electron - hole distance.

In our tight binding basis we only retain, as commonly done, the following interactions

$$e^2 \int \frac{\left| \phi_i\left(\vec{r}_e\right)\right|^2 \left| \phi'_j\left(\vec{r}_h\right)\right|^2}{\left|\vec{r}_e - \vec{r}_h\right|} = \frac{1}{\varepsilon R_{ij}} \qquad i \neq j \qquad (9)$$

$$= U \qquad i \neq j$$

where ϕ_i and ϕ'_j are atomic orbitals centered on atoms i and j whose distance is R_{ij}. The determination of the intraatomic Coulomb term U is detailed in ref [29]. In the calculation we have taken $\varepsilon = 7.4$ which is the value measured for the solvent of ref [26].

Let us now compare our predictions to experimental evidence. The optical absorption energies are given on fig 7 where we see that the theoretical curve gives a good account of the observed values of ref [26]

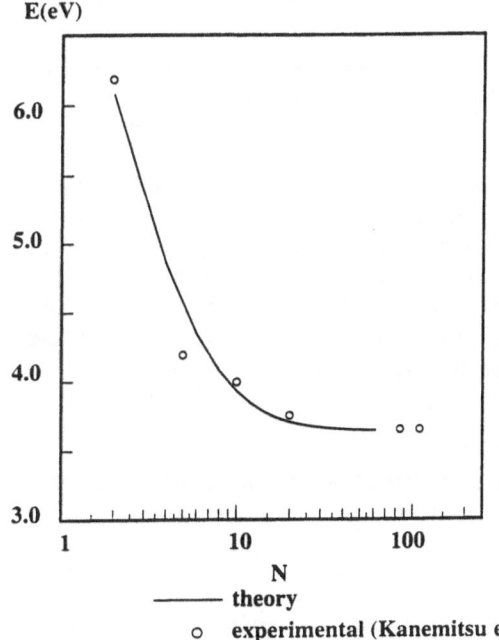

Fig 7 :   Comparison between the calculated and experimental [26] absorption energies of polysilanes

C. Delerue et al.

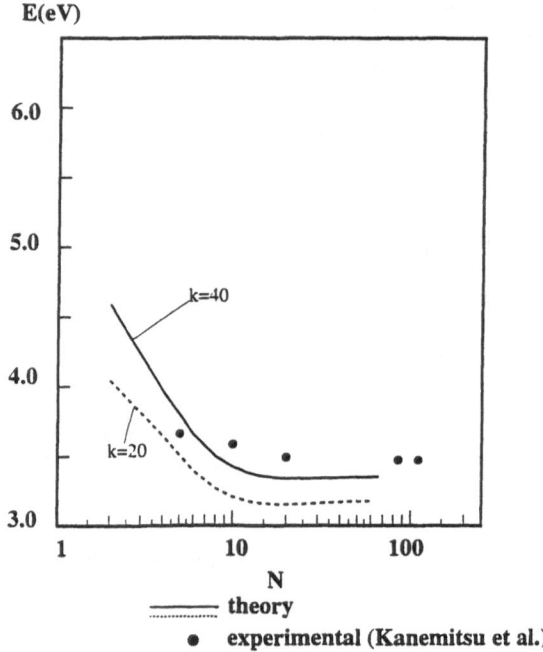

Fig 8 :   Comparison between the calculated and experimental [26] energies of
             luminescence for polysilanes

Finally let us discuss a possible origin of the fast blue band observed in the luminescence of porous silicon [4] in the light of these results. It is clear from fig 8 that linear chains of silicon atoms embeddied in a material with large bandgap (>> 3.6 eV) could be at the origin of this luminescence since they also have a radiative lifetime in the correct range (~ 30 nsec). We have checked that similar results can be obtained for crystallites or wires of small diameter (<~ 10 Å). However this would lead to a large bandgap (> 3.5 eV) and a luminescence in the range 1.8 - 2.8 eV could be obtained only with a large Stokes shift as is exactly the case for polysilanes.

IV. DEFECTS AND IMPURITIES IN CRYSTALLITES

We consider here two cases :

i) Dangling bonds at the surface of a crystallite which can provide an important channel for non radiative recombination
ii) Donor impurities which are expected to behave in a completely different way than in the bulk.

4.1 Surface dangling bonds

Dangling bonds correspond to coordination defects in which the silicon atom has only three equivalent covalent bonds. The best known case is the P_b center at the Si (111) - SiO_2 interface. Such defects are also expected to occur at the surface of crystallites in porous silicon as was indeed demonstrated by electron paramagnetic resonance [34,35]. We examine here the properties of such defects in crystallites in the light of what is known for the P_b center.

The P_b center can exist in three charge states +,0,-. The corresponding ionization energies are $\varepsilon (-,0) = \varepsilon_c - 0.3$ eV and $\varepsilon (0,+) = \varepsilon_v + 0.3$ eV where ε_c and ε_v are the bulk silicon band edges [36]. To transpose these properties to the case of silicon crystallites one can apply the following arguments :

i) Deep gap states are fairly localized in real space and only experience the local situation. Their energy should not be shifted by the quantum size effect.
ii) The valence and conduction states of the crystallite exhibit a blue shift $\Delta E_{c,v}$ due to confinement.

From this, ionization energies of the charged dangling bonds (- or +) which are 0.3 eV for P_b become $\Delta E_c + 0.3$ eV and $\Delta E_v + 0.3$ eV respectively, the wave function remaining unchanged.

This allows us to discuss first the non radiative capture of one electron (or hole) by one neutral dangling bond at the surface of the crystallite. We start again from the known properties of the neutral P_b center which has a measured capture cross section σ in the 10^{-14} - 10^{-15} cm^2 range [36] at 170 K for both types of carriers. Its physical origin is multiphonon capture which can be best understood on the configuration coordinate diagram of Fig 9. On such a diagram the total energy of the system is plotted versus a local lattice coordinate Q to which the electron system is coupled. The curves (i) and (f) denote the initial and final states for the P_b center for which the situation is practically the same for both types of carriers. In the simplest theory, valid for strong coupling and high temperature, the cross section has a thermally activated behavior

$$\sigma \sim \sigma_0 \, \exp\left(- \frac{E_b}{kT} \right) \tag{11}$$

in which E_b is the classical barrier height, equal to the excitation energy necessary to reach the crossing point in the initial state (this is zero on fig 9 for the P_b center but is shown explicitly for curves (i') and (f)). E_b has a simple expression

$$E_b = \frac{\left(E_0 - d_{FC} \right)^2}{4 \, d_{FC}} \tag{12}$$

where E_0 is the ionization energy of the defect and d_{FC} is the Franck-Condon shift, equal to the change in energy of the initial or final state between the values of Q corresponding to the two minima (identical for both states if the two curves are parabolic with the same curvature). For P_b one finds $E_b \approx 0$, i.e $E_0 = d_{FC}$.

One can now extend this well established result to the dangling bond in crystallites. One finds, according to the previous discussion, that the final state remains unchanged while the initial state (electron in the conduction band or hole in the valence band) is blue shifted with respect to (i) by an amount ΔE_c or ΔE_v (on fig 9 we take $\Delta E_c = \Delta E_v = \Delta E$). In the simple theory outlined above one would thus find that between (i') and (f) E_0 changes to $E_0 + \Delta E$ while d_{FC} remains unchanged. From equ (12) this means that E_b is increased from zero to a value equal to $\Delta E^2/4d_{FC}$ which increases quadratically with the blue shift. The consequence is that σ is expected to decrease rapidly with confinement.

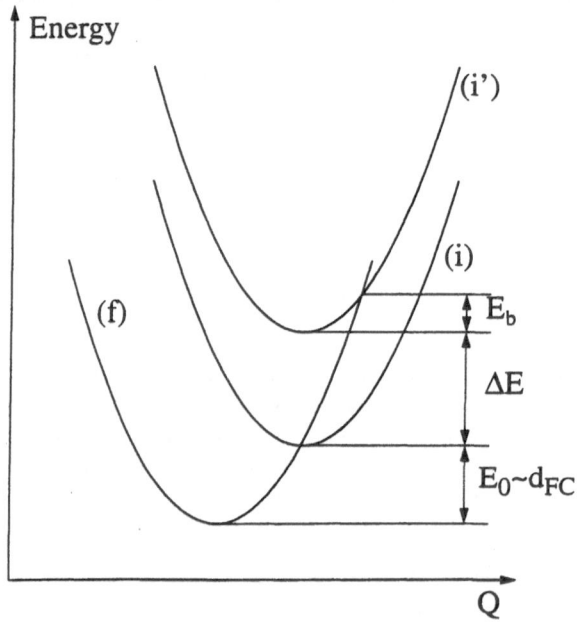

Fig 9 :   Configuration coordinate diagram, representing the variation of the total energy
           versus the atomic displacement for two charge states of the defect
           [initial (i) and final (f)]. Two initial states are indicated, one (i) in bulk
           silicon (ionization energy E_0) and the other (i') in a silicon crystallite
           (ionization energy $E_0 + \Delta E$). the situation in bulk silicon corresponds to
           a negligible barrier for the capture. In silicon crystallites, the increase in ionization
           energy creates a barrier E_b for the recombination (in a classical picture).

In practice we have performed the calculation of the non radiative capture by a neutral dangling bond using the more sophisticated theory of [36] but the conclusions remain qualitatively unchanged (details are given in [2b]). The full results for the capture rate are given on fig 10. It is seen that, for crystallites with a gap smaller than 2.5 eV, non radiative capture by a neutral dangling bond dominates the intrinsic radiative recombination. This means that "the presence of one dangling bond at the surface of a crystallite is enough to kill its luminescence".

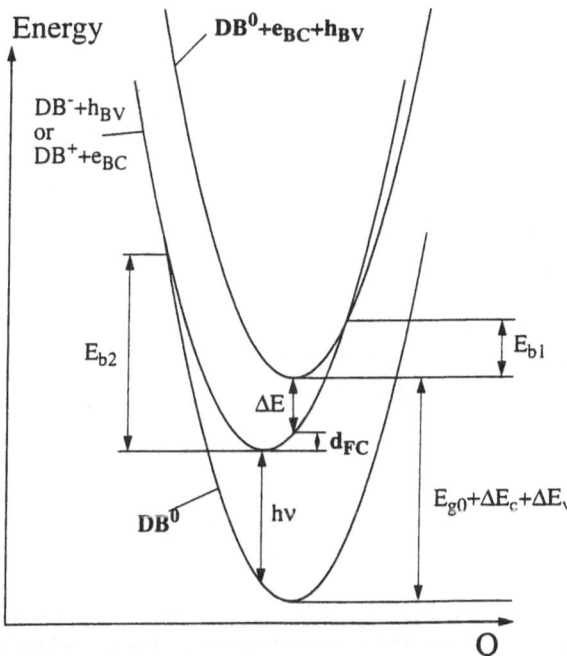

Fig 10 :   Capture rates (T = 300 K) of an electron or a hole in silicon crystallite due to a
nonradiative capture on a single neutral silicon dangling bond plotted with respect
to the excitonic band-gap energy of the crystallites (————); Crosses give
the radiative recombination rates of the electron-hole pairs in the same crystallites.
The other curves are the radiative capture rates of carriers on a neutral dangling
bond (hole capture: - - - ; electron capture: . . . .).

After capture of one carrier by the neutral dangling bond this one becomes charged.   To
complete the recombination process one must then look at capture by the other type of carrier,
i.e by the charged dangling bond. Let us first look at non radiative capture with the help of the
configuration diagram of fig 10 on which we directly see that the barrier for capture E_{b2} by the
charged dangling bond is much larger than E_{b1} for the neutral one.

Its expression is

$$E_{b2} = \frac{\left(E_{go} + \Delta E - 2d_{FC} \right)^2}{4d_{FC}} \tag{13}$$

where E_{go} is the bulk band gap. If we take $\Delta E = 0.3$ eV we find E_{b2} equal to 0.53 eV which is
fairly large and, according to equ. (11) will drastically reduce the capture cross section. Using
the more quantitative theory of [36] one indeed finds for σ a reduction factor of 3×10^{-7} at T =
300 K and 5×10^{-11} at T = 10 K. Even if the prefactor σ_0 is somewhat increased for a charged
defect this means that non radiative capture by charged dangling bonds should be completely
negligible.

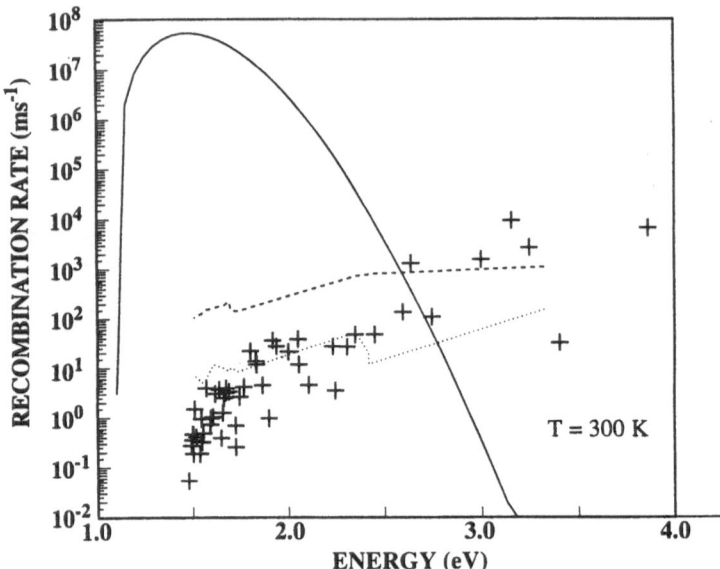

Fig 11 :   Configuration coordinate diagram, representing the variation of the total energy of a
           crystallite with one dangling bond at the surface. The ground state (lower curve)
           corresponds to filled valence states, empty conduction states, and the dangling bond
           in neutral charge state (DB^0). The higher curve [equivalent to curve (i') of Fig 9]
           represents the same system after excitation of an electron in the conduction band
           leaving a hole in the valence band (DB^0 + e$_{BC}$ + h$_{BV}$). The intermediate curve
           [equivalent to curve (f) of Fig 9; the curve (i) of Fig 9 is not reproduced here]
           describes the system after capture of a carrier by the dangling bond. Two situations
           are possible: the capture of the electron, the hole remaining in the valence band
           (DB$^-$ + h$_{BV}$) or the capture of the hole, the electron remaining in the conduction
           band (DB$^+$ + e$_{BC}$). ΔE is the energy shift of the conduction or the valence band due
           to confinement compared to the bulk silicon band structure (conduction for the
           capture of the electron, valence for the capture of the hole).

There remains to see if radiative capture by dangling bonds (charged or not) might be an
efficient process. We extend to this case the method used in section II-3 as detailed in ref
[2b]. The results do not depend on the charge state and are reproduced in fig 11. They fall in
the range 1 to 10 µsec for hole capture and 10 to 100 µs for electron capture. The conclusion
which emerges from fig 11 is that the recombination at dangling bonds proceeds in two steps :
non radiative capture at the neutral defect followed by radiative capture of the other carrier, this
last step being the rate limiting one.

An interesting point to notice is the spectral range of these emitted photons.   The peak
corresponds to hv on fig 10 and is equal to E_{go} + ΔE - 2 d_{FC}. For ΔE = 0.3 eV we thus get hv
= 0.8 eV. Such radiative capture at dangling bonds could thus be at the origin of the observed
infrared emission [6].

4.2 Hydrogenic impurities in silicon crystallites

Donor and acceptor impurities in bulk silicon are known to give rise to shallow levels. Their energies can be predicted by effective mass theory, leading to a series of hydrogenic states. The ground state ionization energy is pretty small, or order 20 to 40 meV so that these impurities are usually ionized at room temperature.

It is interesting to see how this behavior is modified for such an impurity in a crystallite. This has already been studied on the basis of the effective-mass approximation [37] and we have recently considered this problem on the basis of T.B. theory discussed above [38]. We shall here only present some simple considerations leading to the essential results. We first consider a spherical crystallite, with the impurity at the center.

For a crystallite with infinite size this would produce a Coulomb potential.

$$V(r) = \pm \frac{e^2}{\varepsilon r} \tag{14}$$

where ε is the bulk dielectric constant. For a crystallite with radius R this becomes

$$V(r) = \pm e^2 \left[\frac{1}{\varepsilon_{in} r} - \frac{1}{R} \left(\frac{1}{\varepsilon_{in}} - \frac{1}{\varepsilon_{out}} \right) \right] \tag{15}$$

where ε_{in} and ε_{out} are the dielectric constants of the crystallite and the surrounding medium respectively. The energy levels of the crystallite can be calculated as before in the TB approximation, by simply adding this potential to the normal TB hamiltonian. The results are plotted on fig 12 versus crystallite size, showing that the energy level of the extra electron (or hole) remains practically constant. A similar result is obtained by treating V(r) in first order perturbation theory on the intrinsic crystallite states. This is due to the fact that the Bohr radius of the hydrogenic state is much larger than the crystallite size. Another interesting point to notice is that the screened potential of equ (15) takes a simple form when $\varepsilon_{in} \gg \varepsilon_{out}$ which is actually the case. In such a limit it becomes constant, equal to $\pm e^2/\varepsilon_{out} R$. Finally the results remain practically unchanged when the impurity is displaced away from the center of the crystallite, for reasons discussed in [38].

A drastic change which occurs when the impurity is within a crystallite concerns the notion of ionization energy. For the crystallite containing the impurity the above argument shows that all crystallite states are shifted by the same amount ($\sim \pm e^2/\varepsilon_{out} R$) without appreciable change in the wave functions. In such a situation one way of defining the ionization energy would be to transfer the bound carrier to a perfect crystallite of the same size, located far away, and to look at the shift of the level occupied by the extra electron (or hole). This can be deduced directly from fig 12 where it is seen to take large values ($\sim 2eV$) for small crystallites. For large ε_{in} this ionization energy simply becomes $\pm e^2/\varepsilon_{out} R$.

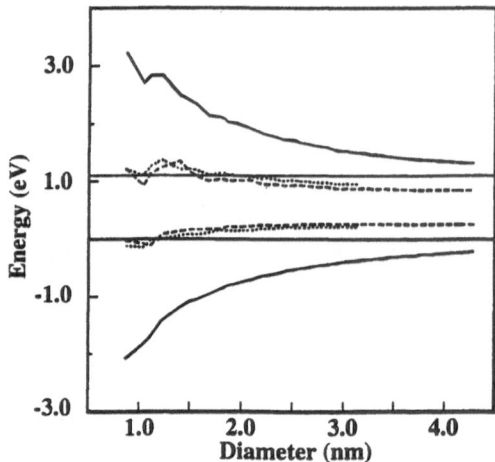

Fig 12 : Energy levels of hydrogenic impurities in porous silicon as a function of the particle diameter. Horizontal lines give the bulk gap limits. The porous silicon bandgap limits are given by the full lines. The dashed (dotted) lines show the energy levels for an hydrogenic impurity at the center of the cluster (located at 0.9 times the radius away from the center).

V. EXCHANGE SPLITTING AND STOKES SHIFT OF CRYSTALLITES

Although porous silicon is heterogeneous at the microscopic scale, some fine structures appear in the excitation spectrum of the luminescence at 2K [4]. In particular an onset of a few meV is observed [4,39] and, in addition, the lifetime of the visible luminescence decreases when going from low (4K) to higher temperatures (~ 100 - 200 K) in parallel with an increase in luminescence intensity [4,40,41]. Both results have been interpreted as consequences of the exchange splitting of the exciton [4,39]. Indeed the electron and hole, with spin 1/2, can be combined into a singlet (S = 0) and a triplet (S = 1 state), the exchange interaction stabilizing the triplet as shown in fig 13. In bulk silicon the exchange splitting is thought to be smaller than 0.15 meV but it was proposed that confinement could bring it in the range of several meV [41,42]. Then the level scheme of fig 13 could explain both types of experiments with, however the difficulty that a different value of the exchange splitting is required in each case. The aim of this section is to explore this point more quantitatively and also to discuss the Stokes shift which can also bring a substantial contribution.

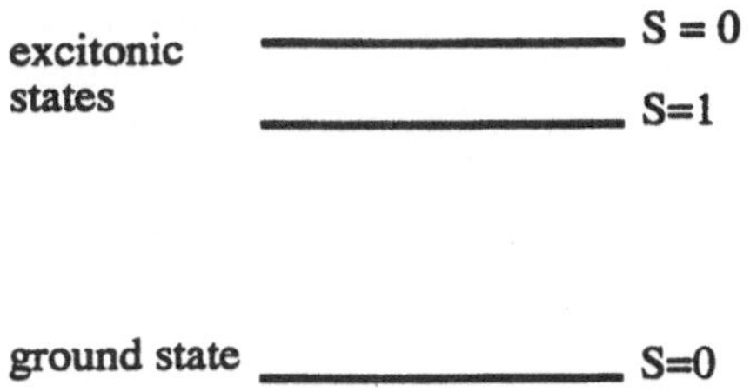

excitonic   ——————————— S = 0
states   ——————————— S=1

ground state   ——————————— S=0

Fig 13 :   Two-level model for the recombination of excitons in porous silicon.
           The lowest excitonic state is split due to exchange interaction between
           the electron and the hole. The upper level has much smaller lifetime than the lower
           one. Thermal equilibrium between the two levels could explain the temperature
           dependence of the radiative lifetime.

5.1 Calculation of the exchange splitting

As discussed before the basis of the simple model of fig 13 is the existence of a lower triplet
excited state S = 1, followed by a singlet S = 0. It was proposed [4,41] that these are slightly
mixed by the spin-orbit coupling so that the lifetime of the lower level ($\sim$ 1 msec) is much
longer than the lifetime of the upper one ($\sim$ 1 μsec).

We have thus calculated the excitonic states and their lifetimes from a tight binding method,
exactly in the same way as for polysilanes (section III). The only difference is that we add to
the exciton hamiltonian two extra contributions : exchange (which can be expressed in a form
similar to direct Coulomb terms) and the spin orbit interaction which is diagonal in the T.B.
basis. We then diagonalize exactly the full excitonic hamiltonian, deduce the wave functions
and calculate radiative recombination rates. The procedure is detailed in ref [43].

Let us first describe the results for spherical crystallites. Fig 14 shows the exciton binding
energy versus diameter. It is pretty large ($\sim$0.1 - 0.2 eV), in agreement with effective mass
models [44] and represents the average electron-hole attraction. On fig 15 we plot a typical
excitonic spectrum, in an energy range of 22 meV from the lowest excitonic state, where the
height of the bars corresponds to the calculated radiative recombination rate. We see that the
spectrum is very complicated and cannot correspond to the two level model of fig 13 with
much smaller recombination rate in the S = 1 lower state. This complexity is due to the high
degeneracy at the bulk band edges (6 for the conduction band, 3 for the valence band) which is
not present in a simple two band model.

If the two levels model of fig 13 is to be valid it cannot be explained by spherical crystallites.
We have thus explored the case of crystallites of lower symmetry. Starting with ellipsoidal
crystallites one finds that the lowered symmetry does not lift completely the degeneracies so
that one still gets a complex behavior, qualitatively different from the simple model. The last
possibility is that the symmetry is lower than axial. We have thus studied the case of deformed
ellipsoids in which an angular distorsion perpendicular to the main axis is imposed. When the
angular distorsion becomes of order 25% we get the typical spectrum of fig 16. Here the lifting
of degeneracy is complete enough so that spin-orbit coupling is practically quenched. The net
result on fig 16 shows that one indeed recovers the simple two-levels model for the lowest

states with the qualitatively correct behavior for the recombination rates. However the predicted exchange splitting between these two states is still too small compared to experiment, as shown on fig 17 which summarizes all the asymetrical cases we have explored. This means that another contribution is required which might possibly be due to the Stokes shift between absorption and luminescence.

5.2 Calculation of the Stokes shift

The situation is similar to the one studied in section II for polysilanes. The exciton is expressed as a linear combination of electron-hole excitations, each one weakening the strength of the bonds. The system, when in the excitonic state, will thus experience a lattice relaxation when compared to the ground state. The average optical transition occurs at the equilibrium lattice configuration of the initial state. The average energy in absorption will then exceed the corresponding value in luminescence by a quantity equal to $2d_{FC}$ where d_{FC} is the Franck-Condon shift (see fig 18). We have estimated d_{FC} by using bulk deformation potentials together with the electron-hole distribution given by our calculation. The calculation is described in ref [43]. In this way we get $d_{FC} = 14$ meV and 1.8 meV for spherical crystallites of diameter 2nm and 4 nm. Although such an estimation remains crude this means that the Stokes shift should provide a substantial contribution to the onset in selectively excited photoluminescence, explaining most of the difference between experiment and theory on fig 17.

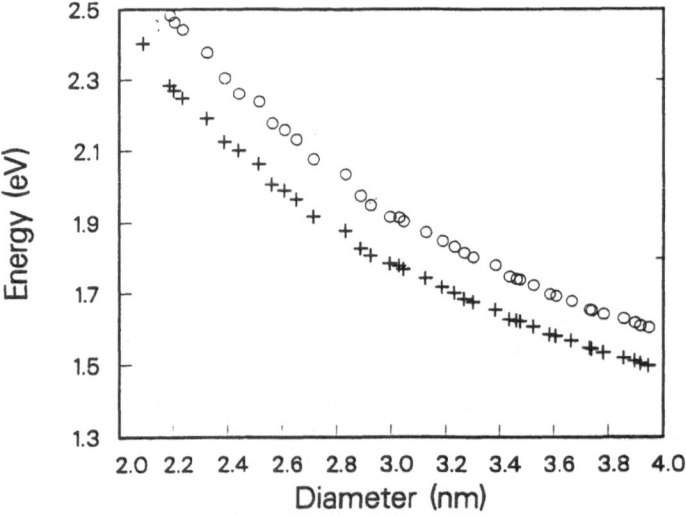

Fig 14 :   One-electron band gap (circles) and excitonic bandgap (crosses) calculated for spherical silicon crystallites with respect to their diameter.

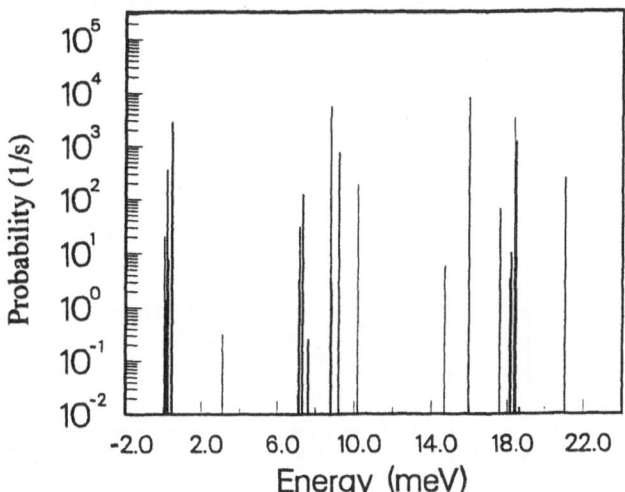

Fig 15 : Calculated excitonic structure of the spherical silicon crystallite of diameter 3.86 nm. The levels are indicated by vertical bars. The zero of energy corresponds to the lowest exciton level. The height of the bars represents the calculated radiative recombination rate (inverse of the radiative lifetime).

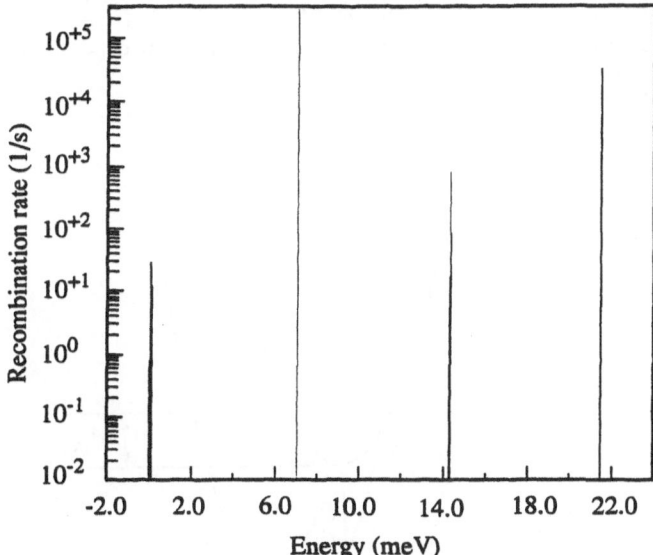

Fig 16 : Calculated excitonic structure of a silicon crystallite with complex shape built from an ellipsoid with long axis of 2.4 nm and short axis of 1.8 nm. The levels are indicated by vertical bars. The zero of energy corresponds to the lowest exciton level. The height of the bars represents the calculated radiative recombination rate (inverse of the radiative lifetime).

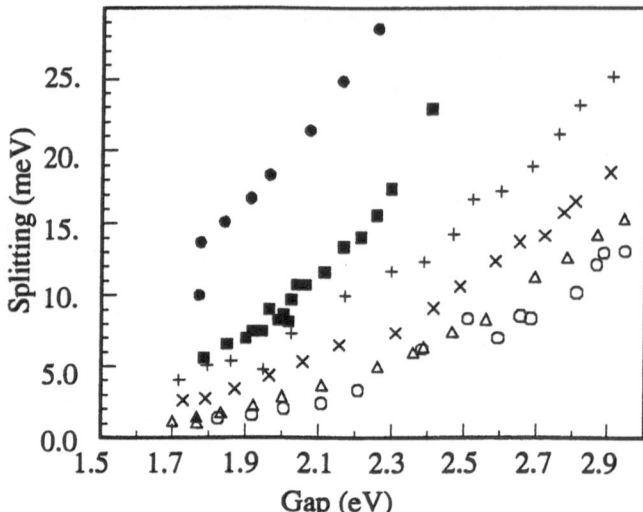

Fig 17 :   Splitting between the two lowest calculated excitonic levels in several silicon
           crystallites with respect to their excitonic bandgap. Crystallites have complex
           shapes but with a longer axis in the 100 direction (open circles), 110 direction
           (open triangles), 111 direction (+). Crosses (x) correspond to the average over all
           the orientations of the longer axis of the crystallite. Black squares are the first
           onsets measured by selectively excited photoluminescence and black dots are the
           energy splittings derived from the fit of the temperature dependence of the
           luminescence lifetime.

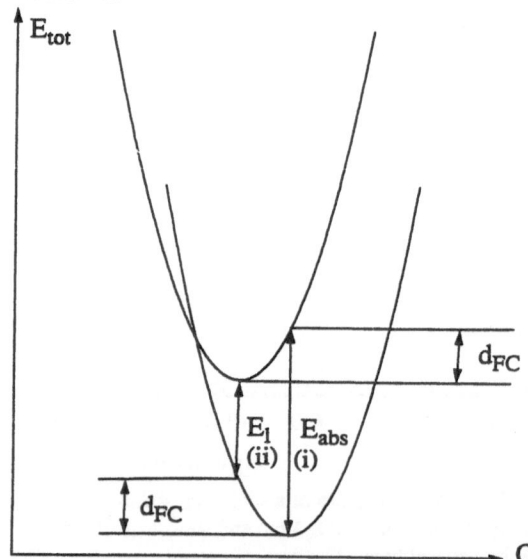

Fig 18 :   Configuration coordinate diagram representing the variation of the total energy
           versus the lattice displacement coordinate Q. The lowest curve corresponds to the
           system in its ground state, the upper curve to the excitonic state.

VI. CONCLUSION

We have reviewed here a number of predictions concerning the optical properties of silicon crystallites with the exception of phonon assisted transitions which are the object of Theory part I. The main results are the following :

- Radiative recombination of silicon crystallites is slow;
- A strong Stokes shift is found in polysilanes, with fast recombination, offering a possibility of interpretation of the fast blue band;
- The presence of one dangling bond in a crystallite is enough to kill its luminescence;
- Donor or acceptor levels are practically independent of particle size;
- The two levels model with exchange splitting is valid only for strongly asymmetric crystallites;
- The Stokes shift of crystallites is non negligible and must be considered in the interpretation of the onset found in selectively excited photoluminescence.

REFERENCES

1. L.T. Canham, Appl. Phys. Lett. **57**, 1046 (1990)
2. a) J.P. Proot, C. Delerue and G. Allan, Appl. Phys. Lett. **61**, 1948 (1992)
   C. Delerue, G. Allan and M. Lannoo, in Optical Properties of Low Dimensional Silicon Structures edited by D.C. Bensahel, L.T. Canham and S. Ossicini, NATO ASI Series, Kluwer Academic Publishers, p 229 (1993)
   b) C. Delerue, G. Allan and M. Lannoo, Phys. Rev. B. **48**, 11024 (1993)
3. A. Bsiesy, J.C. Vial, F. Gaspard, R. Hérino, M. Ligeon, F. Muller, R. Romestain, A. Wasiela, A. Maimaoui, and G. Bomchil, Surface Science **254**, 195 (1991)
4. P.D.J. Calcott, K.J. Nash, L.T. Canham, M.J. Kane, and D. Brumhead, J. Phys. : Condensed Matter **5**, L91 (1993)
5. I. Sagnes, A. Halimaoui, G. Vincent, and P.A. Badoz, Appl. Phys. Lett. **62**, 1155 (1993)
6. B.K. Meyer, D.M. Hofmann, W. Stadler, V. Petrova-Koch, F. Koch, P. Omling and P. Emanuelsson, Appl. Phys. Lett. **63**, 2120 (1993)
7. P. B. Allen, J.Q. Broughton, and A.K. McMahan, Phys. Rev. B **34**, 859 (1986)
8. A. Halimaoui, C. Oules, G. Bomchil, A. Bsiesy, F. Gaspard, R. Herino, M. Ligeon, and F. Muller, Appl. Phys. Lett. **59**, 304 (1991)
9. S. Y. Ren and J. D. Dow, Phys. Rev. B **45**, 6492 (1992)
10. E. F. Steigmeier, B. Delley, and H. Auderset, Phys. Scr. T **45**, 305 (1992)
11. G. D. Sanders and Y. Chang, Phys. Rev. B **45**, 9202 (1992)
12. T. Ohno, K. Shiraishi, and T. Ogawa, Phys. Rev. Lett. **69**, 2400 (1992)
13. A.J. Read, R.J. Needs, K. J. Nash, L. T. Canham, P.D.J. Calcott and A. Qteish, Phys. Rev. Lett. **69**, 1232 (1992)
14. F. Buda, J Kohanoff, and M. Parrinello, Phys. Rev. Lett. **69**, 1272 (1992)
15. T. Takagahara and K. Takeda, Phys. Rev. B **46**, 15 578 (1992)
16. F. Huaxiang, Y. Ling, and X. Xide, J. Phys. Condens. Matter **5**, 1221 (1993)
17. M. V. RamaKrishna and R. A. Friesner, J. Chem. Phys. **96**, 873 (1992)

18. Lin-Wang Wang and A. Zunger, J. Chem. Phys, to be published
19. S. Furukawa and T. Miyasato, Phys. Rev. B **38**, 5726 (1988)
20. S. Schuppler, S.L Friedman, M.A Marcus, D.L. Adler, Y.H. Xie, F.M. Ross, T.D. Harris, W.L. Brown, V.J. Chabal, L.E. Brus and P.H. Citrin, Phys. Rev. Lett. to be published
21. P.M. Fauchet, private communication
22. D.L Dexter, in Solid State Physics, Advances in Research and Applications, edited by F. Seitz and D. Turnbull (Academic, New-york, 1958) Vol 6, p 360
23. M. Hybertsen, Phys. Rev. Lett. **72**, 1514 (1994)
24. V. Lehmann and U. Gosele, Appl. Phys. Lett. **59**, 304 (1991)
25. Y. Kanemitsu, K. Suzuki, H. Uto, Y. Masumoto, T. Matsumoto, S. Kyushin, K. Higushi, and H. Matsumoto, Appl. Phys. Lett. **61**, 2446 (1992)
26. Y. Kanemitsu, K. Suzuki, Y. Nakaoshi, and Y. Masumoto, Phys. Rev B **46**, 3916 (1992)
27. S.M. Prokes, J. Appl. Phys. **73**, 407 (1992); S.M. Prokes, W.E. Carlos, and V.M. Bermudez, Appl. Phys. Lett. **60**, 995 (1992)
28. Y. Takeda, S. Hyodo, N. Suzuki, T. Motohiro, T. Hioki, and S. Noda, J. Appl. Phys. **73**, 1924 (1993)
29. G. Allan, C. Delerue and M. Lannoo, Phys. Rev. B **48**, 7951, (1993)
30. J. Van der Rest and P. Pêcheur, J. Phys. Chem. Solids **45**, 563 (1984)
31. B.J. Min, Y.H. Lee, C.Z. Wang, C.T. Chan, and K.M. Ho, Phys. Rev. B **45**, 6839 (1992)
32. P. Trefonas III , R. West, and R.D. Miller, J. Am. Chem. Soc. **107**, 2737 (1985)
33. M. Lannoo and G. Allan, Phys. Rev. B **25**, 4089 (1992)
34. M.S Brandt and M. Stutzmann, Appl. Phys. Lett. **61**, 2569 (1992)
35. H.J Von Bardeleben, D. Stiévenard, A. Grosman, C. Ortega, and J. Siejka, Phys. Rev. B **47**, 10 899 (1993)
36. D. Goguenheim and M. Lannoo, J. Appl. Phys **68**, 1059 (1990) D. Goguenheim and M. Lannoo, Phys. Rev. B **44**, 1724 (1991)
37. R. Tsu and D. Babic, in Optical properties of low dimensional silicon structures, D.C. Bensahel, L.T. Canham and S. Ossicini Eds NATO, ASI Series (Kluwer Academic Publishers, Dordrecht 1993); see also this volume
38. G. Allan, C. Delerue and M. Lannoo, to be published
39. P.D.J. Calcott, K.J. Nash, L.T. Canham, M.J. Kane, and D. Brumhead, in Microcrystalline Semiconductors - Materials Science and Devices, edited by P.M. Fauchet, C.C. Tsai, L.T. Canham, I. Shimizu and Y. Aoyagi, Mater. Res. Soc. Symp. Proc. **283**, Pittsburg, PA, p 143 (1993)
40. X.L Zheng, W. Wang, H.C Chen, Appl. Phys. Lett. **60**, 986 (1992)
41   J.C. Vial, A. Bsiesy, G. Fishman, F. Gaspard, R. Hérino, M. Ligeon, F. Muller, R. Romestain and R.M. Macfarlane, Mat. Res. Soc. Symp. Proc. Vol **283**, p 241 (1993)
42. G. Fishman, R. Romestain and J.C. Vial, Journal de Physique IV, supplément du Journal de Physique II N° 10. Colloque C5 of the Third International Conference on Optics of Excitons in Confined Systems, Volume 3, p 355 (1993)
43. E. Martin, C. Delerue, G. Allan and M. Lannoo, to be published
44. Y. Kayanuma, Sol. State Comm. **59**, 405 (1986)

Doping of a quantum dot and self-limiting effect in electrochemical etching

R. Tsu and D. Babić

University of North Carolina, Charlotte, NC 28223, U.S.A.

1. INTRODUCTION

The observation of photoluminescence (Canham, 1990) in electrochemically etched porous silicon has launched an intense research activity because this discovery has opened the door for a possible optoelectronic role for silicon, the all important material for electronics. Furthermore, the apparent increase in the fundamental optical absorption energy gap (Lehmann and Gösele, 1991) and the decrease in the Raman phonon frequency with increase of the position of the peak photoluminescence established the role of quantum confinement (Tsu et al., 1992). Fundamentally, quantum confinement pushes up the allowed energies resulting in an increase in the binding energy of shallow impurities such as the cases of quantum well (Bastard, 1981) and superlattice (Ioriatti and Tsu, 1986). Theoretical treatment of the dielectric constant in quantum confined systems (Tsu and Ioriatti, 1986) (Kahen et al., 1985) shows that a significant reduction takes place when the width of the quantum well is reduced to 2 nm and below. In a quantum dot of radius a, the reduction of the size dependent static dielectric constant $\varepsilon(a)$ results in a significant increase of the binding energy (Tsu et al., 1993) of shallow impurities. Since electrochemical etching depends on the current, significant increase in the binding energy can cut-off extrinsic conduction leading to a self-limiting process in the electrochemical etching during the formation of the porous silicon (Tsu et al., 1993). The model used in calculating $\varepsilon(a)$ is the modified Penn model (Penn, 1962) which replaces the continuous electron energies by the discrete energy states of a quantum dot. The calculated $\varepsilon(a)$ agrees (Tsu and Ioriatti, unpublished) with $\varepsilon(q)$ in the results of Walter and Cohen (1970) when q is replaced by π/a. Having obtained $\varepsilon(a)$, we are in a position to compute the binding energy of the shallow impurity, E_b, in a quantum dot. Preliminary results have been presented (Tsu and Babić, 1993) showing a surprising development, that the reduction in the static dielectric constant represents only a small role in the increase of E_b for a small nanoscale silicon particle. The bulk of the increase is due to the induced polarization charges at the boundary of the dielectric discontinuity. The simple physical picture is as follows: (1) the reduction of the static dielectric constant plays a small role in increasing the binding energy of a donor or acceptor via a reduction in dielectric screening: (2) a far more significant role is due to the induced charges at the dielectric interface between the quantum dot and the matrix in which the particle is embedded. With ε_1 and ε_2 denoting the dielectric constant of the particle and the matrix and for $\varepsilon_1 > \varepsilon_2$, the induced charge of the donor is of the same sign resulting in an attractive interaction with the electron of the dot, pushing deeper the ground state energy of the donor resulting in an appreciable increase in E_b. For $\varepsilon_1 < \varepsilon_2$, the opposite is true, E_b is much reduced allowing possible extrinsic conductivity at room temperatures. Discussions of totally different behaviors of porous silicon in air and water were raised by V. Lehmann and J.C. Vial at the Grenoble Workshop (1993), suggesting that the different behaviors of porous silicon in an aqueous solution and air may be attributed precisely to the difference in the binding energies. In retrospect, the good result of electroluminescence device of Nippon Steel fabricated by Matsumoto (1993) using a SiC/pSi/Si pn junction (pSi stands for porous silicon), may very well be due to the matching of the dielectric constant of SiC to the reduced dielectric constant of the porous silicon dot, resulting in the elimination of the induced charges at the dielectric discontinuity. Therefore, an understanding of the binding energy of a quantum dot donor embedded in a matrix is not only of fundamental importance, but also of technological significance in the understanding of extrinsic conduction and doping, and may even play a vital role in future generations of optoelectronic devices such as modulators and light switches.

2. CALCULATION AND RESULTS

The calculation of the binding energy depends on several assumptions: (1) the dielectric constant of the quantum dot is represented by $\varepsilon_1 = \varepsilon(a)$ (Tsu et al., 1993) (Penn, 1962) (Tsu and Ioriatti, unpublished), and the matrix, by ε_2, (2) ideal interface between the dot and the matrix, and (3) the electronic part is described within the effective mass approximation using an isotropic effective mass. Validity of the effective mass approximation for a quantum dot depends on the range of the Bloch function wave vectors necessary to construct the envelope wave function for the dot. This range has to be much smaller than the width of the Brillouin zone. For GaAs wells the effective mass approximation is valid down to well widths of the order of 2 nm (Priester et al., 1983). Due to nearly identical lattice constants of Si and GaAs and thus nearly identical sizes of the Brillouin zones, the effective mass approximation should provide a meaningful results for the silicon quantum dots of similar sizes. The binding energy E_b(bulk) = 24.56 meV, obtained from the hydrogenic model for $\varepsilon_1 = 12$ and the isotropic effective mass. This value is much lower than the experimental value of 45 meV for phosphorus doping. Luttinger and Kohn (1955) attempted to include anisotropic mass but to no avail. It was precisely the introduction of $\varepsilon(q)$ allowing a reasonable agreement with experiments (Pantelides, 1978). Using a wave-vector-independent dielectric constant $\varepsilon(a)$ for ε_1, the Hamiltonian of our problem may be readily written down,

$$H = -\frac{\hbar^2}{2m_e} \nabla^2 + V(r) + \phi_C(r) + \phi_P(r) + \phi_S(r) \tag{1}$$

with

$$V(r) = \begin{cases} 0 & r<a \\ \infty & r \geq a \end{cases},$$

and the direct Coulomb potential ϕ_C between the donor and the electron is

$$\phi_C = - \frac{q^2}{4\pi\varepsilon_0\varepsilon_1 r} \tag{2}$$

and the self-polarization between the electron and its "image" is (Babić, et al., 1992)

$$\phi_s = \frac{1}{2} \sum_l \frac{q^2(\varepsilon_1-\varepsilon_2)\,(l+1)\,r^{2l}}{4\pi\varepsilon_0\varepsilon_1[\varepsilon_2 + l\,(\varepsilon_1+\varepsilon_2)]\,a^{2l+1}} \tag{3}$$

and taking the s-state for the spherically symmetrical ground state, the polarization term between the electron and the induced polarization of the donor ϕ_p is (Babić et al., 1992)

$$\phi_p = - \frac{q^2(\varepsilon_1-\varepsilon_2)}{4\pi\varepsilon_0\varepsilon_1\varepsilon_2 a} \tag{4}$$

Note that we have excluded the self-polarization term between the donor and its induced polarization, because this interaction contributes to the donor formation energy when the donor is introduced into the quantum particle. The ground state energy of the donor, E_0 is obtained by a variational technique of minimizing E_0 with respect to c in the ground state s-type trial function

$$\psi(r) = [\,1 - (\tfrac{r}{a})^2\,]\,e^{-\frac{r}{c}} \tag{5}$$

Note that this trial wave function satisfies the boundary condition of $\psi(a) = 0$. Figure 1 gives the size dependence of the static dielectric constant versus the radius of the dot in angstrom (Tsu and Ioriatti, unpublished). For a given radius a, $\varepsilon_1 = \varepsilon(a)$ is used to obtain the ground state E_0 numerically. Figure 2 shows the computed E_0 for $\varepsilon_2 = 1$ with and without polarization terms. Note that there is a difference of about 0.4 eV at a radius of 15 Å, considered to represent the typical size of porous silicon. Figure 3 shows how we define the binding energy $E_b = E_1 - E_0$ where E_1 is the lowest allowed state without the positively charged donor, but including the self-polarization effects. Obviously our definition makes sense only when the donor density is such that majority of silicon particles contain no donor. As defined, the binding energy measures the energy necessary to ionize the donor in one quantum particle and promote the electron to the lowest state in another neighboring dot with no donor. The electron is thus made available for transport through porous silicon, most likely by tunneling. Figure

4(a) shows a particle with a donor at its center surrounded by particles having no donor. The matrix is ε_2.

However what we have treated is a simplified problem as shown in Fig. 4(b). This simplification has been made to allow much easier computation of the relevant energies.

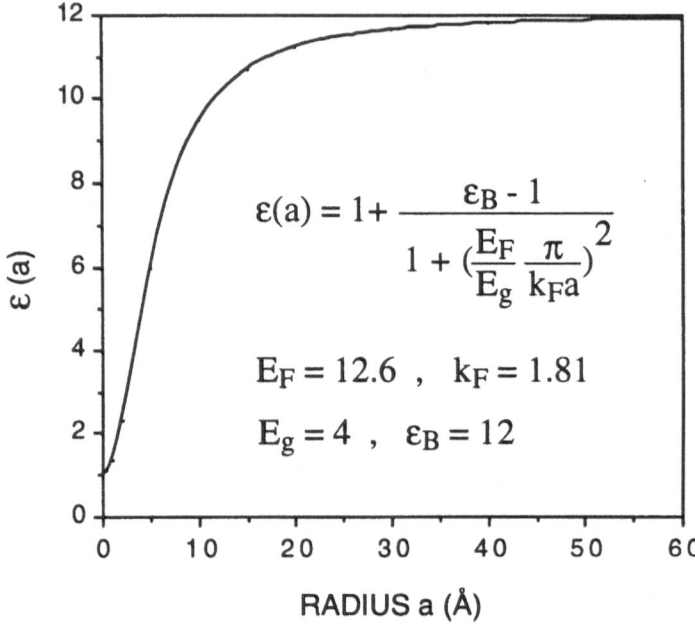

Fig. 1   The calculated static dielectric constant versus the radius of a Si sphere with the modified Penn model.

Charged donor induces polarization surface charge at the dielectric interface as well as the surrounding particles. Since the Coulomb potential of a donor charge decreases with increasing distance from the donor, appreciably less polarization charge is induced on the surrounding particles if they are located further away from the silicon particle with the donor. Therefore the presence of the surrounding particles may be approximated by an effective matrix. On the other hand, if the surrounding particles are very close to the particle with the donor, dielectric screening effects of a thin matrix layer are much reduced so that the dielectric discontinuity is almost absent. This qualitative discussion brings out a very important point: the actual surface polarization effects are bracketed by the case with no polarization at all ($\varepsilon_1 = \varepsilon_2$) and the case of the homogeneous infinite matrix. Figure 5 shows the computed binding energy E_b versus the dot radius for four values of ε_2 representing the cases of the matrix having the same permittivity ($\varepsilon_1 = \varepsilon_2$) and the matrix being air and water. Two very different dielectric constants for water ($\varepsilon_2 = 6$ and 80) have been used (Peter, 1984). A discussion of the proper dielectric constants of water for various mechanisms can be found in Bockris and

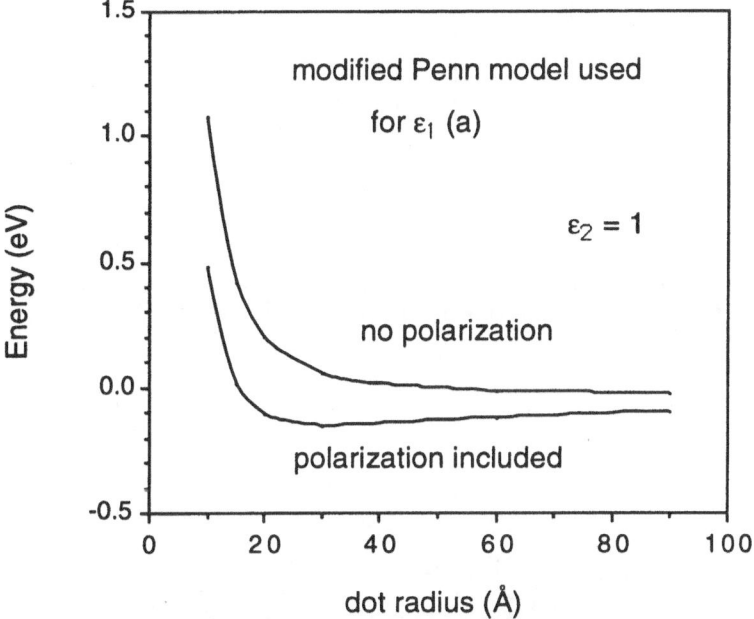

Fig. 2   The ground state energy of a donor at the center of a sphere in air with and without induced polarization.

Reddy (1973). The value of 6 belongs to the thin layer of water in contact with silicon (or some ion). In this layer, referred to as the primary solvation sheath, the dipoles are bound and cannot be oriented by external fields. Therefore the value for the dielectric constant of water in the immediate vicinity of the silicon particle should be much lower than the bulk value. Measurements suggest 6 as the dielectric constant for the primary solvation sheath. The sum of the kinetic and Coulomb energy terms of E_b for air and water is almost the same as for the case $\varepsilon_1 = \varepsilon_2$. The polarization effects thus introduce primarily a rigid shift whose magnitude scales as $1/\varepsilon_2$ being therefore much larger for air than for water. Note that at the typical radius of 15 Å the binding energy in water E_b varies from ~150 meV ($\varepsilon = 80$) to ~300 meV ($\varepsilon = 6$) reducing the carrier concentration by 3 to 6 orders of magnitude compared to the bulk conditions at room temperature. For these circumstances extrinsic conduction ceases. The situation is even more dramatic in air, with a donor binding energy of ~1.1 eV, these nano-particles are practically insulators.

R. Tsu and D. Babić

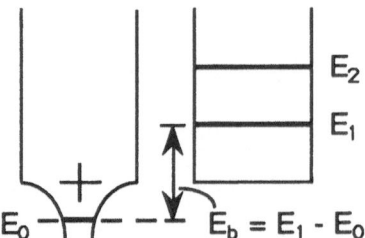

Fig. 3   Our definition of the binding energy E_b in terms of the difference between the lowest allowed
state in a neighboring particle without the donor and the ground state of a donor.

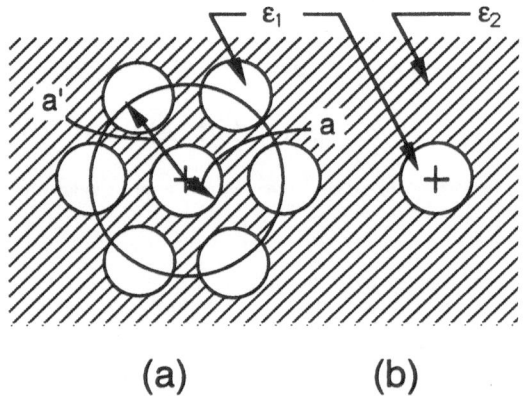

Fig. 4   (a) Actual situation where a donor at the center of a sphere is surrounded by other spheres and
(b), our simplified model where the sphere with the donor is immersed in a uniform matrix.

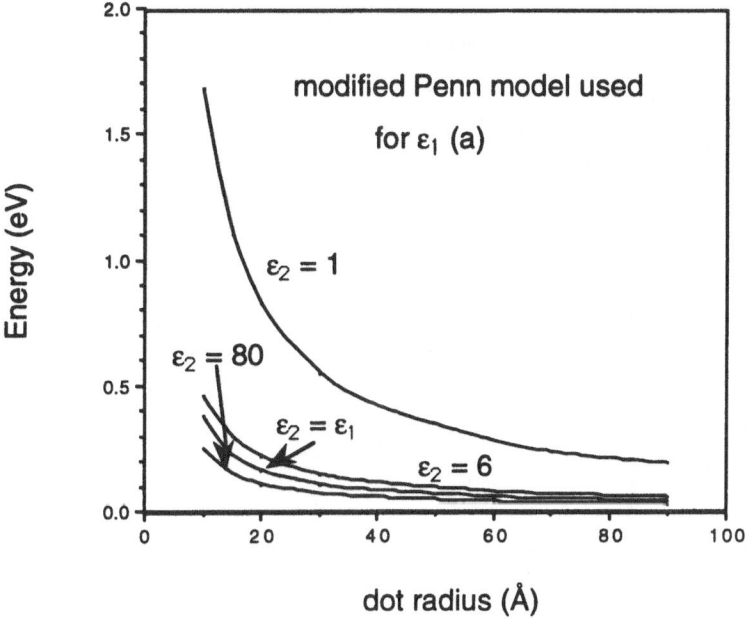

Fig. 5   The donor binding energy E_b versus several values of the dielectric constant for the matrix.

3. SELF-LIMITING ETCHING IN POROUS SILICON

Porous silicon is usually formed by anodic etching of p-type silicon. Although our calculations have been carried out for n-type silicon, the conclusions are immediately applicable to p-type. The reason is the dominance of the electrostatic energy terms which are the same for either donors or acceptors. Dealing with donors has allowed a major simplification of the kinetic energy term compared to the case of acceptors where one has to treat light and heavy hole degeneracy and use much more complicated Luttinger-Kohn Hamiltonian (Luttinger and Kohn, 1955). The dramatic increase of the acceptor binding energy due to dielectric mismatch and quantum confinement offers an explanation for the self-limiting etching of porous silicon. In the beginning of etching, acceptor binding energy is low, with the same value as in the bulk. The positive voltage applied to the p-type silicon produces an accumulation of mobile holes at the silicon electrolyte interface enabling etching. Figure 6(a) shows the energy bands in silicon and redox states in the solution in this case. As the etching progresses the dimensions of unetched silicon are reduced and the binding energy of acceptors increases sharply. Concentration of free holes decreases making silicon to appear intrinsic. Figure 6(b) show the energy bands in nano-silicon with respect to the redox states in the solution for this case. Without the accumulation of holes at the interface electrochemical etching cannot proceed. Although some holes can tunnel from acceptors to the solution, this does not constitute etching since no silicon bond at the interface is involved.

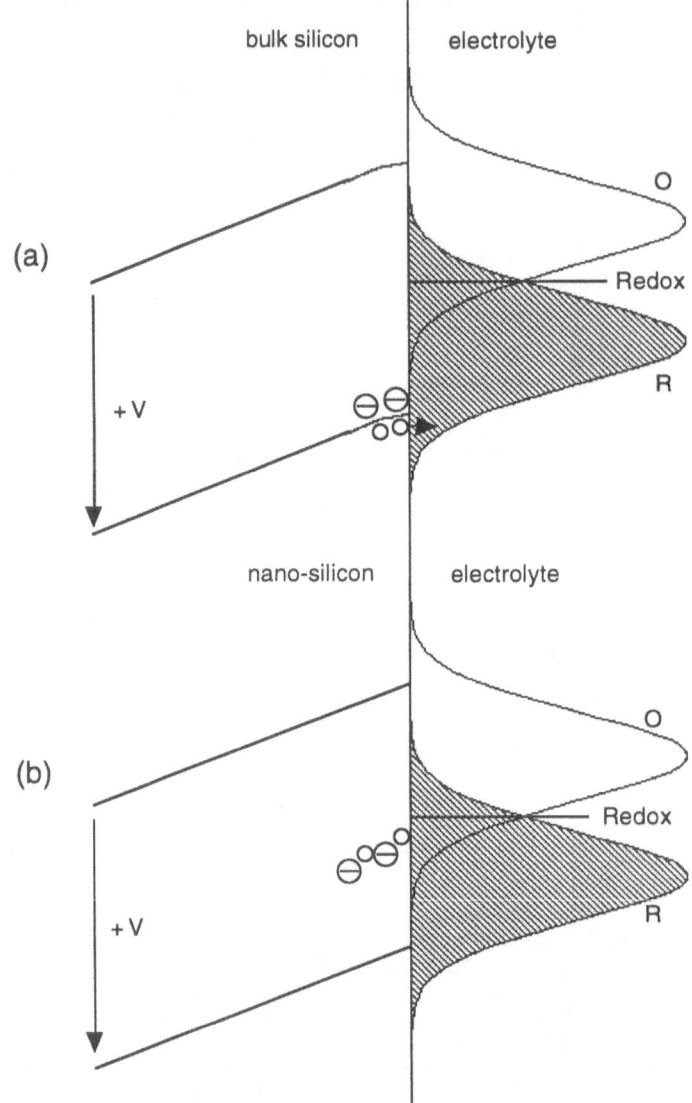

Fig. 6   Energy band diagram in silicon with respect to the redox states in solution under the applied positive anodic voltage for: (a) bulk silicon when etching can proceed; and (b) silicon nanoparticles with no etching.

4. FUTURE WORK

As discussed earlier, referring to Figure 4, the calculated binding energy E_b is that of Figure 4b where a particle of ε_1 is immersed in a medium of ε_2 of infinite extent. What we need to consider is schematically represented by Figure 4a. The actual binding energy lies in the range bounded by $\varepsilon_2 = \varepsilon_1$ and $\varepsilon_2 = 1$ curves for air and the $\varepsilon_2 = \varepsilon_1$ and $\varepsilon_2 = 80$ curves for water. In brief, fraction of the induced charges interact with the electron confined in a given particle with a donor. We are developing a method to tackle this difficult situation which involves the use of an effective "electrostatic" radius for the particle being different from the radius for the electron wave function. The rationale is that a Brüggeman type effective medium for ε_2 may be used to replace the situation of Figure 4(a). The details of such a model will be the subject of further investigation. Nevertheless, our results should serve as the basis for a better understanding of doping in nanoscale particles, and the self-limiting process in electrochemical etch involving silicon of nanoscale dimensions.

Acknowledgment

This work is supported by ARO-DAAL03-90-G-0067 and ONR N00014-90-J-1752.

References

Babic D., Tsu R. and Greene R.F., *Phys. Rev.* **B45** (1992) 14150.
Bastard G., *Phys. Rev.* **B24** (1981) 4714.
Bockris J.O'M. and Reddy A.K.N., Modern Electrochemistry (Plenum, New York, 1973) p. 156.
Canham L.T., *Appl. Phys. Lett.* **57** (1990) 1046.
Ioriatti L. and Tsu R., *Surf. Sci.* **174** (1986) 420.
Kahen K.B., Leburton J.P. and Hess K., *Superlattice and Microstructure* **1** (1985) 289.
Lehmann V. and Gösele U., *Appl. Phys. Lett.* **60** (1991) 856.
Luttinger J.M. and Kohn W., *Phys Rev.* **97** (1955) 869.
Matsumoto T., Futagi T., Mimura H. and Kanemistu Y., 1992 Int. Conf. Solid State Devices and Materials (Tsukuba, Japan) p.478., also in Futagi T., Matsumoto T., Katsuno M., Ohta Y., Mimura H. and Kitamura K., *Mat. Res. Soc. Symp. Proc.* **283** (1993) 389.
Pantelides S.T., *Rev. Mod. Phys.* **50** (1978) 797 .
Penn D.R., *Phys. Rev.* **128** (1962) 2093.
Peter L. suggested that a value of 6 should better represent the physics involved during the discussions at the Les Houches Winter School "Luminescence of Porous Silicon and Silicon Nanostructures" from 8 to 12 February 1994.
Priester C., Allan G. and Lannoo M., *Phys. Rev.* **B28** (1983) 7194.
Tsu R. and Ioriatti L., *Superlattice and Microstructure* **1** (1985) 295.
Tsu R., Shen H. and Dutta M., *Appl. Phys. Lett.* **60** (1992) 112.
Tsu R., Ioriatti L., Harvey J.F., Shen H. and Lux R.A., *Mat. Res. Soc. Symp. Proc.* **283** (1993) 437.
Tsu R. and Babić D., "Effects of the Reduction of Dielectric Constant in Nanoscale Silicon", Optical Prop. of Low Dimensional Silicon Structures, CNET, France, 1-3 March 1993, edited by Bensahel D., Canham L.T. and Ossiani S. (Kluwer, Dordrecht/Boston/London, 1993) p. 203.
Tsu R. and Ioriatti L., unpublished.
Walter J.P. and Cohen M.L., *Phys. Rev.* **B2** (1970) 1821.

Electronic and optical properties of semiconductors quantum wells

R. Ferreira

*Laboratoire de Physique de la Matière Condensée,
Ecole Normale Supérieure, 24 Rue Lhomond,
75005 Paris, France*

INTRODUCTION

In this contribution we review some basic concepts concerning the optical properties near fundamental band gap of low dimensional structures. We shall discuss the electronic and optical properties of semiconductor quantum wells. Such structures are the object of intensive studies since many years, and a good understanding of most of its basic properties is nowadays achieved [1-3]. Various effects linked to the quantum confined motion of the carriers through the different semiconductor layers have been experimentally observed and theoretically interpreted. Various theoretical models have been used for the study of the electronic states of such crystaline heterostructures (for a review see for instance ref.[4]). In this work we shall consider the envelope function approximation. This formalism is characterized by its simplicity and flexibility, and has proven to give a coherent explanation for the effects we discuss in the following.

ELECTRONIC STATES

We present in fig.(1) the bulk states near the fundamental band gap for a zinc-blend semiconductor. The fundamental band gap is direct in the k space and placed at the Γ point of the first Brillouin zone. At k=0 the conduction band is twofold degenerated (Γ_6 symetry) and the valence band present four degenerated Γ_8 states and 2 split-off Γ_7 states. The wavefunction for these states are given by $u_n(\mathbf{r})$ n=1,...,8 and have the appropriate Γ_6, Γ_7 or Γ_8 symetries. The band gap energy is around 1 eV. For such wide gap semiconductors we can describe separately the conduction band and valence band states. In the effective mass approximation we use the k=0 wavefunctions as a basis to the wavefunctions at non-zero (but small) wavevector **k** :

$$\Psi_\mathbf{k}(\mathbf{r}) = \Sigma_n \ \alpha_n(\mathbf{k}) \ \exp\{ \ i\mathbf{k}\cdot\mathbf{r} \ \} \ u_n(\mathbf{r}) \tag{1}$$

where n runs over the 2 (6) k=0 states in the conduction (valence) band and the α_n are k-dependent coefficients. This is a Bloch-like function, with rapidly variÿng center-of-zone functions (periodic in the real space) multiplied by slowly variÿng "envelope" functions (periodic in the reciprocal lattice space ; $|\mathbf{k}|<<\pi/a$, where a is the direct lattice period ; $a\approx5\text{Å}$).

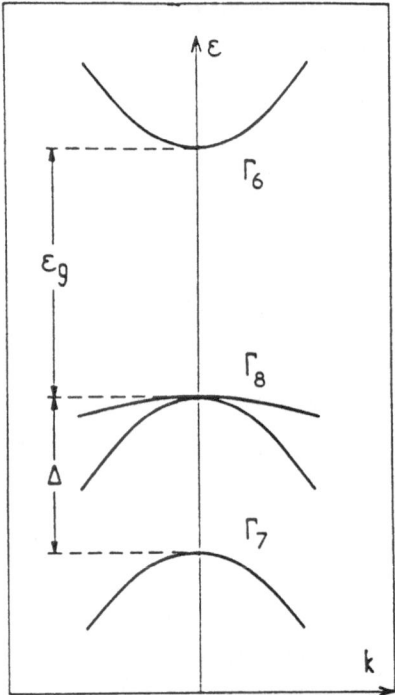

Fig.(1) Schematic representation of the band structure of III-V semiconductors
in the energy region around the fundamental band gap

A quantum well (QW) structure is obtained by the growth (epitaxially, for instance) of a thin layer of semiconductor A between two thick layers of semiconductor B, where A and B have similar structural properties and the material B presents a wider band-gap. The following zinc-blend III-V semiconductors pairs A/B present very low lattice constant mismatch and satisfy the band-gap requirement : A/B = $GaAs/Ga_{1-x}Al_xAs$ (with $x\leq0.3$) ; GaSb/AlSb ; $Ga_{0.47}In_{0.53}As/InP$; $Ga_{0.47}In_{0.53}As/Al_{0.48}In_{0.52}As$. For these A/B pairs the difference in the band-gaps originates sharp conduction band and valence band discontinuities at the QW interfaces such that $\varepsilon(\Gamma_6)_{well} < \varepsilon(\Gamma_6)_{barrier}$ for the botton of the conduction band, and $\varepsilon(\Gamma_8)_{well} > \varepsilon(\Gamma_8)_{barrier}$ for the top of the valence band. These are the so called type I QWs : the central A layer (the lateral B layers) acts as potential "wells" (potential "barriers") for both the electrons and the holes. Different conduction band and valence band alignements do exist for different A/B-based QWs (the reader is refered to ref.[1] for a most detailed description of these different QW structures). In this work we consider only type I QWs.

We consider now the main assumptions of the envelope function formalism to describe the electronic states of QWs. Let z be the QW growth axis. We write a QW total wavefunction in the general form :

$$\Psi_{k\perp}(r) \,|_{QW} = \Sigma_1 \, \Sigma_b \, \alpha_{1,b}(k_\perp) \, \Psi_{b,1,k\perp}(r) \tag{2}$$

where $\Psi_{b,1,k\perp}(r)$ is the bth bulk wavefunction for the 1th material layer forming the QW, as follows. Generally we have a translational invariance in the layer plane ; thus the in-plane component of the wavevector, $k_\perp=(k_x,k_y)$, is a good quantum number. Next we take for each layer "l" all the bulk states "b" associated with the same band extremum (which is the Γ extremum for the valence states of III-V semiconductors and either Γ or X or L for the conduction band states) and with same energy and $k_\perp$ for all the layers, and use these states $\Psi_{b,1,k\perp}(r)$ as a basis for the QW wavefunction. Next we assume that the center-of-zone periodic functions $u_n(r)$ are the same for the A and B material. Since $k_\perp$ and the $u(r)$ are the same for the different layers only the motion along the growth axis remains to be determined. In fact, for each layer there is at least two degenerate bulk states, for the free propagations in opposite directions along the growth axis. Wa take the linear combinations of these plane waves with coefficients $\alpha_{1,b}(k_\perp)$ wich depend on the energy and in-plane wavevector. We call the resulting function f(z) the z-dependent envelope function. In this case we can calculate [1] the effective hamiltonian acting on the slowly varying envelope function f(z) and solve the resulting Schroedinger-like problem with the help of suitable matching conditions at each QW interface (the continuity of f(z) and of the current of probability) and by specifying the limiting (or asymptotic) behaviour of the envelope functions at large distance of the quantum well region (e.g. $|f(z)| \rightarrow 0$ as $|z| \rightarrow \infty$ for the QW bound states (see below)).

For the parabolic conduction band we have the Schroedinger-like problem for the envelope eigenstates :

$$\{ - (\hbar^2/2) [\partial/\partial z \, (1/m^*(z)) \, \partial/\partial z]+ V_c(z) \} \, f(z) = \{ \varepsilon - \hbar^2 k_\perp^2/2m^*(z) \} \, f(z) \tag{3}$$

where $V_c(z)$ is a piecewise function giving the variation along the growth axis of the botton of the conduction band (bulk) energy. Thus, we have a particle-in-a-box unidimensional problem to solve. We might seek for solutions with energy around the two Γ_6 bulk edges in fig.(2). For $\varepsilon < \varepsilon(\Gamma_6)_{barrier}$ we are in the energy gap of the barrier material. Actually, we have two kinds of states in bulk semiconductors : (i) propagative band states, which are true stationnary

R. Ferreira

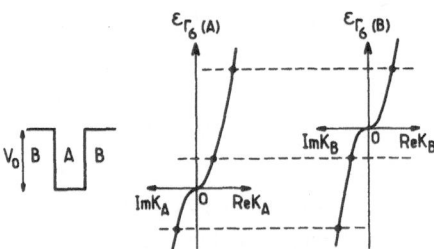

Fig. (2) Bulk energy dispersions of the well (A) and barrier (B) materials
around the Γ_6 conduction band edges (schematic).

solutions, and (ii) states in the forbiden gap region, which are evanescent states and thus non-stationnary solutions of the perfect crystal. Then, in QWs we have three different energy regions to consider. The first one , for $\varepsilon < \varepsilon(\Gamma_6)_{\text{well}}$, present evanescent states in the two materials, and no solution is admited for the QW in this energy region. A second one with $\varepsilon > \varepsilon(\Gamma_6)_{\text{barrier}}$, with propagative modes in the two materials, which will give rise to a continuum of QW states wich are extended along the growth axis. And, finally, the intermediary energy region with propagative bulk states in the well layer and evanescent bulk states in the barriers. The envelope function eigenstates in that energy interval present a discrete spectrum and QW envelope solutions localized in space around the quantum well layer. Finally we have in that last region (we neglect here for simplicity the z-dependence of the Γ_6-related effective mass m* for the in-plane motion ; see also ref.[1]) :

$$\Psi_{\mathbf{k}\perp}(\mathbf{r}) = f_n(z) \ \exp\{ \ i\mathbf{k}_\perp \cdot \mathbf{r}_\perp \ \} \ u_{\Gamma6}(\mathbf{r}) \qquad (4.a)$$
$$\varepsilon_n(\mathbf{k}_\perp) = E_n \ + \ \hbar^2 \mathbf{k}_\perp^2/2m* \qquad (4.b)$$

where the first energy term accounts for the discrete levels associated with the quantum confined motion in the z direction (in fig.(3) we show the envelope functions for the first two bound states of the QW : n=1,2) and the second energy term accounts for the two dimensional dispersions associated with the free propagation in the layer plane. Note the presence of an energy shift for the QW ground state $\varepsilon_1(\mathbf{k}_\perp=0)$, which is due to the quantum confinement imposed by the two barrier layers. That confinement energy depends strongly upon the QW width and barrier height (offset discontinuity $V_0 = \varepsilon(\Gamma_6)_{\text{barrier}} - \varepsilon(\Gamma_6)_{\text{well}} > 0$). We show in fig.(4) the QW eigenstates (at $\mathbf{k}_\perp=0$) as a function of the well thickness (L) for a Ga(In)As-InP QW. The barrier height is $V_0 = 244$ meV. As a general rule any one-dimensional attractive potential binds at least one state (E_1 for 0<L≤50Å). With increasing L more and more states become bound. Note that the confinement energies increase continuously with decreasing well width.

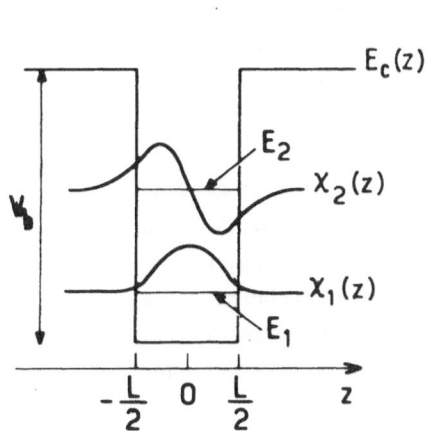

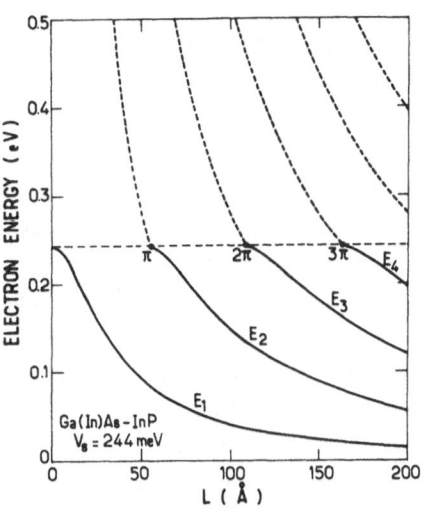

Fig.(3) First two envelope wavefunctions (χ_n=f_n(z)) for the confined motion along the growth direction z in a single quantum well structure (schematic).

Fig.(4) Calculated bound (solid lines) and virtually bound (dashed lines) conduction states in lattice-matched Ga(In)As-InP single quantum wells [3]

For holes in the valence band the situation is generally much more complicated. Actually, the valence states are strongly perturbated by spin-orbit interactions and the in-plane dispersions (or the dependence of the hole energies upon $k_\perp$) are no more given by a simple parabolic law, as for electrons. We refer the reader to ref.[1] for a detailed discussion of the valence band dispersions in quantum wells. In this work we will suppose that the hole dispersions are parabolic. This constitutes the simplest approximation for the hole states and will permit us to obtain some analytical results in the following. On the other hand, it is worth noting that at $k_\perp$=0 we have two different kinds of hole states : the heavy holes and the light holes. Without considering the effective hamiltonian for the hole envelope states, we can justify this statement by considering again the topmost valence Γ_8 bulk states (we are not concerned with the Γ_7 states, which are generally well energy separated of the topmost Γ_8 states, and thus can be neglected when we are interested only in the QW states near the fundamental band gap). We have two different branches in fig.(1), the heavy one and the light one. Since the quantum confined effect depends upon the mass of the confined particle, the heavy hole states and the light hole states will have different confinement energies and thus each bulk dispersion shall contribute with a different kind of hole state in the QW. The $k_\perp$=0 heavy hole and light hole states present the same confinement aspects as found previously for electrons : discrete levels in the energy region between the $k_\perp$=0 Γ_8 extrema of the well and barrier materials, associated to which there are z-dependent envelope functions localized along the growth axis around the quantum well layer. Finally we note that the offset discontinuity is the same for the heavy and light states (but different from the conduction band one) and that the total offset discontinuity (conduction band + valence band) is equal to the difference between the A and B energy gaps.

We discuss now briefly the excitonic states in QW's. In the effective mass approximation an electron in the conduction band and a hole in the valence band interact through a coulomb-like screened potential. Since this potential is attractive, the ground energy for the correlated electron-hole pair is lower than the QW energy gap (which is the ground energy of the uncorrelated electron-hole pair). As usual, the motion of the interacting two-body problem can be replaced by a relative and a center-of-mass motions The relative motion is dominated by the coulombic interaction and give rise to hydrogen-like states whereas the center-of mass propagates freely (Wannier-Mott excitons). To discuss the effects of confinement on the excitonic states we will consider that the electron and the hole are both confined between hard walls. We present schematically in fig.(5) the exciton binding energy (in units of the three dimensional Rydberg R_y*) as a function of the QW width (in units of the three dimensional Bohr radius a_B*). We consider only the $K_{CM,\perp}=0$ excitonic states, where $K_{CM,\perp}$ is the wavevector for the free center-of-mass in-plane motion. We have four different cases. (i) The three dimensional (L=∞) ground bulk exciton has a spherical 1S-like symmetry for the relative motion ; its energy reads $\varepsilon_{3D} = \hbar^2 (K_{CM,z})^2/2M - R_y$*, where R_y* $= \hbar^2/(2\mu a_B$*$^2)$, with μ (M) the reduced (total) mass of the electron hole pair. (ii) For wide wells (a_B*$<<L<∞$) the confinement is weak and does not modify the strongly correlated electron-hole relative motion. Only the z-motion of the exciton as a whole entity is modified by the confinement : we are in the center-of-mass confinement regime. The exciton ground energy reads approximatively : $\varepsilon_{CM} \approx \hbar^2 \pi^2/(2ML^2) - R_y$*. (iii) For thin wells ($0<L<$some few a_B*) the electron and the hole motions are imposed by the strong confinement potential and thus, since these motions are separately quantized, so the relative motion along the growth axis. The exciton energy reads approximativelly $\varepsilon_{rel} \approx \hbar^2 [(\pi/L)^2 - 1/\lambda^2]/2\mu$ where $1/2 < \lambda/a_B$* < 1, where λ is a L-dependent variational parameter. This is the relative motion quantized regime. (iv) Finally, the two dimensional (L=0) ground bulk exciton has an in-plane 1S symmetry for the relative motion and its ground binding energy reads $\varepsilon_{2D} = - 4R_y$* .

In conclusion, the exciton becomes more and more stable, in the sense that the binding of the electron-hole pair increases continuously with increasing confinement of the carriers. This result holds both for the heavy hole and for the light hole excitonic states.

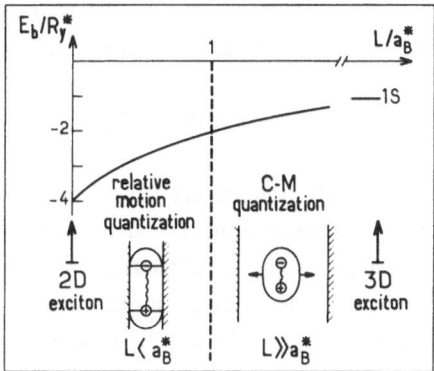

Fig.(5) Schematic representation of the center-of-mass and relative motion quantization regimes
for excitons confined between hard walls.

INTERBAND OPTICAL ABSORPTION

We consider now the interband optical absorption in quantum wells. The absorption probability
(or the absorption coefficient times the quantum well thickness) is given by the Fermi Golden
Rule :

$$\alpha(\omega)\, L \approx (1/S\omega) \sum_{i,f} |<f\,|\,\hat{e}\cdot\mathbf{p}\,|\,i>|^2\ \delta\{\ \varepsilon_f(\mathbf{k}_{\perp f}) - \varepsilon_i(\mathbf{k}_{\perp i}) - \hbar\omega\ \} \tag{5}$$

where S is the in-plane area ; $\hbar\omega = \hbar\omega_{las} - E_{gap}$ is the excess laser energy as regards the direct Γ
fundamental gap of the bulk quantum well material ; the sum is performed over all the initial $|i>$
and final $|f>$ states in the valence and conduction bands respectively ; $\hat{e}$ is the light polarization
vector and $\mathbf{p} = -i\hbar\partial/\partial\mathbf{r}$. We have two different main contributions (we take the population
factor equal to one for the undoped structures we are interested in). The first factor is the
conduction band - to - valence band joint density of states, given by the sum over all the initial
and final states of the energy conservation delta-like expression. For a given laser energy above
the fundamental QW band-gap this joint density of states gives us the answer to the question :
How many (valence-band) $\rightarrow$ (conduction band) transitions are in principle possible in a small
energy interval around this laser energy ? The next question is almost evident : How strong are
they? In the weak absorption regime the strenght of a given possible transition is set by the
dipole matrix element in eq.(5). Thus, it is these matrix elements that give us the selection rules
for the energy permitted optical interband transitions. In the following we discuss somewhat

qualitatively the different selections rules for the interband optical absorption. In the envelope function approximation the optical matrix elements are given by a sum of terms of the form

$$M_{c,v} \Rightarrow <f_c(z) \mid f_v(z)> <u_c(r) \mid \hat{e} \cdot p \mid u_v(r)> \delta(k_{\perp c} - k_{\perp v}) \tag{6}$$

connecting the initial (an electron in the valence band) and the final (an electron in the conduction band) QW states. That simple form is obtained grace to the very different spatial variations of the rapidly oscillating periodic part of the Bloch functions and the slowly varying envelope functions. The first term is the envelope function overlap in the z-direction. If the QW is symmetric as regards the center of the well layer, then the functions f(z) have well defined parity (see fig.(3)). Thus, only parity-conserving transitions are possible. The last term accounts for the translational invariance in the layer plane and show us that only vertical transitions (in the wavevector space) are permitted. Finally, the optical matrix elements give us the selection rules for the polarization and propagation directions of the laser light. Indeed, we have no more an isotropic bulk-like material and the absorption should be different for different propagation directions and/or polarizations.

We can easily estimate the absorption coefficient in a simple model. For infinite conduction band and valence band barriers the envelope functions and energies are the solutions of a particle in a hard wall box. We obtain that : (i) the f(z) are sinus or cosinus functions ; in this case the envelope function overlap between the nth valence and the mth conduction states is simply given by $\delta_{n,m}$; (ii) for a given particle (electron, heavy hole or light hole) the confinement energies increase quadratically with the discrete level number n=1,2,3,..., and the confinement shift is inversely proportional to the particle mass and to the square of the well size : $E_n = n^2 E_1$ with $E_1 = \hbar^2 \pi^2 / (2m^* L^2)$. Also, we take parabolic dispersions for both the conduction and the valence bands. Finally, we have :

$$\alpha(\omega) L \approx |m_{c,v}|^2 \Sigma_n Y[n^2 \hbar^2 \pi^2 / (2\mu L^2) - \hbar\omega] \tag{7}$$

where $m_{c,v}$ is the optical matrix element (second term in eq.(6)) and Y[x] is the step function (Y[x>0]=0 and Y[x<0]=1). Thus, the absorption probability is proportional to the square of the optical matrix element times a sum of constant terms. Fig.(6) shows the absorption coefficient as a function of the excess photon energy. In the upper pannel only two transitions are presente. For a fixed laser energy, more and more transitions become permitted with increasing L, but the absorption step decrease since $\alpha(\omega)$ is proportional to 1/L (middle pannel). As L goes to infinity (botton pannel) we have an infinite number of very weak transitions and we can evaluate exacly the sum in eq.(7) to obtain for the absorption coefficient :

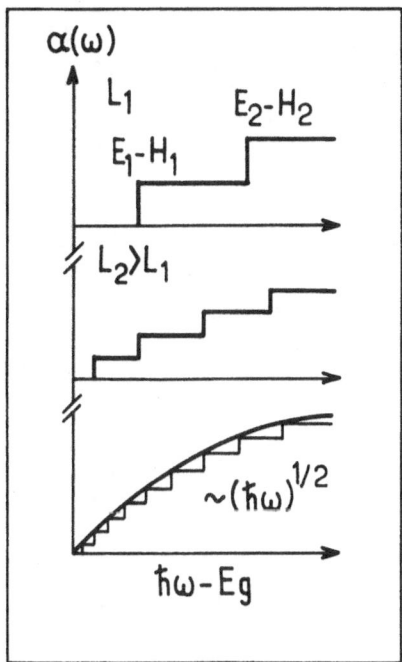

Fig.(6) Schematic representation of the continuous evolution of the band-to-band absorption profile
for a single quantum well when increasing the well width (from top to botton).

$$\alpha(\omega) \approx |m_{c,v}|^2 \ [\ \hbar\omega \ ]^{1/2} \tag{8}$$

In this case $\alpha(\omega) \approx \sqrt{(\hbar\omega_{las} - E_{gap})}$, as expected for the three dimensional bulk absorption. Thus, in short, the use of such a simple model permitted us to follow the increasing modification of the optical response of the QW which appears with decreasing QW size or, in other words, with increasing carriers' confinement.

The question now is : how does the excitonic effects change that simple band-to-band picture? For excitons we have the following selection rules. The only optically active excitonic states are those with "S" symmetry for the relative motion and with vanishing center-of-mass wavevector K_{CM}. Note that K_{CM} is a three dimensional vector for bulk excitons and is an in-plane vector for QW's or for idealized two dimensional systems. We present in fig.(7) schematically the three dimensional and the two dimensional absorption coefficients. In the gap region ($\varepsilon < 0$) the laser light interacts with bound excitonic S-like states and for positive energies ($\varepsilon > 0$) we have the photocreation of dissociated (but correlated) electron-hole pairs. The bound excitons binding energies (E_b) and oscillator strengths (OS) are given by

$$E_b |_{3D} = R_y^*/n^2 \qquad ; \qquad OS |_{3D} \approx [\ 4\pi a_B^{*3} n^3 \ ]^{-1}$$
$$E_b |_{2D} = R_y^*/(n-1/2)^2 \quad ; \qquad OS |_{2D} \approx [\ \pi a_B^{*2}(n-1/2)^3 \ ]^{-1}$$

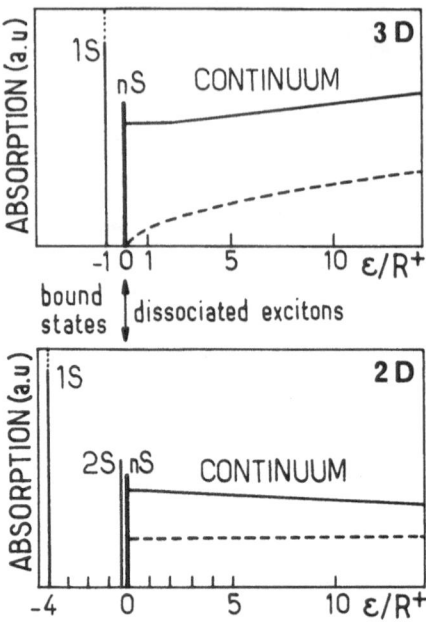

Fig.(7) Schematic absorption profiles for 3D (upper pannel) and purely 2D (lower pannel) materials, with (solid lines) and without (dashed lines) Coulomb interaction.

Note that the OS are inversely proportional to the "volume" of the exciton and decrease rapidly with increasing n for the successif nS states. The dashed lines in fig.(7) are the band-to-band absorptions, plotted for comparison. Thus, finally, we have two main effects due to excitons : the presence of sharp lines in the band gap, due to bound excitons, and an enhancement of the absorption in the continuum part. For quantum wells, very briefly, and as a general rule, the oscillator strenght increases with increasing binding energy, and thus, it should increase with increasing confinement.

EXPERIMENTAL ASPECTS

Fig.(8) shows the photoluminescence excitation (PLE) spectrum of a quantum well structure schematically presented in the inside pannel : a thin GaAs QW (50Å) inside a large $Ga_{0.87}Al_{0.13}As$ QW (200Å wide) which is confined lateraly by thick $Ga_{0.66}Al_{0.34}As$ barrier layers. It is obtained by using two different Al concentrations during the growth. The potential barrier at a $Ga_{1-x}Al_xAs/Ga_{1-y}Al_yAs$ interface (x<y<0.4) increases almost linearly with y-x. We remark various features in fig.(8). (i) The main pics are atributed to 1S-like excitons. Note that excitonic states attached to various (excited) band-to-band transitions are observed. In fact, excitonic effects are expected to dominate the absorption near the energy edge of any band-to-band transition in the QW. (ii) We see two series of accidents : excitonic states associated with heavy holes (E_1H_1, E_1H_3, E_2H_2, ...), and excitonic states associated with light hole states (E_1L_1, E_2L_2, ...). (iii) Note that most of the transitions are parity conserving : $1 \rightarrow 1$, $1 \rightarrow 3$, $2 \rightarrow 2$, ..., buth we observe also a $2 \rightarrow 3$ transition, which is not parity conserving. The reason is that the QW is not exactly symetric : the intermediate $Ga_{0.87}Al_{0.13}As$ barriers are slightly asymmetric (they have different widths). (iv) Also, we have plateaux (almost energy-independent absorption intensity) between two well separated excitonic transitions, due to the

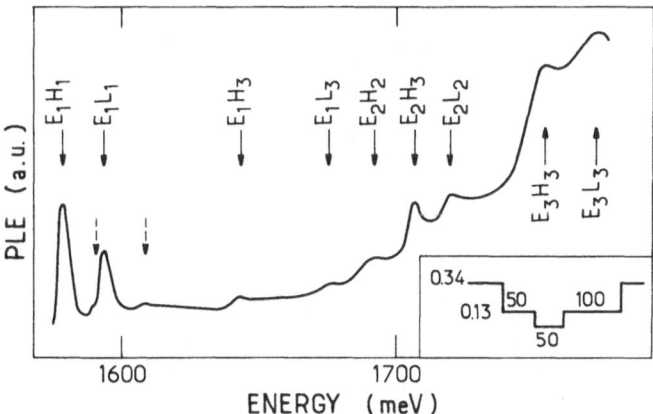

Fig (8) Low temperature (2K) PLE spectrum of a 50Å-thick quantum well (the potential profile is schematically presented in the insert). The two dashed arrows indicate the shoulders assigned to the excited 2S levels of the E_1H_1 and E_1L_1 excitons (from ref.[5])

continuum absorption (see fig.(7)). (v) We note also that the excitonic pics are NOT delta-like pics. In fact the absorption lines are inhomogeneously broadened, due to the presence of defects in the QW region (see below). (vi) Finally, note the presence of small shoulders at the high energy sides of the E_1H_1 and E_1L_1 excitonic pics. They are attributed to the 2S excitonic transitions, which have much smaller oscillator strenghts than the corresponding 1S ones (see fig.(7)). Since they are in fact very close in energy from the onset of the band-to-band transitions, the 1S-2S energy differences shall give us an inferior limit for the excitons binding energies. We see that the two binding energies are not the same and we have approximately 8.1meV (heavy) and 9.6 meV (light), wich is sensibly larger than the bulk value of ≈ 5 meV for the GaAs bulk material. The difference is due to the quantum confinement of the excitons.

These results were obtained at low temperature (T=2K). We present in fig.(9) the absorption coefficient for a multi-QW structure and by comparison the absorption of a thick layer of GaAs [6]. For the QWs we see as before the heavy hole and light hole excitons and a large absorption plateau. The excitonic transitions are clearly blue shifted, due to the quantum confinement effect. At the inside pannel we present the temperature dependence of the width of the excitonic pic. At low T it is inhomogeneously broadened and with increasing temperature the interaction with the optical phonons become the dominant broadening mechanism. The full curve is given by $\Gamma(T) = \Gamma_0 + \Gamma_1 n_{ph}(T)$, where Γ_0 and Γ_1 are best fit (temperature independent) constantes and n_{ph} the T dependent optical phonon population. Thus, QW's excitons are more bound than bulk excitons and thus more stables : we can observe sharp excitonic features in QWs even at room temperature.

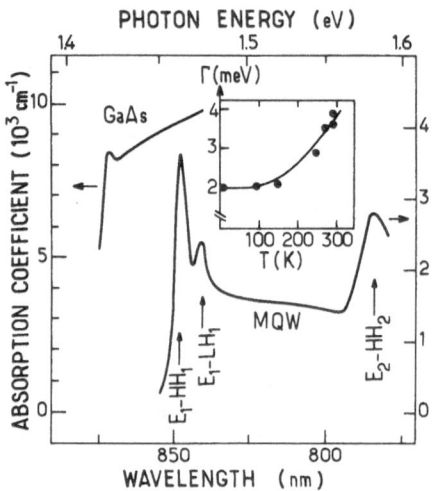

Fig. (9) Room temperature absorption of a GaAs layer and of a GaAs-Ga(Al)As multi-quantum well structure showing the enhancement of excitonic effects [6]. Insert: temperature dependence of the FWHM of the excitonic accident.

We present in fig.(10) the measured (symbols) and the calculated (full lines) size quantization effect for QWs of different material compositions. Δ is the difference between the QW band gap and the bulk bandgap. Note that for the sake of clarity the horizontal axis for the Ga(In)As-InP structures was inverted. We see clearly the existence of a blue shift for the transitions when decreasing L and for large wells a $1/L^2$ law for the energy shifts, as expected for quantum confinement effects.

The previous results were obtained for thin quantum wells, for wich the excitons are always in the regime of relative motion quantization. However, the center-of-mass quantization regime was also clearly evidenced experimentally. We present in fig.(11) the PLE spectra for a series of CdTe/CdZnTe QWs of different sizes, from nearly 200Å up to 1000Å [7]. These are II-VI materials. The PLE spectra are reported relative to the first observed excitonic transition. We have for each spectrum various accidents at high energies : N=1,2,3,... in the figure.

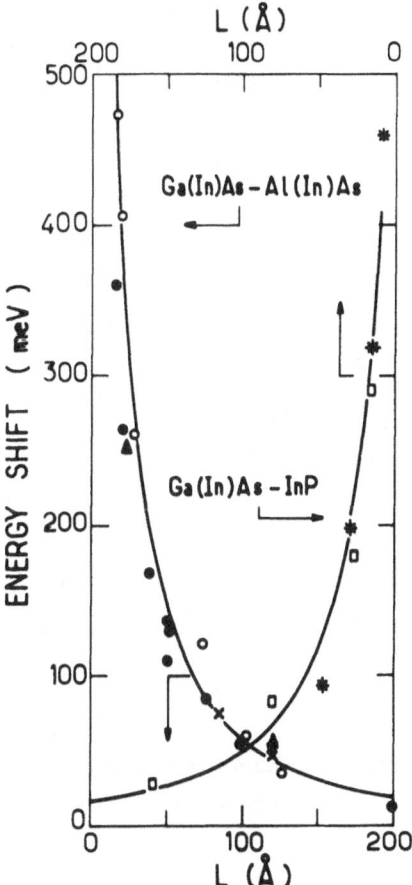

Fig. (10) Calculated energy shift of the band-to-band fundamental absorption edge versus the Ga(In)As slab thickness in Ga(In)As-Al(In)As (lower horizontal scale) and Ga(In)As-InP (upper horizontal scale) quantum wells. The symbols correspond to various experimental data [3].

When we plot all these accidents togheter we have the well thickness dependence for the E_N-E_1 energies presented in fig.(12). Thus, for each accident label we observe a $1/L^2$ variation for the E_N-E_1. These various additional transitions were assigned by the authors to the discrete exciton center-of-mass levels. Note that the full lines are theoretical results which corroborate that interpretation [7].

R. Ferreira

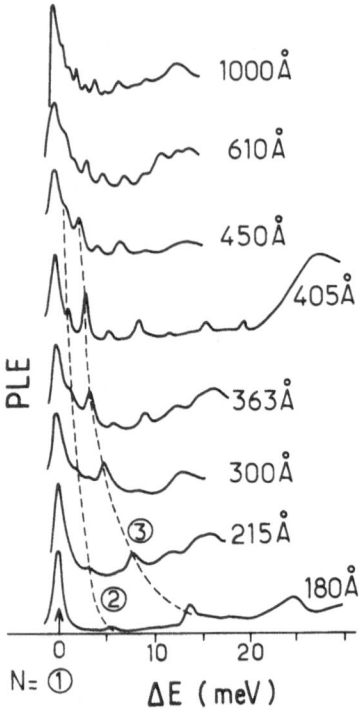

Fig. (11) PLE spectra of various CdTe-CdZnTe multi-quantum wells of different thicknesses. The energy scale
in each case shifted with respect to the energy of the first PLE accident (maximum). Adapted from ref.[7].

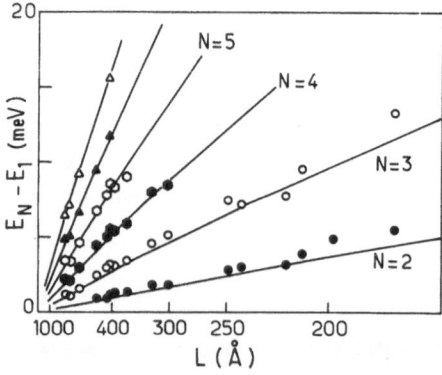

Fig. (12) Energy position of the various maxima E_N (N=2,3,...) in the PLE spectra of various CdTe-CdZnTe
multi-quantum wells of different thicknesses. For a given well width L we plot $\Delta E = E_N - E_1$ with respect to the
first maximum in the PLE spectrum (see also fig.(11)). The solid lines correspond to the various maxima in the
calculated absorption profile. Adapted from ref.[7].

We consider in the following some luminescence (PL) spectrum of quantum wells. Fig.(13) shows the PL and PLE spectra of a thin (70Å) GaAs quantum well, obtained at low temperature [2,5]. We observe in the PLE spectrum the excitonic pics, as before. The PL line in this case is Stoke shifted and is sensibly larger than the first excitonic PLE transition. What is the origin of this Stoke shift or, in other words, why the energy to excite the ground exciton state in the QW is not the same as the mean energy for the radiative excitonic recombination? This energy difference is due to the presence of defects in the quantum well region.

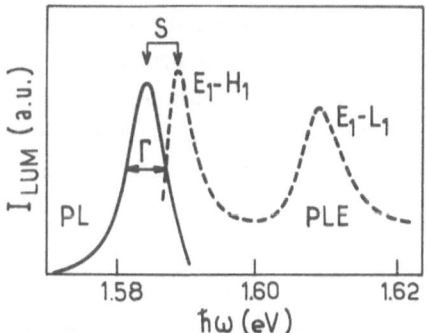

Fig.(13) Low temperature (T=2K) PL (solid line) and PLE (dashed line) spectra of a 70Å-thick moderate quality GaAs quantum well, showing S≈3meV Stokes shift [2,5].

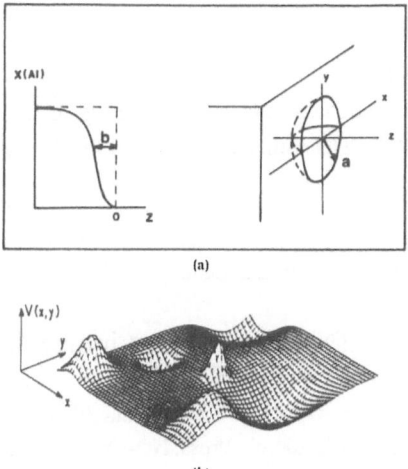

Fig.(14) (a) Schematic representation of a semi-gaussian interface defect. (b) In-plane average potential energy seen by an electron (or a hole) moving in a quantum well whose interfaces display semi-gaussian defects. The average is taken over the ground state for the z motion of the unperturbated well. Attractive and repulsive defects of equal depth b are considered. From ref.[2].

Fig.(14) shows schematically the potential fluctuations in the vicinity of an inverted QW interface (the one obtained by growing the binary over the ternary barrier material). Actually, we can have in the interface region islands of barrier material inside the well region, or vice-versa. In this case, the PL line is due to excitons trapped on some of these intrinsic interface deffects. In fact, some of that islands are attractive for the QW excitons so that the trapped excitons are more tightly bound than the free ones. Since the photoluminescence process favours the lowest energy states, the PL line will be dominated by these bound excitonic states. On the other hand, we saw that the absorption process depends strongly upon the density of states. Since the density of bound states is generally much lower than the twodimensional density of states for free excitons, then the PLE will start at the free exciton energy. That is the origin of the Stoke shift. The consideration of a distribution of such defects accounts for the broadening of the PL line. That interpretation for the Stoke shift and PL broadening was confirmed by a systematic study of QWs of different sizes and quality, and also by considering their temperature dependence.

CONCLUSION

We have considered some aspects of the theoretical description of the electronic states and of the optical response of semiconductor quantum wells near its fundamental band gap. The effect of quantum confinement have been discussed for both the one-particle (electrons and holes) and excitonic states in the effective mass formalism. Finally, we have presented some few experimental results which illustrate the effects of such quantum confinement on the optical response of quantum wells.

ACKNOWLEDGEMENTS

The "Laboratoire de Physique de la Matière Condensée" is "Unité de Recherche Associée au CNRS (France) - URA 1437". The author would like to thank A. Filoramo for the reading of the manuscript.

REFERENCES

1. G. Bastard , Wave Mechanics Applied to Semiconductor Heterostructures, (Les Editions de Physique, Les Ulis, 1988)

2. G. Bastard, C. Delalande, Y. Guldner and P. Voisin, in Advances in Electronics and Electron Physics, (Academic Press, New York, 1988)

3. G. Bastard, J. A. Brum and R. Ferreira, in Solid State Physics, vol 44, ed. by H. Ehrenreich and D. Turnbull , (Academic Press, New York, 1991)

4. D. L. Smith and C. Mailhiot, Rev. Mod. Phys. 62 (1990) 173-234

5. M-H. Meynadier, PhD thesis, Ecole Nationale Supérieure des Télécommunications, 1985

6. D. S. Chemla, in High Excitation and Short Pulse Phenomena, ed. by M. H. Pilkuhn (North Holland, 1985)

7. H. Tuffigo, PhD thesis, Université Joseph Fourrier - Grenoble I, 1990

LECTURE 9

What can be learned from time resolved measurements on porous silicon luminescence

J.C. Vial

*Laboratoire de Spectrométrie Physique , URA 08 du CNRS,
Université Joseph Fourier de Grenoble, BP. 87,
38402 Saint Martin d'Hères cedex, France*

INTRODUCTION

History

The transient behavior of light emission has been known for a long time; the so called phosphorescence of many organic compounds or color centers in inorganic insulators can be easily observed by the naked eye for example. Many people have played with a clock whose needles and numbers are coated with a phosphor and remember the slow vanishing of the emitted light after they have illuminated the clock for a short time. By doing that they have performed a luminescence time resolved experiment and have verified that the luminescence lifetime of this particular phosphor is of the order of an hour. Some of them may have verified that this lifetime is the same for various excitation intensities or excitation times.

As simple as these observations can be, they already point out the major interest of the luminescence lifetime measurement i.e. the luminescence decay is very often insensitive to the excitation parameters and consequently represents intrinsic properties of the luminescent entities.

For a long time ruby ($Al_2O_3:Cr^{3+}$) has been the prototype on which a many dynamically resolved luminescence experiments have been performed. With a luminescent lifetime in the range of ms and an emission in the visible range which is easily excited by flash lamps or by the first efficient pulsed laser (the ruby laser itself) it offers exceptional qualities which render possible the analysis of phenomena as fundamental as the relation between absorption oscillator strength and lifetime, the non-radiative relaxation, the non-radiative transfer of excitation and related transport properties. Nowadays, a large variety of tunable pulsed lasers with time resolution as small as several femtoseconds are commercially avalaible. The time resolved technique can be used for systems as different as organic or inorganic compounds, direct or indirect semiconductors and amorphous or crystalline materials, hopefully, the fundamental mechanisms which govern the dynamics are very similar.

Porous silicon is an efficient phosphor with long luminescence lifetimes at room temperature. They appear to be strongly dependent on various external parameters as different as the temperature or the level of passivation. Therefore porous silicon is well suited for investigations in the time domain.

The objectives

The main objective is to use the dynamical properties to access the intrinsic properties of a material. For this purpose several questions have to be addressed. The first one concerns the origin of the excited state population decay; is it radiative or non radiative ? Depending on the answer, the models used to interpret the decay shapes are totally different.

Lifetime measurements also give directly information about transition probabilities and the interaction of the electronic system with photons, phonons or other carriers. The decay shapes can also represent the statistics of carriers and excitation transport (carrier transport) and can be useful when classic electrical measurements are impossible (when it is impossible to make contacts for example).

The vocabulary

Depending on the sub-field different words can be used to represent the same phenomena. For example "luminescence" and "phosphorescence" have the same meaning but phosphorescence is always long lived and luminescence is more general. "Geminate luminescence", a concept often used by the amorphous silicon community is closely related to the term "exciton recombination". On the other hand the term "non geminate luminescence " is equivalent to "delayed luminescence" and the term "radiative recombination" has a meaning very close to the term "fusion of excitation" used by the molecular spectroscopist.

The orders of magnitude

Lifetimes depend strongly on the nature of the Hamiltonian H responsible of the transition between an initial state i and the final state f and on the selection rules specific to the Hamiltonian and to the wavefunction symmetry (due to spins, carrier momentum, photon and phonon momentum...) This is expressed by the Fermi rule :

$$W_{if} \propto |\langle i|H|f \rangle|^2 \, g(\omega_{if})$$

where $g(\omega)$, which is the density of final states at the energy of the transition, can have very different values for different transition mechanisms.

For a Radiative lifetime, the interaction is with the electromagnetic radiation and the rates are governed by $g(\omega) \propto c^{-3}$ where c is the speed of the light, consequently the orders of magnitude range **from seconds** (forbidden transition : indirect transition in semiconductors, singlet to triplet transition in organic systems) **to ns** (allowed transition, direct band gap transition, singlet to singlet transition in organic systems)

For Non-radiative transitions, induced by phonons $g(\omega) \propto v^{-3}$ where v is the sound velocity, so the decays can be as fast as **fs.** But phonons have limited energies and for transition energies higher than the Debye energy a higher order in a perturbation calculation (multiphonon transition) is needed and rates are slower. The Auger process is another efficient non radiative process. In that case the electron-hole pair energy is transferred to a third carrier. Depending on the carrier density, Auger processes can be as fast as **ns.**

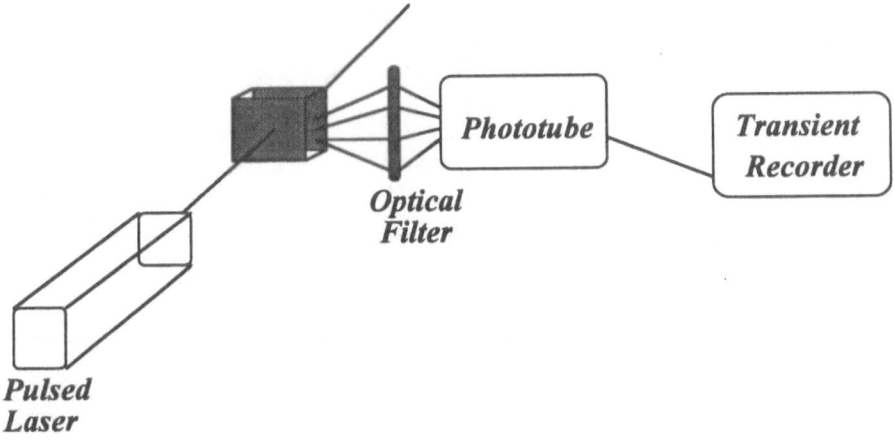

Figure 1

The classic experimental setup for time resolved measurements on the photoluminescence .

EXPERIMENTAL TECHNIQUES

The experimental setup shown in Figure 1 is well adapted to the measurement of decay times as short as several 100 ps (for shorter time measurements, refer to the paper by P. Fauchet in this book).

The excitation is provided by a pulsed laser associated with a dye laser in case selective excitation is needed. Pulses are obtained either by switching a CW laser (argon laser for example) or by a nanosecond pulsed laser (Nitrogen laser, Q-Switched Yag laser...). The luminescence filtered by a monochromator or a simple optical filter, is then detected by a fast phototube and averaged (usually 10 to 1000 shots) by a multichannel analyzer.

Following are some general considerations which will help us to understand what can be the best excitation technique for a given purpose.

comparison with the characteristic decay time of the system under study.

Short pulse and long pulse excitation

The meanings of the words short or long are not absolute and have to be considered in Let us consider the simplest system : the so called two level system schematically represented on the left in Figure 3. In a small flux approximation the equation which represents the pumping and decay processes is :

$$\frac{dn^*}{dt} = kIn - Wn^*$$ where kI is the pumping rate

The optical response is given by the solution of this equation and shown on the right in Figure 3 for short and long pulse excitations.

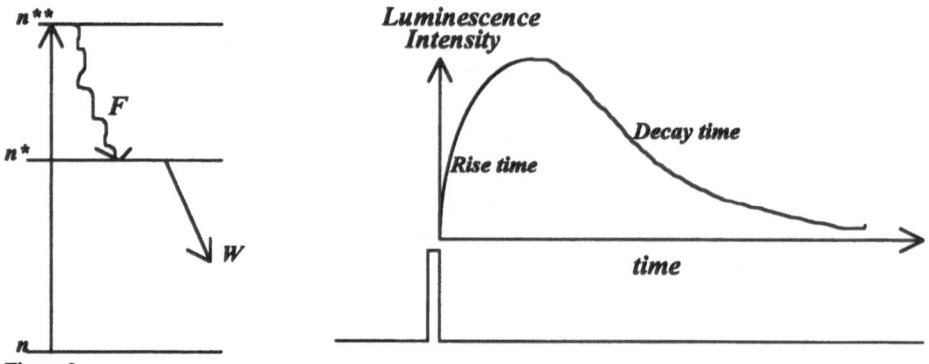

Figure 2

Photoluminescence time response to a short pulse excitation of a three level system

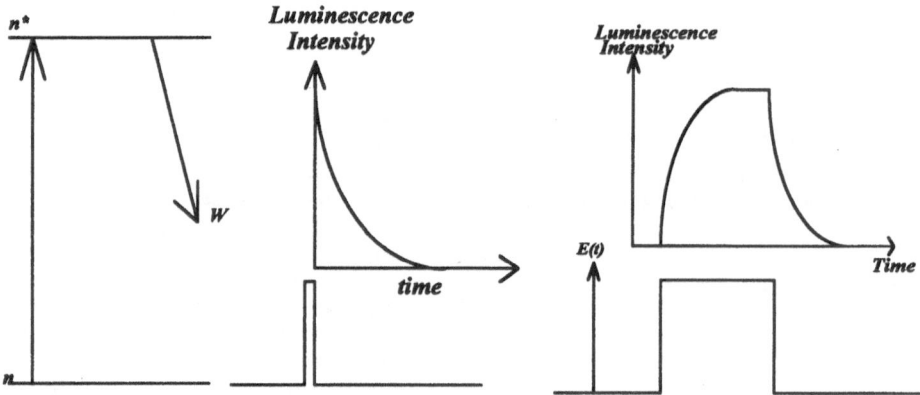

Figure 3

Photoluminescence time response to short and long pulse excitation of a two level system

The information which can be extracted from these two responses are the same. Note for example that the luminescence onset shown on the right does not give information about a new relaxation process. However when an excited state population is established after several successive relaxations, as shown for example on the left side of Figure 2 (for a non resonant excitation), the optical response will now be informative of the different steps. The time dependent equations are as follows :

$$\frac{dn^{**}}{dt} = kIn - Fn^{**} \text{ and } \frac{dn^{*}}{dt} = -Wn^{*} + Fn^{**} \Rightarrow n = A(e^{-Wt} - e^{-Ft})$$

Note that even in that simple case where only one intermediate step has been considered, the time evolution becomes more complex than before . For long pulse excitation the rise time and the decay time are both physically informative.

In general the problem is much more complicated : the luminescence comes from various species which are not all equivalent and various steps are involved; some are relaxation between the levels of each species others are relaxation between species (excitation transfer).

A GENERAL PRESENTATION OF THE PROBLEM :

Porous silicon, seen as an ensemble of interconnected crystallites, is a multi-disordered system, due to :
- A size distribution with its consequences on transition frequencies and radiative recombination W_r.
- A passivation distribution and consequently a distribution of non radiative rates W_{nr}.
- A proximity distribution between the various crystallites and therefore a distribution of transfer rates W_{ij} between all the crystallites i and j.
- A distribution of initial states after excitation.
- Different types of excitations : electrons, holes, "excitons".
Figure 4 summarizes all these contributions.

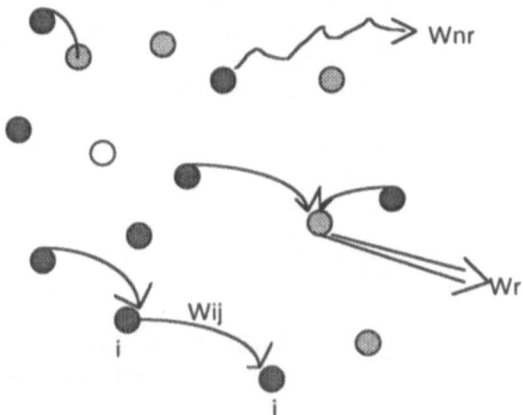

Figure 4

A schematic view of some processes involved in the excitation re-distribution following a pulse excitation. The porous silicon is represented as a spatial random distribution of confined zones where the radiative and non radiative recombination takes place (with the rates Wr and Wnr), in addition carriers localized at i can escape to reach j.

The master equation (in the small flux approximation)

Consider the equations describing the time evolution of $P_i^{e,h}(t)$, the probability for crystallite i to be filled with an electron hole pair at time t. They can be divided in two groups :

The losses :
$$\frac{dP_i^{e,h}(t)}{dt} = -P_i^{e,h}(t)\left[W_{r,i} + \sum_j W_{i\to j} + W_{nr,i}\right]$$

and the feeding terms $+\sum_j W^e_{j\to i}P^e_j(t)P^h_i(t)+\sum_j W^h_{j\to i}P^h_j(t)P^e_i(t)$

After a short pulse excitation (initial injection) the time evolution of filled quantum crystallites is governed by **losses** (radiative , non-radiative and transfer) and **feeding** (delayed injection). The feeding gives a **bimolecular** character to the evolution.
Depending on the relative importance of these two terms the time evolution will more represent the intrinsic dynamical properties of quantum crystallites or the hopping processes of carriers.
In fact the final solution $S(\lambda,t)$, the signal observed at λ, represents all the various crystallites (with various passivations, various feedings...) which are emitting at λ so

$$S(\lambda,t)=\sum_i P^{e,h}_i(t)$$

The master equation, even over simplified as above, cannot be solved except for some limiting cases which will be considered now.

The case of Strongly Localized Excitation (No Delayed Injection)

The time evolution is then governed by the losses and an exact solution $\Phi(\lambda,t)$ can be obtained as shown in expressions (1) where W_i is the rate of the total losses for the crystallite i and $D(W,\lambda)$ the distribution of rates for crystallites emitting at λ.

$$\frac{dP^{e,h}_i(t)}{dt}=-P^{e,h}_i(t)\left[W_{r,i}+\sum_j W_{i\to j}+W_{nr,i}\right]\Rightarrow P^{e,h}_i(t)\propto exp(-W_i t)$$

$$\Downarrow \qquad\qquad\qquad\qquad (1)$$

$$\phi(\lambda,t)=\frac{S(\lambda,t)}{S(\lambda,0)}=\int D(W,\lambda)exp(-Wt)dW$$

The important consequence of the strong localization of the excitation are the following :
- The emission intensity is linearly dependent on the excitation (before saturation of the absorption).
- The time evolution of the signal is the Laplace transform of the distribution of decay rates as shown in the last expression.
- An **average lifetime** τ defined by the **area** under the decay curve coincides with an exact definition of the average as shown in the last equation in expression (2). This can be useful because $D(W,\lambda)$ is not often known.

$$\tau=\int\phi(\lambda,t)dt$$
$$\phi(\lambda,t)=\int D(W,\lambda)exp(-Wt)dW \qquad (2)$$
$$\int dt\int D(W,\lambda)exp(-Wt)dW=\int D(\tau,\lambda)\tau\,d\tau$$

In some particular cases $\Phi(\lambda,t)$ can be analytically expressed as for luminescence in the presence of excitation traps via the so called Inokuti-Hirayama model[1] .

The Inokuti-Hirayama model:

In this case, the population losses are due to a direct energy transfer from luminescent species (the donors) to acceptors without any back transfer (acceptors = traps). The analytical solution can be obtained because :

- The law $W_{ij}(r)$ is known since the energy transfer mechanisms have Dipole-Dipole, Dipole-Quadrupole or Quadupole-Quadrupole origins with an r^{-n} dependence with n=6, n=8, n=10 respectively. In addition the spatial distribution of donors and traps is known so D(r) is a known function of donors and trap concentration. Then a simple application of the general formula shown before gives:

$$s(t) = exp(-w_r t) \int drD(r)exp- ta / r^n$$

$$s(t) = exp(-w_r t)exp(-bt^{3/n})$$

This expression is instructive because it shows that the signal is a stretched exponential decay (at least when the radiative rate is slow compared to the nonradiative rates). It is interesting to point out that this behavior is a consequence of the **absence** of diffusion of excitation among the donors. What happens now if, on the contrary, the diffusion of excitation becomes very fast ?

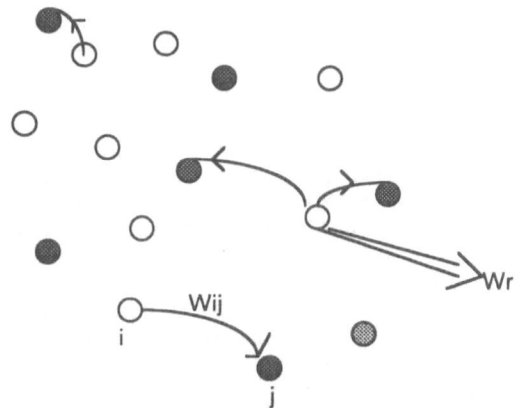

Figure 5

A simplified excitation redistribution for the Inokuti-Hirayama model : the only excitation transfer considered is from the optical center to traps.

The Yokota and Tanimoto model[2] (1967)

The diffusion homogenizes the donors : all donors see the same averaged trap distribution, i.e. all the donors are equivalent. The decay shape which is a sum of identical exponential become itself exponential !

Experimentally a transition between localized to delocalized excitation along with a non exponential to exponential transition has been seen :

-When the temperature allows quasi resonant transfer among the donors,

-When the average donor-donor distance decreases (concentration dependence).

But this conclusion has to be revised for the equations which are supposed to represent the case of porous silicon.

Delayed injection is dominant (a qualitative description)

Then the important terms to look at are

$$\sum_j W^e_{j\to i} P^e_j(t) P^h_i(t) + \sum_j W^h_{j\to i} P^h_j(t) P^e_i(t)$$

The decay shape will represent the probability that electrons and holes will meet themselves during a random walk. This process is strongly dependent on the density of carriers which itself decreases with time and therefore tends to **stretch** the exponential decay. In addition the bimolecular reaction introduces a quadratic term , so the emission intensity should be super linearly dependent on the initial excitation.

When the disorder introduces a distribution in the transition energies (inhomogeneously broadened spectral line) the delayed injection can introduce a spectral drift to lower energies which is a consequence of a phonon assisted transport from "blue species" to "red species".

The criteria that allow the distinction between strongly localized and delocalized excitation

The physics behind the localized and delocalized models are so different that one of the first questions that the experimentalist has to address concerns the distinction between these two situations. Hopefully temporal behavior of the photoluminescence is not the only criterion that can be used to make this distinction.

a) The spectral criteria

The nature of the broadening of an optical transition affects strongly the possibilities of excitation transfer. When the optical transition is homogeneously broadened all the optical entities which contribute to the line have the same behavior. On the contrary an inhomogeneously broadened line is the sum of individual transitions at different frequencies which represent different optical entities. The selective excitation and/or the selective detection of a given subset of these entities gives information on the flow of excitation between them. Fluorescence Line Narrowing (F.L.N.) and the Spectral Holeburning are techniques well suited for that purpose, especially if they are time resolved.

b) The temporal criteria

The nature of the lifetime, i.e. determined by radiative or non radiative processes, is another important question to address because the physical backgrounds of these two processes are very different. It will be shown later that measurements of the quantum efficiency and the lifetimes as a function of several parameters are able to answer this important question.

The distinction between the "first step emission" or the delayed emission is also an important task. Observing a rise time is an important step for that purpose. Finally the analysis of the temporal shape can be very informative but as pointed out before it cannot be considered alone.

c) Additional perturbations

As in many domains of physics, the addition of perturbations can be a very useful technique. Examples include temperature effects which modify the localization / delocalization. The same idea holds for the concentration dependence (which modifies the mean distance between species). Finally the introduction of additional species (as excitation donors or acceptors or killers) can be used to reduce or enhance the transport properties.

GENERAL PROPERTIES OF THE LUMINESCENCE DECAYS OF POROUS SILICON

The line shape, the "selective detection"

Figure 6 shows a typical CW luminescence spectrum for a typical porous silicon layer. It has often been suggested that the luminescence lineshape represents the inhomogeneous distribution of crystallite sizes multiplied the quantum efficiency (QE) of each crystallite. In fact the experimental observations of Calcott et al (in this book) indicate that the line is not totally inhomogeneous, one also has to take into account an homogeneous contribution whose origin is, at present, not clear. Nevertheless the model in which the observed spectrum is associated with energy confinement remains valid i.e. the more confined the carriers are, the bluer is the emission. This is essential to understand the following.

The various parameters which affect the decays

The right part of Figure 6 represents a typical luminescence (detected at 700nm) decay following a nanosecond pulse excitation in the UV. An unusual time axis scale has been chosen in order to emphasize the two different components. The fast component, in the nanosecond range, has an intensity which is strongly dependent on the level of oxidation. It has been studied in detail[3] and presented by Fauchet in this book and can be attributed to a color center in the oxide. We will not focus on it here. On the contrary the slow component is characteristic of the porous silicon itself and is our main subject of interest. It is dependent on various parameters as summarized below.

The decay times are :

- Long (μs to ms).
- Strongly affected by the passivation
- Strongly temperature dependent
- Strongly wavelength dependent
- No dependent on the intensity of excitation

- No particularly dependent on the porous silicon nature (p, p-, n, n-) or type of passivation (hydrogen or oxygen)

The temporal shape is :

- Non-exponential
- No dependent on the excitation power.
- The Shape is "universal" (the same for all wavelengths)

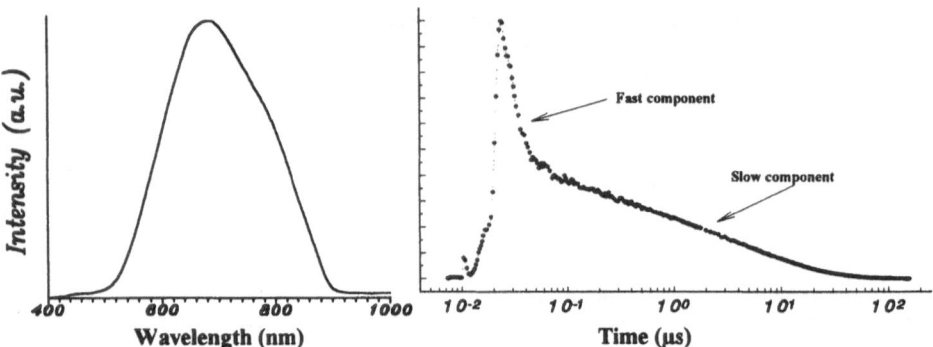

Figure 6

Left : A typical luminescence spectrum of a high porosity porous layer observed at room temperature excited in the UV.
Right : A typical photoluminescence decay for a high porosity porous layer at room temperature.

NON RADIATIVE PROCESSES OF POROUS SILICON LUMINESCENCE

The first question to address concerns the nature of the emission : what does the lifetime represent ? To answer this question we have to remember that the luminescence Intensity I, proportional to $W_r/(W_r+W_{nr})$, and the lifetime $\tau =1/(W_r+W_{nr})$ are not independent quantities. Figure 7 represents the evolution of the luminescence intensity and lifetime as a function of an important parameter : the temperature. In a shaded area which corresponds to room and above room temperatures, the lifetime and the intensity clearly show parallel evolution. The meaning is clear : In this region W_r is constant and for room temperature (and above) $W_{nr}>>W_r$, that is to say that $\tau=1/W_{nr}$. Therefore the lifetime measurements, in this temperature regime give access to the nonradiative processes.

The tunneling model[4]

Two interesting conclusions can be drawn from the fluorescence lifetime measurements.
(i) At room temperature, the long non-radiative lifetimes measured, together with the quantum efficiency of 3%, yield a radiative lifetime of order 1 msec. This does not support the commonly made assumption concerning the breakdown of translational invariance in the porous silicon structure. If this were the case, radiative lifetimes would be much shorter, approaching that of a direct transition, that is to say in the range of nanoseconds to microseconds.
(ii) A second interesting result is the relatively low efficiency of non-radiative processes evidenced by the rather high quantum efficiency. Although standard silicon wafers have been used, the measured non-radiative rates are comparable with those obtained on highly pure bulk silicon with a highly passivated surface. We believe that this interesting behavior originates from the micro-structure of porous silicon and mainly from the restricted volume available to the carriers. Crystallites of nanometer dimensions have indeed such small volumes and surface area (10^{-19} cm^3 and 10^{-12} cm^2 respectively) that the probability of finding non-radiative centers inside or at the surface is very small even for silicon of modest purity. For example the p-type bulk silicon used in experiments reported here, has doping levels as low as 10^{16} cm^{-3} so even if all the dopants were non-radiative centers (which is unlikely) the

proportion of quenched crystallite should be as low as 10^{-3}. For these reasons, we propose that the mechanism for the remaining non-radiative decay comes from an escape of the carriers from the confined zone to more extended or less passivated regions where non-radiative recombination can occur. This is schematized on the diagram in Figure 7.

This escape can be described by a carrier tunneling and expressed by the following expression

$$W_{nr} \propto exp- \frac{4\pi a}{h} \sqrt{2m(Vo-E)}$$

An important consequence of this model is the expected strong dependence of W_{nr} on the passivation via the barrier parameters "a" and "V_0" and on the confinement via the confinement energy and therefore via the luminescence energy $h\nu$.

W_{nr} Dependence on passivation

The dependence of W_{nr} on the passivation or on the other hand on the porous silicon degradation is analyzed within the framework of the tunneling model by Hérino in this book. The same idea has been extended to describe the evolution of the electroluminescence which appears during the anodic oxidation of porous silicon.

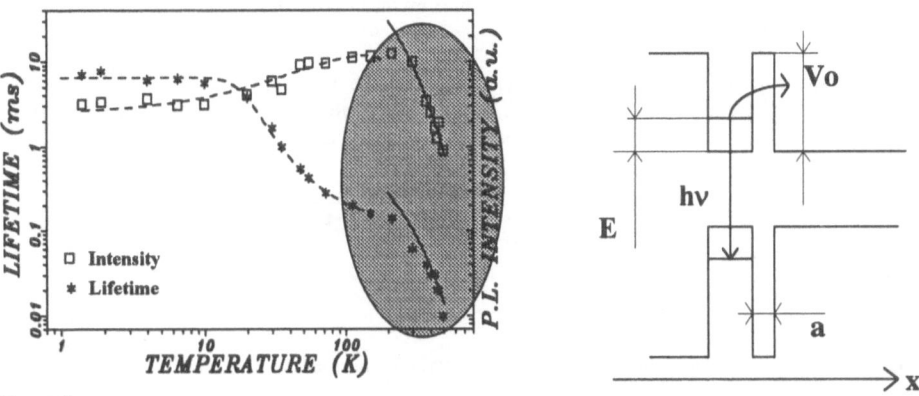

Figure 7

Left : Log-log plot of simultaneously recorded temperature dependencies of the lifetime and intensity of luminescence (detected at 700nm) for a 65% porosity porous silicon sample anodically oxidized up to Q_0. Note the tendency for parallel evolution at high temperature. The lines joining the experimental points are just guides for the eye. The shaded area is the range were the intensity and lifetime evolution are dominated by the nonradiative processes

Right : An energy diagram for silicon confined zone schematically represented by a quantum well and quantum barriers. The vertical arrow represent a radiative recombination and the curved arrow an escape of the electron by tunneling through the barrier

W_{nr} dependence on confinement

Figure 8 demonstrates the strong dependence of Wnr on the luminescence energy. Qualitatively, this behavior is explained by the confinement effect which rises the energy levels

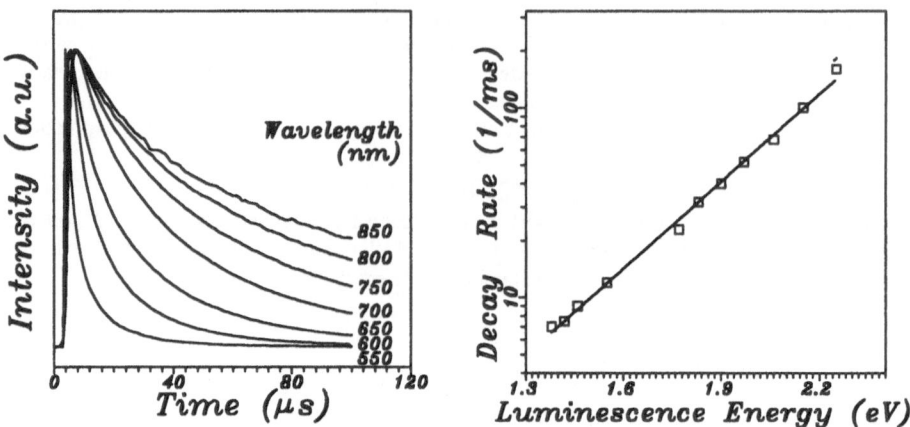

Figure 8

Left : Photoluminescence time evolution following short pulse excitation in the UV for a high porosity porous layer at room temperature and for various detection wavelengths.

Right : The average decay rate (deduced from the time evolution presented on the left) as a function of luminescence energy.

close to the top of the barriers and then enhances the non radiative escape of carriers. Quantitatively the exponential dependence of the decay rates on the luminescence energy is also understood with use of the tunneling model because the expression

$$W_{nr} \propto exp- \frac{4\pi a}{h} \sqrt{2m(V_0-E)}$$

can be approximated by

$$W_{nr}= Aexp(E/\alpha)$$

when the barriers are much higher than the confinement energies.

The Decay shapes

As already mentioned the decay curves $\phi(E,t)$ observed at various energy E, are non-exponential, nevertheless they have some particular features that we will try to describe in a pure phenomenological approach.

First remark : the decay shapes at various wavelengths can be deduced one to the others just by a linear stretching (or contraction) of the time axis. In other words we can define a **unique shape $\phi(E,T)$** for all the luminescence energies with

$$T = t \, exp(E/\alpha)$$

Where $exp(E/\alpha)$ is the law describing the variation of the average decay rates with the luminescence energy as shown before. In fact this observation is an other way to introduce an average decay rate as the scaling factor which connect the various decays. Expressions (3) and

Figure 9 show that a unique distribution of confinement energies or passivations inducing a distribution of decay times via $W_{nr} = A exp(E/\alpha)$ explains the scaling law on the shape. We have verified that the experimental shape is not strongly dependent on the passivation (Hydrogen, oxygen) so the distribution of A is ruled out. On the contrary, the observation of a narrowing of $D(W)$ when α decreases (by lowering the temperature) favors the idea of a distribution of E. A distribution of confinement energy is not a new idea for porous silicon, it can take its origin in a distribution of crystallite shapes for example, Figure 10 shows that the vibronic nature of the optical transitions can indeed be an other origin for the distribution of confinement energies.

$$\phi(t) = \int D(W) exp(-Wt) dW \quad \text{and} \quad W = A\,exp(E/\alpha) = A\,exp\frac{(E_0 + \Delta E)}{\alpha}$$

$$\Downarrow \quad\quad (3)$$

$$\phi_{E_0}(t) = \int D(\Delta E) exp(-Ate^{E_0/\alpha} e^{\Delta E/\alpha}) d\Delta E \quad \text{or} \quad \phi_E(t) = \int D(A) exp(-Ate^{E/\alpha}) dA$$

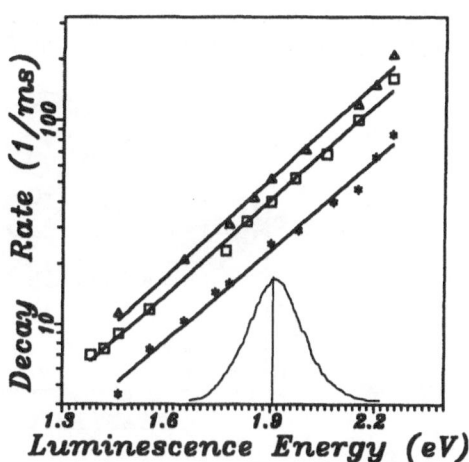

Figure 9

A plot of the decay rate evolution as a function of luminescence energy and for three different passivation levels. The bell shaped curve is not an experimental result it shows the consequences that can have an energy distribution on the decay rate distribution.

Taking the inverse Laplace transform of the decay shapes and assuming that the origin of the decay distribution is only due to an energy distribution, give access to the curves in Figure 10. They show that the distributions are very similar for all the wavelength except for the far blue side. It is difficult to affirm that the shape of this distribution is only due to a phonon distribution, as shown in the left part of Figure 10 but this hypothesis cannot be excluded.

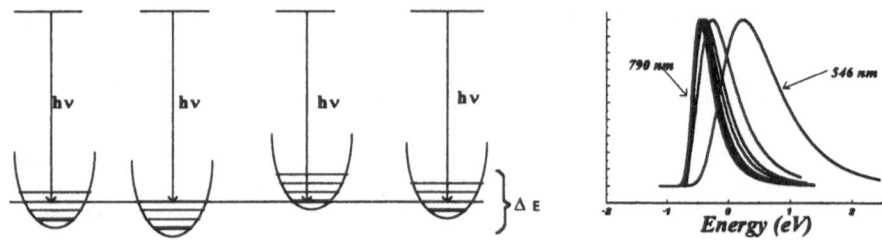

Figure 10

Left : a schematic view of vibronic transition, selected at hν, within an inhomogeneously broadened line, showing the distribution ΔE of electronic transitions.
Right : the energy distribution deduced from the Laplace transforms (and expression (3)) of decay shapes at various wavelengths

RADIATIVE PROCESSES[5]

Figure 11 shows the temperature dependence of the photo-luminescence intensity, I(T) recorded over a broad temperature range and detected at 700 nm. Similar curves but with

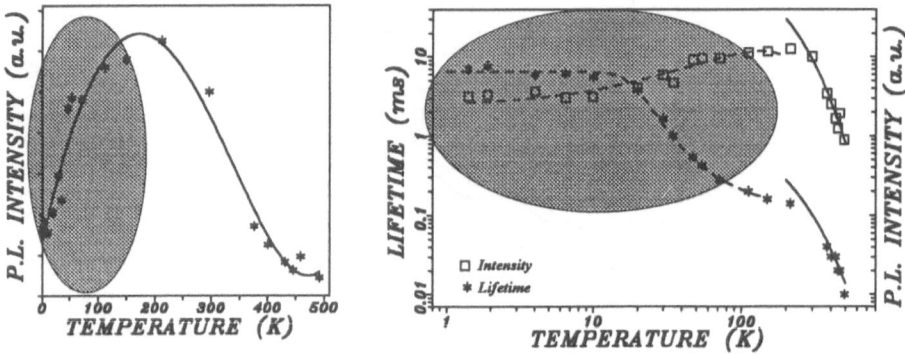

Figure 11

Left : temperature dependence of the photo-luminescence intensity (detected at 700 nm) for a 65% porosity porous silicon sample anodically oxidized up to Q_0 .
Right : Log-log plot of simultaneously recorded temperature dependencies of the total lifetime and intensity of luminescence (detected at 700nm) for a 65% porosity porous silicon sample anodically oxidized up to Q_0. The shaded area emphasizes the temperature domain where radiative process variation is important. The lines joining the experimental points are just guides for the eye.

slightly different peak positions can also be obtained for different detected wavelengths. The main feature is the so called "Anomalous temperature dependencies of photoluminescence", specially below 200K, where the intensity decreases with decreasing temperature while non-radiative processes usually decrease. As pointed out such an anomalous behavior has also been observed for amorphous silicon and this has sometime been used as an argument in favor of an amorphous phase as the origin for visible light emission from porous silicon. In fact, the luminescence intensity dependence on temperature alone is not sufficient to allow any clear conclusion because it depends simultaneously on the radiative recombination rate W_r and non-radiative recombination rate W_{nr} by the law:

$$I(T) \propto W_r(T)/(W_r(T)+W_{nr}(T)).$$

The simultaneously recorded temperature evolution of the luminescence lifetimes together with the intensity can provide a more definite conclusion because the lifetime follows the law

$$1/\tau(T) = W_r(T)+W_{nr}(T)$$

In that sense Figure 11 is the key which allows a separation between the radiative and non-radiative contributions. It shows clearly two different temperature regimes.

i) Above room temperature, the intensity and the decay times both decrease with increasing temperature and show parallel evolution; this can be explained if the radiative processes stay constant while the non-radiative ones accelerate with increasing temperature.

ii) On the contrary, below room temperature (and mainly below 200K), for decreasing temperature, the intensity slightly decreases while the decay times show a dramatic increase. This can be explained if we suppose that now the radiative processes become strongly temperature dependent and decrease with decreasing temperatures. An analysis of the radiative processes therefore becomes possible. But while it is always possible to obtain the non-radiative contribution, simply by taking bad samples for example, it not so easy to extract the radiative part because decay times and intensity, even at low temperature, contain both the radiative and non-radiative contributions. We can write the quantum efficiency $QE(\lambda,T)$ at wavelength λ and temperature T as :

$$\eta W_r(\lambda,T)/(W_r(\lambda,T)+W_{nr}(\lambda,T))$$

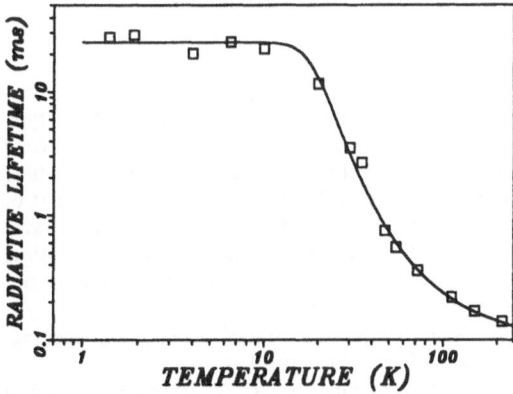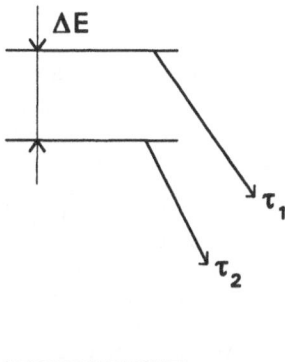

Figure 12

Left : Temperature dependence of the radiative lifetime deduced from the total lifetime evolution of Figure 11. The solid line is the best fit of the experimental data following the scheme shown on the right with $\Delta E=10$ meV, $\tau_2 =25$ ms and $\tau_1= 22\mu$s.

Right : Energy level diagram for an exciton having an excited state split by ΔE and with τ_1 and τ_2 for radiative lifetimes from the two components.

(where η is the proportion of emitting crystallites) and the lifetime for the same wavelength and temperature as :

$$\tau(\lambda,T)=1/(W_r(\lambda,T)+W_{nr}(\lambda,T))$$

Then the radiative recombination rate is given by $W_r(\lambda,T)=QE(\lambda,T)/\tau(\lambda,T)/\eta$. Unfortunately, due to the inhomogeneous spectral nature of the line, the QE is not known for all wavelengths (but in first approximation the 3% average value at room temperature will be kept for λ =700nm and T=300K). In addition, η is not exactly known and can range between 0.1 and 1 (the lower limit has been used for Figure 12) this makes it difficult to determine the precise absolute value of the radiative lifetime but its dependence on temperature is retained and this is our primary concern. Consequently it has been possible to use the data of Figure 11 to extract the temperature dependence of the radiative lifetime, at least for a domain of temperature where the non-radiative rates are not excessively dominant. The result is shown in Figure 12

This strong temperature dependence for radiative processes, showing an acceleration for increasing temperature and plateau at high and low temperatures has been encountered in a variety of light emitting systems. In most cases it originates from the evolution of a Boltzman distribution of the population among excited levels which have very different radiative relaxation rates.

In Figure 12 is presented a particular system with an excited multiplet split into only two components. The experimental data can be explained by a radiative relaxation from these two excited levels (1 and 2) having populations which satisfy the Boltzman equilibrium; the transition probability from the lowest level being much smaller than from the excited one. For the case of porous silicon, several authors in this book (P. Calcott, M. Lannoo, M. Hybertsen) show that the two-spin-1/2 electron-hole exchange interaction[6] can be at the origin of the excited state splitting; level 1 which has the fastest decay is then identified as the singlet while the triplet is associated with level 2

OTHER PRACTICAL USES OF TIME RESOLVED MEASUREMENTS

Voltage quenching of the photoluminescence[7]

An efficient voltage-selective quenching of the photoluminescence is observed for n type porous silicon under polarization **in aqueous solution**. Starting at low voltage a luminescence quenching from the red side of the photoluminescence spectrum shifts continuously to the blue side for increasing cathodic voltage until a total quenching is obtained (see on the upper part of Figure 13). The origin of this quenching can be found in an increase of the non-radiative rates or in a decrease of the radiative efficiency or in an induced transparency. Decay time measurements have shown (upper part of Figure 13) that this efficient quenching is not accompanied by a reduction of the lifetime.

On the contrary when the voltage is applied via **Solid contacts**[8] a voltage quenching, along with an increase of the non-radiative rates, is observed (Figure 13).

Without doing any physics we can conclude that time resolved measurements, on two phenomena which seem very similar, reveal two different physical origins.

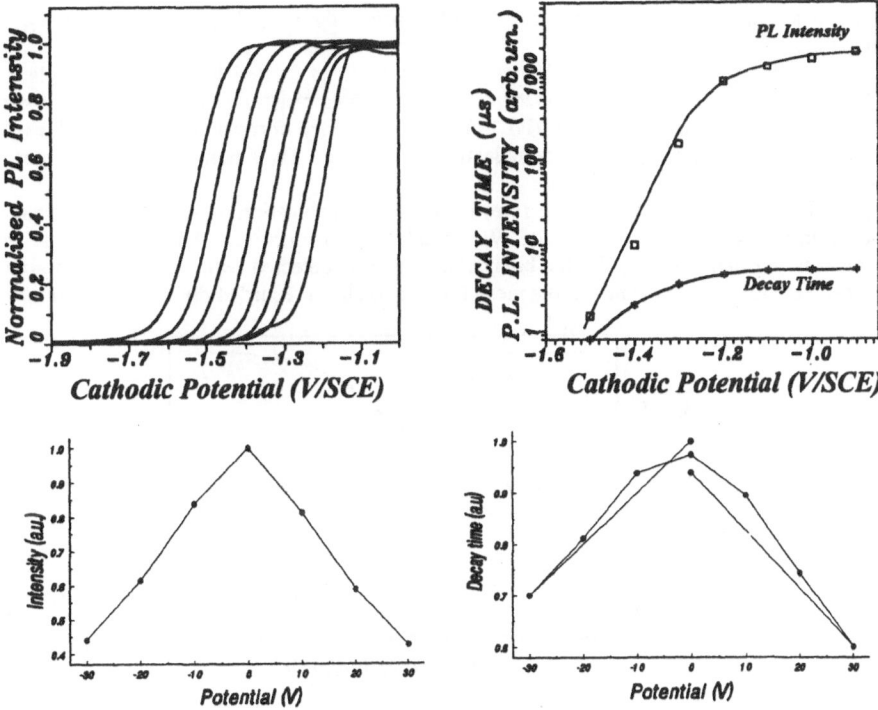

Figure 13

Upper figures : The voltage quenching in aqueous solution
Left : P.L. intensity as a function of the applied cathodic voltage for various luminescence energies (from left to right : 2 eV, 1.88 eV, 1.77, eV, 1.68 eV, 1.59 eV, 1.51 eV, 1.44 eV and 1.35 eV).
Right : Simultaneous measurement of the photoluminescence intensity and decay time recorded at 700nm as a function of the cathodic potential. The continuous lines are just guides for the eyes. Note the dramatic intensity quenching while the decay time is not affected.
Lower figures : The voltage quenching with solid contacts
Photoluminescence intensity and decay time variations as a function of the voltage applied by solid contact on the porous silicon layer.

The Saturation of the Excitation of the Photoluminescence[9]

Although for low intensities the PL of a porous silicon layer has a linear dependence on the excitation intensity we have found that at high intensities this dependence becomes sublinear and tends to saturate (cf Figure 14).

In order to avoid thermal effects, we have used a N2 laser excitation with a pulse width of 3ns and a repetition rate of 20Hz. In our laser focus conditions and for the highest excitation level we estimate that the sample temperature does not increase at the end of the pulse by more than 30K. This is in good agreement with the slight diminution of the lifetime, as measured in our experiment, but is not sufficient to explain the PL saturation.

These experimental facts can be interpreted consistently according to the following two competing models :

-1) A transparency due to population storage in long lived excited levels. But to explain the still linear dependence of the layer absorption the absorbing centers have to be divided into two classes as luminescent (L) or not (NL). The NL category corresponds to crystallites with defects for which non radiative recombination is very fast. Their number is much larger and they dominate the optical absorption.

-2) An increase in the non-radiative rates by Auger recombination as soon as more than one excitation is injected into a crystallite. We calculate the Auger recombination rate and show that it is larger than 10^8 (s^{-1}). Such a high value cannot be detected with our experimental setup and can explain why no change in the mean lifetime has been recorded during the saturation.

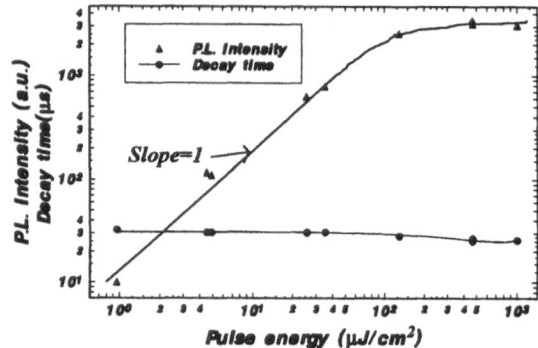

Figure 14

Photoluminescence intensity and decay time evolution as a function of the excitation intensity.

We believe that the PL saturation correlates with the electrical quenching of the PL (with liquid contacts) that presents the same phenomenological variations: a strong quenching of the PL intensity while the lifetimes remain almost constant.

GENERAL CONCLUSION

Bright visible light emission from porous silicon after photo- or electro-excitation is a very exciting phenomenon from a fundamental point of view and for its promise in applications. At room temperature and above, the measurement of the photoluminescence quantum efficiency and decay times show that non-radiative recombination dominates, while at lower temperature the decay time is mainly determined by radiative processes. The observation of an exponential dependence of the non-radiative rates with confinement energies, which increases with increasing confinement, is explained in terms of a model in which carriers escape from the crystallite in which they were confined. A model of escape by tunneling accounts well for the experimental results. It provides an other illustration of quantum size effects complementing the quantum size effect on energy levels. An extension of this model in term of a distribution of confinement is proposed and can explain the non exponential decay shapes. Time resolved luminescence measurements have indeed proved to be a powerful tool for such investigations. They are also useful in separating the various processes involved in the new phenomena such as the voltage quenching of the photoluminescence or the saturation of its excitation.

[1] M. Inokuti and F. J. Hirayama J. Chem. Phys. **43** (1965) 1978.
[2] M. Yokota and 0. J. Tanimoto J. Phys. Soc.Japan **22** (1967) 779.
[3] J.C. Vial and I. Mihalcescu in <u>Optical Properties of Low Dimensional Silicon Structures</u>, page 117, edited by D. Bensahel, L.T. Canham and S. Ossicini, Nato ASI Series, Series E : applied Sciences Vol **244**, 1993, Kluwer Academic Publishers.
[4] J.C. Vial, A. Bsiesy, F. Gaspard, R. Hérino, M. Ligeon, F. Muller, R. Romestain and R.M. Macfarlane
Phys Rev.B **45** (1992) 14171.
[5] J.C. Vial, A. Bsiesy, G. Fishman, F. Gaspard, R. Hérino, M. Ligeon, F. Muller, R. Romestain and R.M. Macfarlane in <u>Light Emission from Silicon</u>, edited by P .M. Fauchet, Y. Aoyagi, C. C. Tsai, MRS Symposia proceedings **Vol 283**(Materials Research Society, Pittsburgh, 1993).
[6] P.D.J. Calcott, K.J. Nash, L.T. Canham, M.J. Kane and D. Brumhead , J. Phys. Condens. Matter. **5** (1993).L91.
[7] A. Bsiesy, F. Muller, M. Ligeon, F. Gaspard, R. Hérino, R. Romestain and J.C. Vial in <u>Light Emission from Silicon</u> page 29, edited by J.C. Vial, L.T. Canham, W. Lang, J. Lum vol **57** N° 1-6 (1993), North Holland Publisher.
[8] H. Koyama, T. Oguro, N.Koshida, Appl. Phys.Lett. **62** (1993). 3177.
[9] I. Mihalcescu, R. Romestain, J.C. Vial, F. Madéore, F. Muller, E. Martin, C. Delerue, M. Lannoo and G. Allan, to be published in the proceeding of the 1994 Spring E.M.R.S. conference.

Ion beam analysis of thin films.
Applications to porous silicon

C. Ortega, A. Grosman and V. Morazzani

*Groupe de Physique des Solides, URA 17 du CNRS,
Universités Paris 7 et 6, Tour 23, 2 place Jussieu,
75251 Paris cedex 05, France*

INTRODUCTION

The aim of this paper is twofold : 1)- to present a summary of the fundamental interactions between ion beam (such as proton, deuteron or helium) of MeV energy and solids, interactions that are used in material analysis techniques such as Rutherford Backscattering Spectrometry (RBS), Elastic Recoil Detection Analysis (ERDA) and Nuclear Reaction Analysis (NRA), and 2)- to illustrate the use of these techniques to determine the composition of the surface and outer microns of material. Some examples will be given concerning porous silicon layers.

Rutherford Backscattering Spectrometry (RBS) consists in measuring the intensity and energy of elastically backscattered incident particles by heavier target atoms. As it will be shown these two parameters are characteristic of the scattering atoms providing the identification and the amount of atoms.

RBS is also used to study the structure of crystals. When a major symmetry direction or plane of a single crystal is aligned with the ion beam, the yield of the backscattered incident particles is considerably reduced compared to that observed when the target is randomly oriented with regard to the beam direction. The incident ions are steered along channels between atomic rows or planes. This phenomenon, called Channeling is used to study the crystal defects.

Elastic Recoil Detection Analysis (ERDA), also called Forward Recoil Spectrometry is similar to RBS but instead of detecting the backscattered incident particles, one detects the recoil target atoms, when the incident particle interacts elastically with a target atom. This technique is used to measure depth profile of light elements in solids such as hydrogen and deuterium.

Nuclear Reaction Analysis (NRA) is a quite different technique. When an incident particle with energy comparable to Coulomb barrier of the target nucleus comes very closed to the nucleus it may be captured by the nucleus to form a compound in an excited state which deexcites by emitting protons, neutrons, α particles, γ-rays etc... The nature and the energy of the emitted particles are characteristic of the target nucleus. Detecting and counting these emitted particles allows one to identify the atoms and to measure its amount.

One interesting feature of nuclear interactions is the existence of narrow resonance in the yield of many nuclear reactions as a function of the energy of the incident particle. By varying the incident energy one explores the target and determines the atom profile.

When a particle penetrates into a target it slows down and its kinetic energy decreases. The energy lost per unit length, called stopping power, depends on the charge of the particle, on its energy and on the target. The energy E and the number of detected particles of energy E depends on the energy of the incident particles just before the collision and on the energy loss of the detected particle during the outcoming path. To determine what part of the target the detected particles come from and consequently what is the concentration of the target atoms it is hence necessary to know the stopping power of the incident and of the detected particles. Together with the physics underlying the techniques quoted above we will present a brief picture of energy loss of light ions, such as protons, deuterons, helium of MeV energy in solids.

I. RUTHERFORD BACKSCATTERING SPECTROMETRY

A - Basic physical concepts

Rutherford Backscattering Spectrometry is based on classical elastic scattering of incident positive ion by the central force field of a target nucleus. During this elastic two body collision, the incident particle tranfers a part of its energy to the target nucleus. This leads to the concept of kinematic factor which is first developed in this paragraph. Then we will determine the path of the incident particle and the scattering cross section. For more details [1-3].

1) Kinematic factors

As it is illustrated in fig.1, during an elastic collision between two isolated particles, an incident particle of mass M_1, velocity V_0 and energy E_0, transfers an amount E_2 of its initial kinetic energy E_0 to a target nucleus of mass M_2 which is initially at rest. The target mass is scattered with a recoil angle ϕ and velocity V_2. The projectile is scattered with an angle θ, a velocity V_1 and an energy E_1.

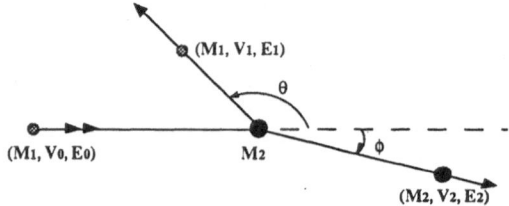

(M_1, V_1, E_1)

θ

(M_1, V_0, E_0) M_2

ϕ

(M_2, V_2, E_2)

Fig. 1: Schematic representation of an elastic collision between a particle (1) of energy E_0 and a particle (2), initially at rest.

In the laboratory system for a non-relativistic case, the conservation of energy and momentum are expressed by the equations:

$$\frac{1}{2} M_1 V_0^2 = \frac{1}{2} M_1 V_1^2 + \frac{1}{2} M_2 V_2^2$$
$$M_1 V_0 = M_1 V_1 \cos\theta + M_2 V_2 \cos\phi$$
$$0 = M_1 V_1 \sin\theta - M_2 V_2 \sin\phi$$

from which one finds the ratio of the energies of the projectile and of the target mass after the collision to the incident energy of the projectile, called the kinematic factors, respectively $K(\theta)$ and $K'(\phi)$. For $M_2 > M_1$:

$$\frac{E_1}{E_0} = K(\theta) = \left[\frac{(M_2^2 - M_1^2 \sin\theta)^{1/2} + M_1 \cos\theta}{M_2 + M_1} \right]^2 \qquad (1)$$

Example : For ^{4}He beam and Si and O target atoms at $\theta = 165°$
$K_{Si} = 0.568$ ($M_2 = 28$) and $K_O = 0.366$ ($M_2 = 16$).

$$\frac{E_2}{E_0} = K'(\phi) = \frac{(4M_1M_2 \cos^2\phi)}{(M_1 + M_2)^2} \quad (\phi < \pi/2) \tag{2}$$

Example : For ^{4}He beam and hydrogen target atoms : $K'_H = 0.48$ at $\phi = 30°$.

Equation (1) shows that for a given incident beam and a given detection angle θ, the energy E_1 of the backscattered projectile is completely determined by the mass M_2 of the target atom and then constitutes the signature of this atom.

When a sample contains two types of atoms with masses M_2 and M'_2, the largest value of the scattered projectile energies E_1-E'_1 is obtained with $\theta = 180°$. In practice the detector is placed at $\theta = 170°$ or $165°$ because of its size. This experimental arrangement has given the method its name of backscattering spectrometry.

2) Two-body problem

This is a classical problem which is treated in many books (see for example [3]). We have just summarized here the main stages of the calculation which allow one to determine the path of the particles.

Let r_1 and r_2 be the positions in the laboratory system of two isolated particles with charges Z_1e and Z_2e and masses M_1 and M_2 interacting through the Coulomb force:

$$F(r) = k \frac{Z_1Z_2e^2}{r^2}$$

where $r = |r_1 - r_2|$ (see fig. 2a), and $k = \frac{1}{4\pi\epsilon_0}$.

Before the collision, the energy of the incident particle (M_1, Z_1) is $\frac{1}{2}M_1V_0^2$ and the target particle (M_2, Z_2) is at rest. As the system of the two particles is isolated, the total energy E is constant and equals to :

$$E = \frac{1}{2}M_1V_1^2 + \frac{1}{2}M_2V_2^2 + Ep(r) = \frac{1}{2}M_1V_0^2 \tag{3}$$

where V_1, V_2 are the velocities and $Ep(r) = k\frac{Z_1Z_2e^2}{r}$ is the potential energy of the two particles.

The change of variables $r = |r_1 - r_2|$ and $R = \frac{(M_1r_1 + M_2r_2)}{(M_1 + M_2)}$ leads to :

$$E = \frac{1}{2}(M_1 + M_2)V^2 + \frac{1}{2}\mu v^2 + E_p(r) \tag{4}$$

where $V = \dfrac{dR}{dt} = \dfrac{M_1 V_1 + M_2 V_2}{M_1 + M_2}$ is the velocity of the center of mass in the laboratory system, $v = \dfrac{dr}{dt}$ and the reduced mass μ is defined as $\dfrac{1}{\mu} = \dfrac{1}{M_1} + \dfrac{1}{M_2}$.

The conservation of the total momentum $M_1 V_1 + M_2 V_2 = (M_1 + M_2)V$ leads to :

$$\frac{1}{2}(M_1 + M_2)V^2 = E_R = \text{constante}$$

and equation (4) may be written :

$$\frac{1}{2}\mu v^2 + E_p(r) = \text{constante} = \frac{1}{2}\mu V_0^2 \tag{5}$$

where V_0 is the velocity of the incident particle (M_1, Z_1) before colliding the charge at rest (M_2, Z_2).

The total angular momentum is also a constant of the motion :

$$L = l_1 + l_2 = r_1 \wedge M_1 V_1 + r_2 \wedge M_2 V_2 = \text{constant}$$

which may be written :

$$L = R \wedge (M_1 + M_2)V + r \wedge \mu v \tag{6}$$

The first term of equation (6) is constant so that it may be rewritten :

$$l = r \wedge \mu v = \text{constante} = l_0 \tag{7}$$

where $|l_0| = |r \wedge \mu V_0| = \mu |V_0| b$, b is called the impact parameter of the particle (see fig.2b).

Equation (7) shows that $r(t)$, the vector joining the two charges, is in a plane. Equations (5) and (7) depend only on the relative position $r(t)$ of the two particles. The path $r(t)$ schematically represented in fig.2b shows the relative motion of the two particles. It may be obtained by solving equations (5) and (7) in plane polar coordinates (r, Φ) (fig. 2b).

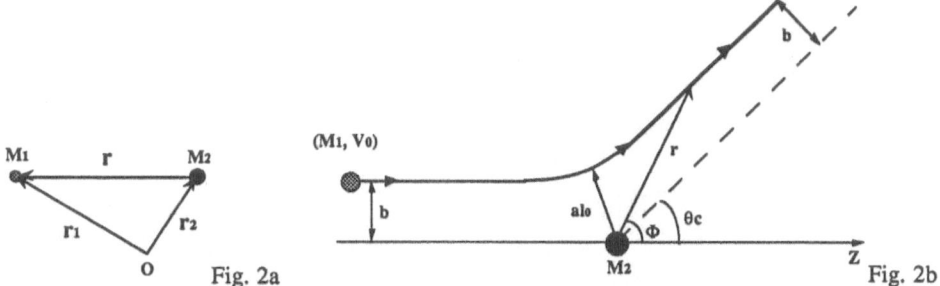

Fig. 2 : (a) Position of two particles in the laboratory system. (b) Schematic trajectory of an incident projectile M_1 of velocity V_0, relatively to the particle M_2 at rest. The interaction between the two particles is a repulsive Coulomb force.

From equation (7) one can deduce:

$$\frac{d\Phi}{dt} = \frac{l_0}{\mu r^2}$$

Equation (5) becomes :

$$\frac{1}{2}\mu\left[\left(\frac{dr}{dt}\right)^2 + \frac{l_0^2}{\mu^2 r^2}\right] + E_p(r) = \frac{1}{2}\mu V_0^2 = E_0 \qquad (8)$$

from which we can deduce $\frac{dr}{dt}$ and

$$\frac{dr}{d\Phi} = \frac{dr}{dt}\frac{dt}{d\Phi} = \frac{\mu r^2}{l_0}\sqrt{\frac{2}{\mu}\left(E_0 - E_p(r) - \frac{l_0^2}{2\mu r^2}\right)} \qquad (9)$$

By integrating this equation one finds the path r(t) of the relative motion which is a hyperbola. The asymptotical direction θ_c of the scattered particle is given by :

$$\text{tg}\,\frac{\theta_c}{2} = \left(k\,\frac{Z_1 Z_2 e^2}{2E_0 b}\right) \qquad (10)$$

As it is shown by eq. (10) and illustrated in fig.3, for a given target nucleus (Z_2) and incident energy E_0 the lower is the impact parameter b, the higher is the scattering angle θ.

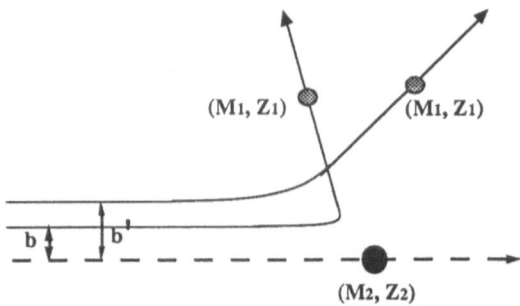

Fig. 3: Coulomb interaction between projectile (M_1, Z_1) with a particle (M_2, Z_2): schematic representation of two different trajectories of a projectile (M_1, Z_1), relatively to the target atom for two different impact parameters b and b'.

θ_c is the angle of the straight lines joining the two particles before and after the collision. Therefore θ_c is also the scattering angle in the system of the center of mass.

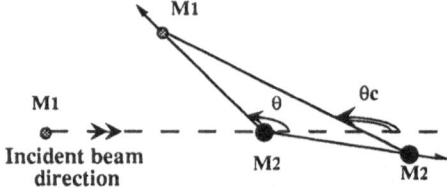

Fig. 4: Scattering angles in both laboratory (θ) and center of mass (θ_c) frames of reference.

The relationship between the scattering angle θ in the laboratory system and the scattering angle in the center of mass is determined using the definition of the center of mass given above (fig. 4).

$$\text{tg } \theta = \frac{\sin\theta_c}{\cos\theta_c + M_1/M_2}$$

The distance of closest approach between the two particles, noted al_0, is determined by solving the equation $\frac{dr}{dt} = 0$. The solution is :

$$al_0 = \frac{a_0}{2} + \sqrt{\left(\frac{a_0}{2}\right)^2 + b^2} \qquad (11)$$

where $a_0 = k \dfrac{Z_1 Z_2 e^2}{E_0}$ is the distance of closest approach when b=0 which corresponds to the case of a frontal collision.

3) Rutherford scattering cross section

Every incident particle passing at a distance between b and b + db from a target nucleus is scattered at angle between θ_c and $\theta c + d\theta_c$, the relation between db and $d\theta_c$ being deduced from eq. (10) (see fig.5).

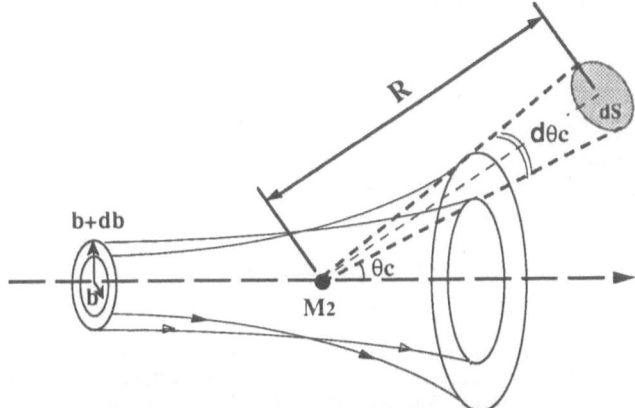

Fig. 5: Schematic path of two projectiles, with b and b+db impact parameters, scattered at angles θ_c and $\theta_c+d\theta_c$. The detection of these particles are performed with a detector of surface area dS, placed at the angle θ_c and a distance R from the target.

By definition, the scattering cross section is the ratio of the "scattering surface" $2\pi b\ db$ to the corresponding solid angle $2\pi\ \sin\theta_c\ d\theta_c$:

$$\sigma(\theta_c) = \frac{2\pi b\ db}{2\pi\ \sin\theta_c\ d\theta_c} = \left(\frac{Z_1 Z_2 e^2}{4E_0}\right)^2 \frac{1}{\sin^4\theta_c/2} \qquad (12)$$

This is Rutherford scattering cross section in the system of center of mass. The transformation to the laboratory frame of reference gives:

$$\sigma_{Ruth}(\theta) = \left(\frac{Z_1 Z_2 e^2}{4E_0}\right)^2 \frac{1}{\sin^4\theta/2} \frac{\left\{\left[1 - \left(\frac{M_1}{M_2}\sin\theta\right)^2\right]^{1/2} + \cos\theta\right\}^2}{\left[1 - \left(\frac{M_1}{M_2}\sin\theta\right)^2\right]^{1/2}}$$

For $M_1 \ll M_2$:

$$\sigma_{Ruth}(\theta) = \left(\frac{Z_1 Z_2 e^2}{4E_0}\right)^2 \left[\frac{1}{\sin^4\theta/2} - 2\left(\frac{M_1}{M_2}\right)^2 + O\left(\frac{M_1}{M_2}\right)^4\right]$$

RBS experiment consists in sending Q incident particles of energy E_0 and detecting the particles scattered in direction θ by target atoms, with a detector of surface area dS placed at a distance R from the target. The probability for these particles to be detected is proportionnal to the detector solid angle $\Omega = dS/R^2$. If the target containing N atoms/cm^2 (M_2, Z_2) is sufficiently thin to neglect the energy loss of the incident particles, the incident particles will be backscattered with an energy $K(\theta) \times E_0$ where $K(\theta)$ is the kinematic factor given by equation (1) and the number of backscattered particles detected will be :

$$Y = Q.N.\Omega.\sigma(\theta, E_0)$$

Order of magnitude:
For $E_0 = 2$ MeV , $Z_1 = 2$ (He) and $Z_2 = 14$ (Si)
$\Rightarrow \sigma_{Ruth}(170°) = 0.25 \ 10^{-24}$ cm^2 = 0.25 barn

<u>4) Deviations from Rutherford scattering cross section</u>

The obtention of Rutherford cross section is based on the assumption of a pure Coulomb interaction between the incident particle Z_1 and a target nucleus Z_2, neglecting the screening of the target nucleus charge by electrons. This assumption is well justified when the distance of closest approach, al_0, is much lower than the extension of the electronic orbitals. As shown by equations (10) and (11) this is the case for large angle detection in RBS experiments which selects collisions with small impact parameter when the energy is sufficiently high.
At lower energy and smaller impact parameter the screening by the electrons may result in large deviations from Rutherford scattering cross section. The magnitude of this screening correction has been measured by Lecuyer et al.[4] who found:

$$\sigma(\theta) = \sigma_{Ruth}(\theta) \times \left[1 - \frac{0.049 Z_1 Z_2^{4/3}}{E_0} \right]$$

For a 1 MeV ^{4}He$^+$ ion on a Si atom, the predicted deviation is only 0.3%.
At higher energy, the distance of closest approach al_0 may be of the order of magnitude of the nuclear radius. Large deviation from Rutherford scattering may occur due to nuclear interactions. For ^{4}He ion this energy is about 7 MeV for C and O atoms and 9 Mev for Si atoms. However, there is some exceptions to these estimations. For O and C, for example, there are strong resonances in ^{4}He elastic scattering cross section at respectively 3.045 and 4 MeV [5,6]. These resonances are used to increase the sensitivity of detection by RBS of oxygen and carbon.

B - Rutherford Backscattering Spectrometry

Rutherford Backscattering Spectrometry is a technique that allows to determine quantitatively atomic profiles in a layer. To determine such atomic profiles, one compares generally the experimental spectra to the spectra obtained from computer program [7, 8] in which one has injected the given profiles. The problem is to determine the energy and the number of detected particles backscattered by atoms (M_2, Z_2) located at depth x in the target. We have seen above (eq.(1)) that the energy of the backscattered particle is proportional to the energy E of the incident particle just before the collision, the factor of proportionality depending on the mass of the scattering atom. We have also seen that the scattering cross section is proportional to $(Z_2/E)^2$ where Z_2 is the atomic number of the scattering atom. To determine the energy and the number of detected particles backscattered by atoms (M_2, Z_2) at a depth x in the target it is hence necessary to know the energy lost by the particle before the collision during the incoming path and after the collision during the outcoming path.

In this chapter, we describe the experimental setup used to register backscattered particle spectra, and we present a brief summary of energy losses of light ions in materials. Then we show how to obtain calculated backscattered particle spectra and how to read them.

1) Experimental setup

Fig. 6 shows a schematic representation of the experimental setup for Rutherford Backscattering or Nuclear Reaction Analysis. A beam of MeV charged particles such as protons, deuterons or helium etc..., is send on a target. A small fraction of the incident ions which undergoes small impact parameter collisions are backscattered in θ direction.

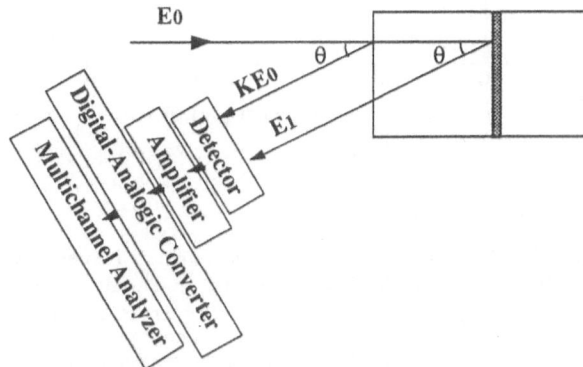

Fig. 6: Schematic setup for RBS experiments.

The detector is a Shottky barrier (Au-Si) with a depletion zone near the surface the thickness of which may be controlled by a reverse biased polarisation. The backscattered particles loose their energy in the depletion zone creating electron-hole pairs the number of which is proportional to the energy loss. The electron-hole pairs are collected to produce a V(t) signal the amplitude of which is proportional to the energy of the detected particles. This V(t) signal is sent to a digital-analogic converter and the corresponding value to a multichannel analyser. RBS or NRA spectra represents the number of detected particles (counts) as a function of their energy (channel number).

2) Energy loss. Energy straggling

When an ion moves in a solid it looses energy through elastic collisions with the nuclei, more or less screened by the electrons, and through inelastic collisions with electrons which are raised to excited states or ejected from atoms. In RBS experiments, we are interested in the processes that control the slowing down of particles before and after backscattering. As most of elastic collisions with the nuclei, before and after backscattering, occur with large impact parameter; the energy loss through these elastic collisions is negligible in the slowing down process with regard to the energy lost through inelastic collisions with electrons, at least at energy which are considered in RBS experiments.

The energy loss per unit length S(E) ≡ dE/dx of a particle in a medium is called the stopping power. Fig. 7 shows for example the stopping power S(E) of ^{4}He in Si and O (gas) expressed in $eV/10^{15}$ at/cm^2 as a function of the energy.

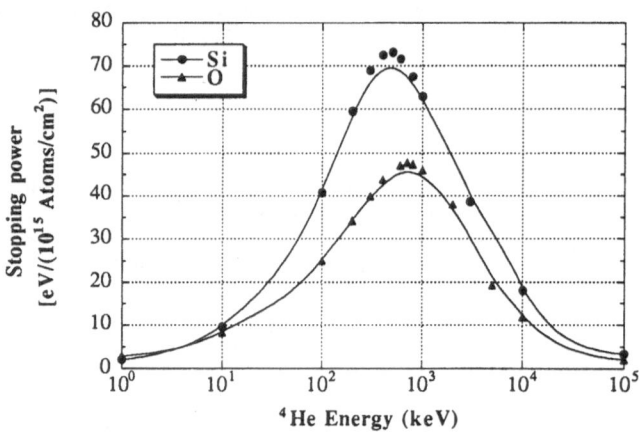

Fig. 7: Stopping power of ^{4}He in silicon and oxygen, as a function of energy.

The treatment of inelastic collisions with electrons considers two energy regions. In the high energy region, typically above a few MeV for ^{4}He, the ion is totaly stripped of its electrons. In these conditions the stopping power S(E) decreases with increasing particle velocity V as $1/V^2$ [9, 10]. As the energy of the particle decreases from this energy region, the probability that an electron is captured by the moving particle becomes important and the influence on the slowing down process of the inner electrons of the target atoms declines. The stopping power no more varies as $1/V^2$, increases less rapidly with decreasing velocity, reaches a maximum and decreases. In the low energy region, the stopping power is roughly proportional to the particle velocity V [11, 12]. For ^{4}He the maximum of the stopping power occurs typically at 0.6 to 1 MeV depending on the target.

For compounds A_xB_y, one applies the Bragg's rule, which states that the stopping power $S_{AxBy} = xS_A + yS_B$. The experimental stopping power of light ions, in different monoatomic targets, has been measured by a great number of independent investigators and assembled by J. F. Ziegler and al.[13]. These empirical stopping powers are used in computer simulations (see below).

The processes through which the particles that move in a material loose energy are subject to statistical fluctuations. The probability that a particle of energy E$_0$ has an energy between E and E + dE at depth x in a target is approximately gaussian when the energy loss is small compared to E$_0$:

$$P(E_0, E, x)\, dE = \frac{1}{\sqrt{2\pi\, \sigma^2}}\ \exp{-\frac{[E - E(x)]^2}{2\sigma^2}}\, dE$$

where $E(x) = E_0 - \int_0^x S(E) \, dx$ is the mean value of the particle energies at depth x and σ^2 is the

variance. σ is called the energy straggling. According to Bohr's model [14] the energy straggling equals to $\sigma = \sqrt{4\pi(Z_1 \, e^2)^2 \, N \, Z_2 \, x}$, in the case of monoatomic sample, where N is the number of target atoms per unit volume, and Z_1 and Z_2 are respectively the atomic number of the projectile and of the target atom. The main features of the Bohr formula is that the energy straggling is independent of particle energy and is proportional to the square root of the film thickness.

Energy straggling contributes to the energy resolution together with the resolution of the detectors (10-15 keV). For example the straggling for ^{4}He in Si equals to 6 keV at a depth of 5000 Å. The energy straggling and the resolution of detectors are taken into account in the different computer programs which simulate RBS spectra, nuclear spectra and excitation curves.

3) Simulated backscattering spectrum

As shown in fig. 8, an incident particle of energy E_0 is backscattered at a depth x by an atom (M_2, Z_2), the concentration of which is C(x). The energy of the particle just before the collision is:

$$E(x) = E_0 - \int_0^x S(E) \, dx$$

where $S(E) \equiv \dfrac{dE}{dx}$ is the stopping power of the particle in the target.

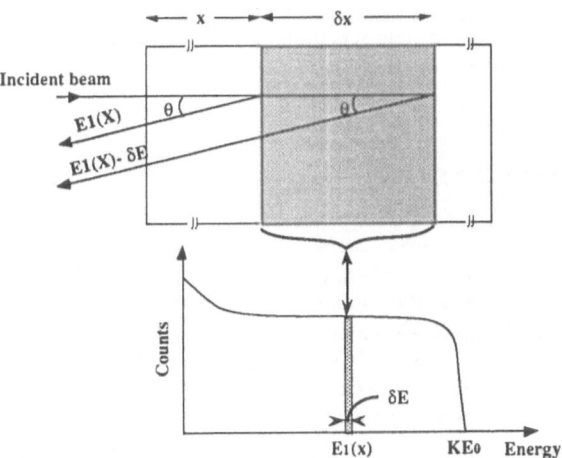

Fig. 8: Schematic ^{4}He backscattering spectrum on a monoatomic target containing. This figure shows the correspondance between the thickness δx of a slice inside the film and the channel width δE.

After scattering in θ direction, the energy of the particle is $K\,E(x)$ where K is the kinematic factor given by equation (1). Finally, it is detected with an energy :

$$E_1(x) = K\,E(x) - \int_x^0 S(E)\,\frac{dx}{\cos\theta}$$

Let be δE the channel width of the multichannel analyser. The particles, with energy $E_1(x)-\delta E$, scattered by atoms (M_2, Z_2), come from a part of the target between x and $x + \delta x$ the relation between δE and δx being:

$$\delta E = \delta x \left\{ \left[K\,S(E)_{E(x)} + \left[\frac{1}{|\cos\theta|} S(E) \right]_{KE(x)} \right\} \right. \tag{13}$$

The number of particles in the channel corresponding to energy E_1 is hence equal to:

$$Y(E_1) = Q\,\Omega\,\sigma[\theta, Z_2, E(x)]\,C(x)\,\delta x$$

where Q is the number of particles sent on the target and σ is the scattering cross section. This is the contribution to the spectrum of atoms (M_2, Z_2) located at depth x in the target. The particles of energy E_1 may come from other scattering atoms different from (M_2, Z_2) located in other part of the target. For example they may come from heavier atoms located in deeper regions of the target or lighter atoms located near the surface. In these conditions $Y(E1)$ will be the sum of all these contributions. By making the above calculation for every part of the target one obtains a simulation which is compared to the experimental spectrum. Backscattering spectra and profile can be obtain from computer programs [7,8].

4) How to read a RBS spectrum

Fig. 9 represents a spectrum of backscattered α particles from SiO_2 thermal oxide formed on Si. The beam energy was 2 MeV. We will interpretate this spectrum by starting from the higher to the lower energies. The Si edge corresponding to α particles backscattered by the Si atoms at the SiO_2 surface is a $2 \times 0.568 = 1.136$ MeV, where 0.568 is the value of the kinematic factor corresponding to the mass 28 (see eq. (1)). The plateau which begins from this edge and which ends at the second edge, called Si substrate in the figure, corresponds to α particles backscattered by the Si atoms in SiO_2. The second edge corresponds to α particles backscattered by the Si atoms at the Si/SiO_2 interface. The distance, in energy scale, between these two edges corresponds to the energy lost by the α particles in SiO_2 during the incoming path and the outcoming path.

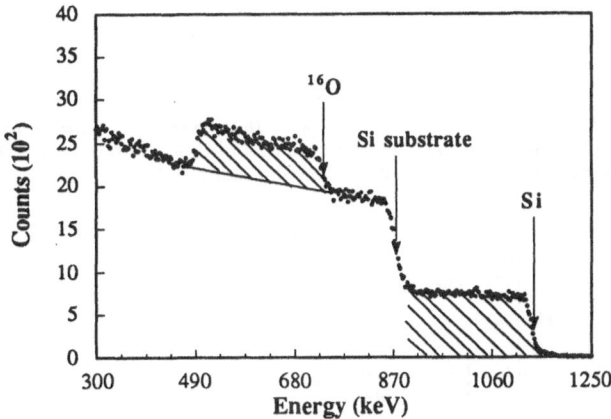

Fig. 9: RBS spectrum from 5400 Å thick SiO_2 on Si. The arrows indicate the energies of α particles backscattered by: 1)- Si atoms at the oxide surface (Si), 2)- Si atoms at the interface SiO_2/Si (Si substrate) and 3)- oxygen atoms at the oxide surface (^{16}O). The hatched regions correspond to silicon and oxygen atoms in SiO_2. The non hatched region corresponds to silicon atoms in Si substrate.

The second plateau until the ^{16}O edge corresponds only to α particles backscattered by the Si atoms in the substrate. The height of the second plateau is higher than that of the first one. This is due to the fact that the stopping power in Si is different from that in SiO_2 (we compare the heights of the two plateaus just at the interface Si/SiO_2 to rule out the influence on the yield of the variation of the Rutherford cross section with energy). The most simple to understand this is to express the film thickness in number of Si atoms/cm^2 instead of in nm, which avoid to take into consideration the density of the different layers (this is very convenient for porous silicon). In this case the stopping power $S(E) \equiv dE/dx$ is higher in SiO_2 than in Si : $S(E)_{SiO2} = S(E)_{Si} + 2\ S(E)_{O}$. The relation (13) shows that the thickness δx of a slice inside the film corresponding to the channel width δE is lower in SiO_2 than in Si. This is the reason why the height of the plateau corresponding to SiO_2 is lower than that corresponding to Si. This is also the reason why the plateau corresponding to Si atoms in porous silicon is lower than that the one corresponding to compact Si : the presence of impurities in porous silicon increases the stopping power expressed in unit energy per number of Si atoms/cm^2, hence decreases the thickness δx of the slices of the films corresponding to the channel width δE.

The ^{16}O edge corresponds to α particles backscattered by the ^{16}O atoms at the SiO_2 surface. The energy of these α particles is equal to 2 MeV x 0.366 = 0.73 MeV, where 0.366 is the value of the kinematic factor corresponding to the mass 16 (see eq. (1)). From the ^{16}O edge the spectrum is the superposition of the α particles backscattered by ^{16}O and that backscattered by the Si atoms in the Si substrate.

C - Ion Channeling in crystals [15]

When the material under study is a crystal, we have to care for the orientation of the crystal with regard to the beam direction. As a matter of fact, when a symmetry axis of the crystal is aligned with the incident beam, the backscattered particle spectrum is very different from the one obtained when the crystal is randomly oriented with regard to the beam direction (Fig. 10).

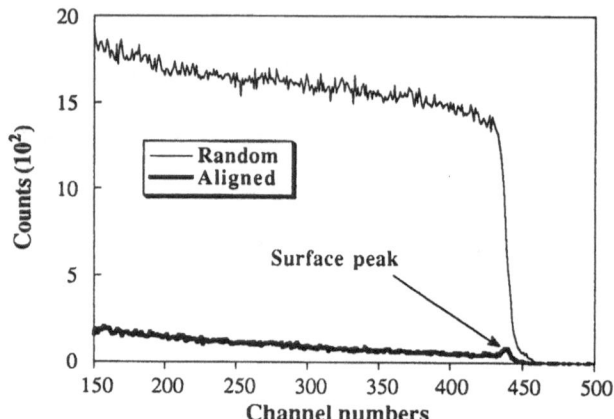

Fig. 10 : ^{4}He RBS spectra from monocrystalline silicon, in channeling and random geometries. E_{4He} = 2.2 MeV, detection angle = 165°.

Fig. 11 represents a schematic view of trajectories of particle entering the crystal. It reveals a major effect of channeling phenomenon : due to the deflection of incident ions by the first layer of target atoms, there is a forbidden region called the shadow cone. Without any thermal vibration this would prevent the incident ions to be backscattered by the second and the subsequent layers of atoms.

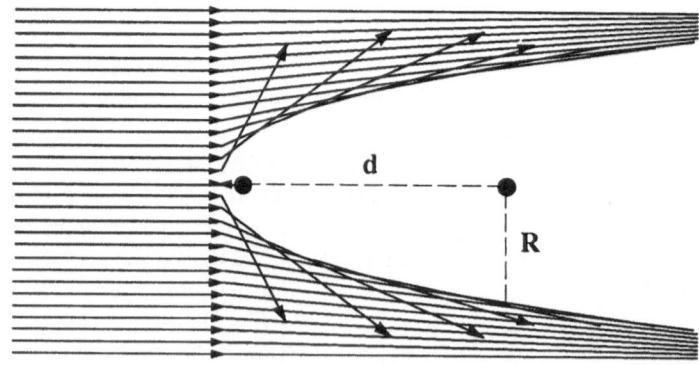

Fig. 11: Schematic representation of the shadowing effect, due to surface atoms, on the underlying atoms.

For example, for 1 MeV ^{4}He in Si (100), the distance of closest approach to the second atom, also called the shadow cone radius R is $\approx$ 0.1 Å which is very high with regard to the impact parameter corresponding to a backscattering event (0.5 10^{-4} Å for a detection angle 165°). In fact the relevant parameter is the ratio of the thermal vibration amplitude normal to the beam to the shadow cone radius. It can be shown that only the first layers of atoms contribute to the surface backscattering i.e. to the surface peak represented in Fig. 10. The surface peak integral that one would obtain for a perfect crystal can be determined theoritically at any temperature. The deviation from this value of the experimental surface peak provides informations on the surface structure.

After passing the first layers of target atoms, most of ions ($\approx$95%) have trajectories with small angle relative to crystallographic axis, smaller than a critical angle Ψ_C and travels far from the atomic rows. In these conditions, these ions do not feel individual atom potential but rather the average continuous potential of the atomic rows by which they are steered. As a first approximation, within the crystal, these ions can never get close enough to the target atoms to undergo large scattering angle and have an "oscillatory" motion between rows or planes of atoms : they are channeled.

Only a few percent of particles having a deflection angle higher than Ψ_C after passing the first layers of atoms can be eventually backscattered by atoms further inside the crystal. These particles are responsible for the yield of RBS spectrum behind the surface peak (Fig. 11). The ratio χ_{min} of the yields under aligned and random geometries at energy just below that of the surface peak is called the minimum yield and gives information on the quality of the crystal.

Experimentally the critical angle Ψ_C and the minimum yield χ_{min} are determined by measuring the backscattering yield just behind the surface peak as a function of angle between the beam and the crystal axis (see fig. 12). The value of the angle at the half minimum is taken as a measure of the

critical angle Ψ_C, and χ_{min} = B/A. For a given temperature, Ψ_C varies linearly with $\sqrt{\dfrac{Z_1Z_2e^2}{E\,d}}$
where Z_1 and Z_2 are the atomic numbers of the incident particle and the target atom, d is the spacing of atoms along the rows and E the energy of the particle. For compounds, Z_2 and d are average values. For 1 MeV ^{4}He along (100) direction in Si at room temperature Ψ_C = 0.6°.

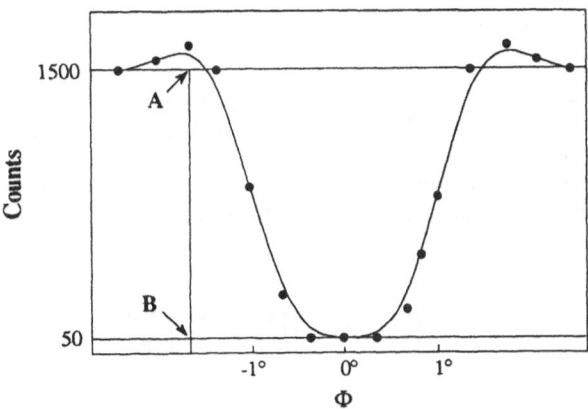

Fig. 12: Typical variation of the yield near the surface as a function of Φ the angle between the beam and a symmetry direction of the crystal. χ_{min} = B/A.

D - Elastic Recoil Detection Analysis (ERDA)

1) Experimental setup

This method, illustrated in fig.13, consists in detecting light target nuclei (^{1}H, ^{2}H) which are scattered in the forward direction during elastic collisions with more massive incident ions (^{4}He, C, O). The target is tilted at angles around θ_i = 75° using the convention θi = 0 for normal incidence and the detection angle θ is typically 30°. A mylar film is put in front of the detector to stop the incident particles (^{4}He, C, O) backscattered by the target atoms. For 2 MeV ^{4}He the thickness of the mylar foil is typically 10 μm.

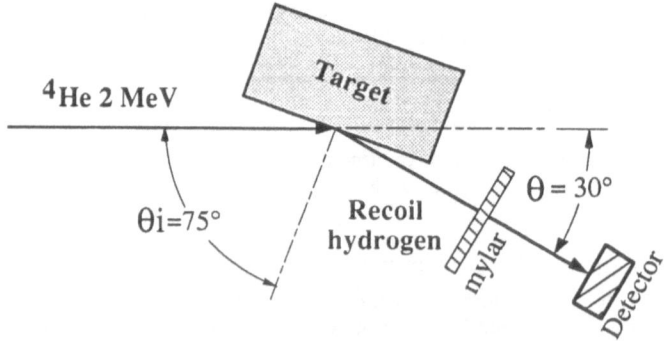

Fig. 13: Usual geometry for ERDA experiments.

2) Simulated recoil spectrum.

The principle of the calculation is the same as for backscattering spectrum, except that the detected particle is the recoil target atom and one has to know the stopping power of the incident particle and of the recoil atom.

Let be C(x) the hydrogen concentration at depth x in the target tilted at θ_i. The energy of the incident particle, ^{4}He for example, at depth x just before the collision is :

$$E(x) = E_0 - \int_0^x [S(E)]_{He} \frac{dx}{\cos\theta_i}$$

where : $[S(E)]_{He} \equiv [\frac{dE}{dx}]_H$ is the stopping power of ^{4}He incident particles.

After scattering, the energy of the recoil atom is K' E(x) where K' is the kinematic factor given by Eq. 2 The recoil atom is detected at θ with an energy E'$_1$ given by :

$$E'_1 = K' E(x) - \int_x^0 [\frac{dE}{dx}]_H \frac{dx}{|\cos(\theta_i + \theta)|}$$

where : $[S(E)]_H \equiv [\frac{dE}{dx}]_H$ is here the stopping power of the recoil proton.

If δE is the channel width of the multichannel analyzer, the recoil atoms with energy between E'_1 and $(E'_1-\delta E)$ come from a part of the target between x and δx, the relation between δE and δx being:

$$\delta E = \delta x \left\{ K' [S(E)]_{He} \frac{1}{\cos \theta i} + [S(E)]_H \frac{1}{|\cos(\theta i + \theta)|} \right\}$$

where $[S(E)]_{He}$ is the stopping power of He at energy $E(x)$ and $[S(E)]_H$ is the stopping power of H at energy $K'E(x)$.

The number of recoil particles in the channel corresponding to energy E_1 is hence equal to :

$$Y(E_1) = Q \ \Omega \ \sigma_{recoil}[\theta, E(x)] \ C(x) \ \delta x$$

3) Cross section for H and ^{2}H elastic recoil using MeV ^{4}He

The cross sections for H and D elastic recoil using MeV ^{4}He have been measured by many authors (see for example [16-17]). Large deviation of the experimental cross sections from Rutherford cross section have been measured in the range of 1 to 2.5 MeV. This is due to nuclear interactions.

For H there are two broad resonances near 8.05 (width 7.2 MeV) and 20 MeV in the scattering of ^{4}He by H which correspond to levels of the ^{5}Li nucleus. For D there is a resonance near 2.13 MeV of width 70 keV and probably other resonances at higher energies which corresponds to levels of ^{6}Li nucleus. In both cases the fitting parameters depend on detection angle.

Tabulated experimental data or analytical expressions which have been obtained by fitting experimental data such as that quoted above, are used in special programs such as SENRAS [16] or RBX [8] to determine H or D depth profiles.

II. NUCLEAR REACTION ANALYSIS

Incident particles with impact parameters of the order of magnitude of the target nucleus radius and with energy sufficiently high to get close enough to the nucleus may induce nuclear reactions i.e. a processus which leads to the formation of different nuclei.

If X is the target nucleus and a the light projectile we can write :

$$X + a \rightarrow Y + b \qquad \text{or} \qquad X(a,b)Y$$

where Y and b are the new nuclei formed.

Nuclear reactions obey the conservation of nucleons and the conservation of charge.

A - Q values and kinetic energies

The Q value of a nuclear reaction $X + a \rightarrow Y + b$ is defined by:

$$(M_a + M_X) c^2 = (M_b + M_Y) c^2 + Q$$

where the M's are the masses.

The conservation of total energy and momentum gives in the non-relativistic case the energy E_b of the particle b detected at angle θ.

$$\sqrt{E_b} = A \pm \sqrt{A^2 + B} \qquad\qquad (12)$$

where :

$$A = \frac{\sqrt{M_a M_b E_a}}{M_b} + M_Y \cos\theta \qquad \text{and} \qquad B = \frac{M_Y Q + E_a(M_Y - M_a)}{M_b + M_Y}$$

These relations show that the energy E_b of the detected particle is characteristic of the nuclear reaction for given θ and E_a.

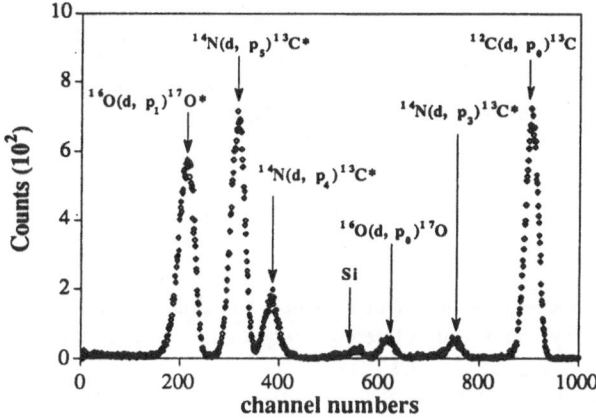

Fig. 14: Energy spectrum from deuteron beam at 850 keV on a p+ psl, 1 μm thick after thermal oxidation in ($^{16}O_2$, 12 mbar), at 300°C during 1 hr followed by a thermal treatment in ($^{14}NH_3$, 12 mbar) at 1000 °C during 8 min.

B - Nuclear reaction cross sections

Nuclear reaction cross sections are not generally given by analytical functions and their determination is mainly experimental.

Figure 15 shows some nuclear cross sections. The $^{18}O(p, \alpha)^{15}N$ presents two properties that are used in Nuclear Reaction Analysis. It varies smoothly in some energy regions, around 700 keV for example, and presents a narrow resonance ($\Gamma = 2$ keV) at 629 keV.

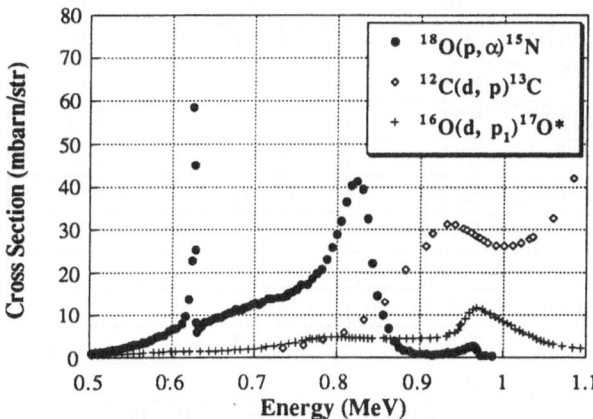

Fig. 15: Variations of the cross sections as a function of energy for the following reactions: $^{16}O(d,p_1)^{17}O*$ [19], $^{12}C(d,p)^{13}C$ [20], $^{18}O(p,\alpha)^{15}N$ [21].

The region of energy where the cross section varies slowly, is used to determine the total amount of nuclei. When the layer is thin enough, i.e., when the energy loss of the incident particles in the layer is low enough to neglect the variation of the cross section, the number N of the emitted particles is practically proportional to the number of ^{18}O atoms in the layer, independently of their distribution $C(x)$ in this layer :

$$N = Q.\Omega. \ \sigma_{nucl}(E_0) \int_0^x C(x) \ dx$$

where E_0 is the energy of the incident particle and Q is the dose. By comparing the counts from the unknown layer and from a reference target one can determine the number of nuclei /cm^2 without knowing the value of Q, Ω and σ_{nucl}.
When the layer is too thick to neglect the variation of the cross section the number of the emitted particles is :

$$N = Q.\Omega. \int_0^x C(x) \ \sigma_{nucl}[E(x)] \ dx$$

In this case, one uses a computer program which simulates the nuclear reaction spectra [18].
The presence of strong resonances in the cross section at energy $E = E_R$ is used to determine the depth profile of the nuclei. The energy E_0 of the incident particle is increased step by step from $E_0 = E_R$. Due to energy losses, the incident ions reach the energy E_R at a depth x. The number $N(E_0)$ of detected particles which depends strongly on $C(x)$ is thus an image of $C(x)$, and from the curve obtained one can deduce the nuclei depth profile.

Reactions		Q (MeV)	E_0 (keV)	E_R (keV)
$^{12}C(d, \ p)^{13}C$	[20]	2.719	970	
$^{16}O(d, \ p_0)^{17}O$	[19]	1.919	860	
$^{16}O(d, \ p_1)^{17}O *$	[19]		860	
$^{19}F(p, \ \alpha_0)^{16}O$	[27]	8.119	1250	
$^{11}B(p, \ \alpha)^8Be$	[25]	8.582	660	
$^{14}N(d, \ \alpha_0)^{12}C$	[28]	13.579	1530	
$^{16}O(d, \ \alpha)^{14}N$	[19]	3.11	1000	
$^{16}O(^3He, \ \alpha)^{15}O$	[22]	4.914		2370 (Γ=70 keV)
$^{15}N(p, \ \gamma\alpha)^{12}C$	[23, 24, 26]	4.964		429 (Γ=120 eV)
$^{19}F(p, \ \gamma\alpha)^{16}O$	[27]	8.119		872 (Γ=4.7 keV)
$^{18}O(p, \ \alpha)^{15}N$	[19, 21]	3.97	750	629 (Γ=2 keV)

Table I : Some nuclear reactions with the Q values. E_0 is the usual bombarding energy for measuring the total amount of nuclei. E_R and Γ are the energy and the width of the nuclear resonance.

Some of nuclear reactions currently used and their corresponding Q values are given in Table I. We indicate the energy region E_0 where the cross sections vary smoothly and where the measurement of total amount of nuclei are routinely performed. We also indicate the energy position (E_R) and the width (Γ) of the narrow resonance used to determine nuclei depth profile.

C - Experimental setup

Typical values of $\sigma_{nucl}(E)$ are about ten mbarn wheareas typical values of $\sigma_{Ruth}(E)$ are of the order of few hundred mbarn. It is hence necessary to put mylar foil of convenient thickness in front of the detectors to stop the elastically scattered beam particles and to prevent count-rate saturation of the detector and electronic systems. The detectors used in NRA present sensitive surface with surface area of about 300 mm^2 and are placed at 7 cm from the sample which leads to a detection solid angle $\Omega \approx 60$ msteradian.

III. APPLICATIONS TO THE CASE OF POROUS SILICON

A - Impurity content in porous silicon

The visible photoluminescence of porous silicon layers (psl) has been attributed by some authors to silicon based molecules such as siloxene, siloxene derivatives with fluorine substitution, hydride species. To test the validity of this idea, we have performed quantitative analysis of the major impurities in psl, i.e. hydrogen, oxygen, carbon and fluorine [29, 30]. The substrates used were (100) p+ Si and (100) n+ Si. The porous layers were 10 μm thick. Two kinds of layers have been formed : layers of medium porosity (47%) which were non-photoluminescent and layers of high porosity which were photoluminescent. The time between the end of the sample preparation and the measurements was about 30 min.

Oxygen content has been measured using $^{16}O(d, p_1)^{17}O^*$ nuclear reaction at $E_d = 860$ keV. For oxygen reference we used anodic tantalum oxide calibrated by coulometric measurements (3%). For carbon measurement we used the $^{12}C(d, p)^{13}C$ nuclear reaction at Ed = 970 keV. Since no satisfactory well characterized carbon reference is available one uses anodic tantalum oxide and the measured ratio of the cross sections :

$$\frac{\sigma[^{12}C(d, p)^{13}C]_{970\ keV}}{\sigma[^{16}O(d, p_1)^{17}O^*]_{850\ keV}} = 5.56 \pm 5\%$$

Porous silicon layers 10 μm thick of 47 % porosity contain 26.5 10^{18} Si at /cm^2 (the density of Si is 5 10^{22} at /cm^3). For psl of 80% porosity the Si content is about 10 10^{18} at /cm^3. The energy loss of deuterons in these layers is between 100-300 keV about and the variation of the cross section cannot be neglected. We used a simulation program [18] to detemine the O/Si and C/Si ratios.

The fluorine content has been determined using the $^{19}F(p, \alpha_0)^{16}O$ nuclear reaction at a proton energy of 1250 keV. The stationary part of the cross section extends down to 1100 keV. The depth corresponding to the quasi plateau is typically 4 μm in silicon, i.e. 8 μm in psl of 50 % porosity or 20 μm in psl of 80% porosity. Except in psl of medium porosity where a slight correction was needed to take into account the decrease of the cross section at depth between 8 and 10 μm, the fluorine content could be deduced simply by comparing the counts from the psl and from a reference which was a CaF_2 layer on Ta.

Hydrogen profile was determined using elastic recoil detection analysis. The samples were tilted at an angle of 75° and the recoil protons were detected at 30°. The reference used was a thin layer of polystyrene C_8H_8 on Si. Fig. 16 shows typical elastic recoil proton spectra and the computer simulations (solid lines).

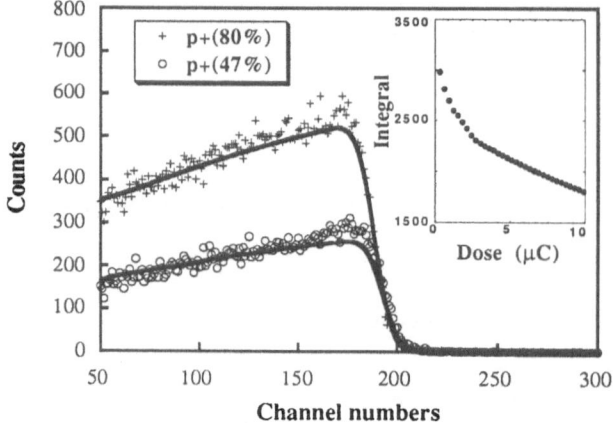

Fig. 16: Recoil proton spectra and the simulations (solid lines) corresponding to p+ psl of 80% and 47% porosity. In the inset we show the proton integral recorded during consecutive small doses for p+ layer of 80% porosity.

The ratio H/Si deduced from these simulations had to be corrected to take into account the hydrogen losses we observed during the measurements. These H losses are due to breaking of bonds by the ion beam and also by secondary electrons emitted during beam irradiation. In the inset, we have represented the proton integral recorded during consecutive small doses for the p+ layer of 80% porosity. The correcting factor was determined from the extrapolation of the curve at the origin.

RBS measurements have been performed on these samples. Fig. 17 shows experimental spectra together with the simulations corresponding to the composition found by nuclear analysis. RBS results confirm the nuclear analysis and show that the impurity concentrations are practically uniform in the freshly prepared psl.

Samples	O 10^{17}at/cm^2	F 10^{17}at/cm^2	C 10^{17}at/cm^2	H 10^{17}at/cm^2	N$_{Si}$ 10^{17}at/cm^2
p+ (47%,9μm)	0.34	0.22	2.2	7.9	26.5
p+ (80%,10.7μm)	0.22	0.15	1.5	6.6	10
n+ (53%,10μm)	0.45	0.32	2.3	9.9	23.5
n+ (80%,10μm)	0.17	0.17	1.2	5.7	10

Table II. Impurity and Si contents in 1μm thick psl.

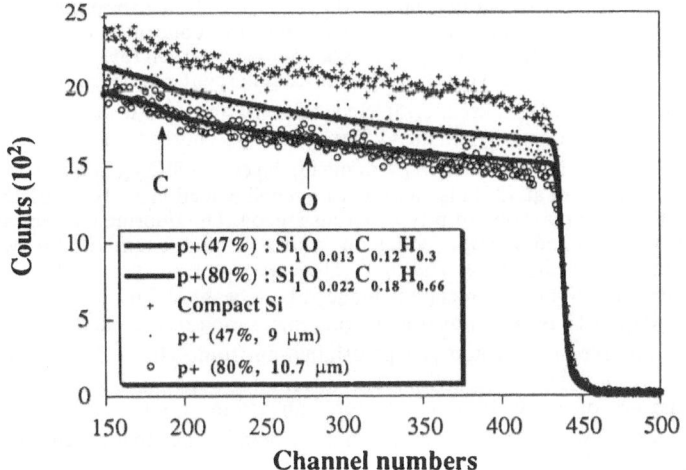

Fig. 17: ^{4}He RBS spectra from compact Si and as-prepared psl together with their simulations.

In all these calculations, the stopping power is an important parameter which is known once the composition is known. Then the determination of psl composition is recurrent.

In Table II, we present the impurity content in 1 μm thick psl.
In Table II, we present the ratios of the impurity contents to the Si content in the different psl.

Samples	O/Si	F/Si	C/Si	H/Si
p+ (47%,9μm)	0.013	0.009	0.13	0.3
p+ (80%,10.7μm)	0.022	0.015	0.18	0.66
n+ (53%,10μm)	0.019	0.016	0.10	0.42
n+ (80%,10μm)	0.017	0.017	0.12	0.57

Table IIi. Ratio of impurity contents to Si content in psl.

The impurity contents are systematically slightly lower in the psl of higher porosity. This may be due to the fact that the specific surface area is lower in psl of higher porosity. In any case these results seem to indicate that the photoluminesence is not due to molecules adsorbed at the pore surface. By dividing the impurity content by the actual surface area found by Herino et al. 200 m^2/cm^3, we found that 10-20 % of the internal surface is covered by a mixture of SiO_2 and O-Si-H species, the remaining surface beeing covered by SiH_x species as it has been shown by IR absorption. There is an excess of hydrogen which is probably bounded to carbon atoms coming from the atmosphere : in average 2 hydrogen atoms for one carbon atom. The quantity of such organic impurity is equivalent to one monolayer adsorbed at the pore surface.

Other groups measured impurity content in porous silicon using nuclear microanalysis.

L. G. Earwaker et al. [31] used nuclear microanalysis to study porous silicon formed on p-type substrate, with (100) orientation and resistivities in the range 0.01-700 Ω.cm. They used the $^{19}F(p, \alpha\gamma)^{16}O$ resonance at 872 keV (Γ = 4.2 keV), the $^{16}O(d, \alpha)^{14}N$ and $^{12}C(d, p)^{13}C$ nuclear reactions to depth profile fluorine and oxygen and to measure the carbon content. Proton recoil analysis was carried out to profile hydrogen. They also performed RBS experiments. It was found in as-prepared samples, that reasonably uniform concentration were present of oxygen (5%), fluorine (1-2%), hydrogen (up to 50% atomic) and carbon (3%) atomic.

P. Steiner et al. [32] investigated electroluminescent nanoporous silicon by accelerator techniques. Rutherford Backscattering Spectrometry and $^{16}O(d, \alpha)^{14}N$ reaction were used to determine the oxygen content and the O/Si ratio in n-type porous silicon which showed strong red photoluminescence. They found $SiO_{1.8}$ in the top layer and $SiO_{0.8}$ in 6 μm depth.

R. Sabet-Dariani et al. [33] also used elastic recoil detection analysis (ERDA), RBS and NRA to determine the composition of p-type porous silicon. The fluorine concentration was estimated using particle-induced gamma emission with a proton beam of 2.6 MeV. The fluorine concentration was estimated at 265 ppm. The carbon and oxygen content were obtained from measurements of the nuclear reactions induced by a beam of 2.45 MeV $^3He^+$ ions. They found $SiO_{0.48}C_{0.5}H_{0.48}$. It must be noted that, when measuring oxygen and carbon content with ^3He beam, one observes two proton groups originating from ^3He-^{12}C reactions and one α group originating from ^3He-^{16}O reaction. The different groups are not well separated. As shown in [22], the use of a thin barrier detector with a thin enough depletion zone allows a full discrimination between the α-particles and the protons, hence to measure oxygen content with a better accuracy.

B - Boron content in p+ porous layers

The electrical resistance of porous silicon layers is several order of magnitude higher than that of the substrate in which it has been formed. The porous layers behave as an intrinsic material. The reason of such a behaviour is not yet well known. Dubin [34] for example, proposed a pore formation mechanism in which pores selectively initiate at surface dopant atoms. The result would be the selective removal of dopant atoms within the Si lattice. Other explanations have been proposed according to which the interpore regions are devoid of mobile carriers due to overlapping of depletion regions. In order to decide between these two explanations, we report, in the present study, measurements, by Nuclear Reaction Analysis, of boron concentration in porous layers formed in p+ silicon substrate with regard to the one of the p+ substrate.

The boron content of p+ porous layers, 10 μm thick, of 50% and 80% porosity, was measured using the nuclear reaction $^{11}B(p,\alpha)2\alpha$ at 660 keV. At this energy the cross section reaches a maximum, then it decreases with decreasing energy until negligible values (a few % of the maximum) at 300 keV [25]. At 660 keV, a proton energy loss of 360 keV corresponds to about 5.6 μm thick compact silicon layer. To be sure to probe only the porous part, excluding the substrate, the samples were tilted at convenient angle.

In fig.18, we have represented the α spectra from p+ Si substrate and from p+ porous layers of 47% and 80% porosity. The α integrals (N_α) corresponding to the p+ porous layers of 47% and 80% porosity are respectivelly 1.2 and 1.7 times higher than the integral corresponding to compact Si. These values are mean values of many measurements.

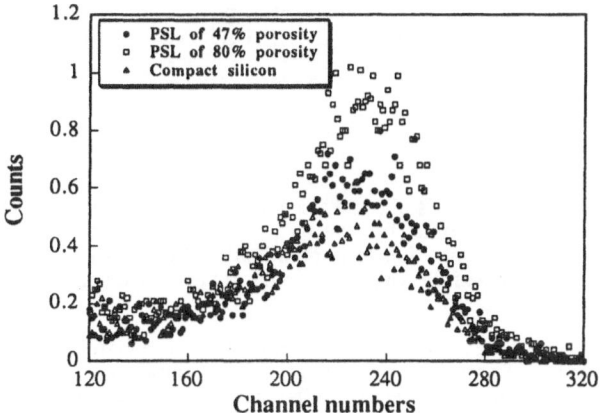

Fig. 18: α spectra from $^{11}B(p,\alpha)2\alpha$ on psl of 47% and 80% porosity and for compact Si. Ep = 660 keV, θ = 150°.

To determine the B/Si ratio in the two p^+ psl with regard to that of the Si substrate it is necessary to determine also, for the three samples, the thickness of the region - expressed in number of Si at/cm^2- analysed by the proton beam. The porous layers contain impurities at the pore surface such as oxygen, carbon, hydrogen and the global formula of such layers may be written $SiO_xC_yH_z$. The stopping power of the psl expressed in unit energy per number of silicon atom/cm^2 is $S_{psl}= S_{Si} + xS_O + yS_C + zS_H$ where the S's are the stopping power of each element. We estimated the stopping powers of protons in the energy range 660-300 keV for p^+ psl of 47% and 80% porosity to be about 1.1 and 1.15 times higher than that of p^+ compact Si respectively.

The number of α particles which come from the target may be expressed by :

$$N_\alpha = Q\,\Omega \int\limits_{0}^{\infty} C(x)\; \sigma[E(x)]\; dx$$

We have verified by SIMS that the boron depth profile is constant in the psl. By making the change of variable $dE = \dfrac{dE}{dx}\, dx \equiv S[E(x)]\, dx$, one obtains :

$$N_\alpha = Q\,\Omega\,C \int\limits_{E_0}^{0} \sigma(E)\; \frac{dE}{S(E)}$$

Since $S_{psl}(E) = K \, S_{Si}(E)$ with $K = 1.1$ for 47% p+ psl and 1.15 for 80% p+ psl, one can compare simply the boron concentration from the α integrals in the three samples :

$$\left[\frac{B}{Si}\right]_{psl} = k \, \frac{N\alpha(psl)}{N\alpha(Si)} \, \left[\frac{B}{Si}\right]_{Si \text{ monocrystal}}$$

We conclude that the B/Si ratio is about $1.2 \times 1.1 = 1.3$ times higher in the p+ porous layer of 47% porosity than in the substrate. For the p+ psl of 80% porosity this ratio is $1.15 \times 1.7 = 2$ times higher. This excess is not due to contamination by air. At first, the samples we have analysed were freshly prepared. Secondly we could not evidenced any boron contamination in 2 years old n+ porous silicon.

The first conclusion of these results is that the dopant atoms are still present in the porous layer. As we have shown by infrared absorption that the dopant atoms are not passivated by hydrogen, these results seem to indicate that there is no free carriers in the psl, i.e. Fermi level in p+ porous Si is no longer pinned by the acceptor impurities but has moved to the midgap position.
The second conclusion of these results is that together with the formation of pores perpendicular to the substrate there is a lateral chemical or electrochemical process which dissolves selectively silicon region in which there is no boron atoms.

C - Channeling experiments

In order to have some informations onto the microstructure of p+ psl we have made some channeling measurements by using α particles with an incident energy of 2 MeV. The RBS spectra, in random and channeling geometry, coresponding to the Si substrate and to the p+ psl are shown in fig. 19. Similar results were obtained for the n+ porous layers. For the Si substrate the χ_{min} value, which is the ratio of the aligned yield to the random one near the surface, is equal to 2.9% . In the case of p+ porous layer of 47% porosity we found 20% which indicates that the crystalline parts are only lightly disoriented with regard to the Si substrate. We have shown, using cross section TEM, that p+ or n+ PSL of medium porosity, consists in long pores of polygonal section, the mean size of which is about 20 nm, oriented perpendicularly to the substrate, with no major rupture between the substrate and the top of the layer, except small "buds" on the side of the pores. The pores are separated by silicon walls 100 nm thick about. The above channeling experiments show that this kind of microstructure preserves quite correctly the orientation of the substrate. It has been confirmed by electron diffraction that the single crystal character of the substrate is very well conserved in the porous layer.

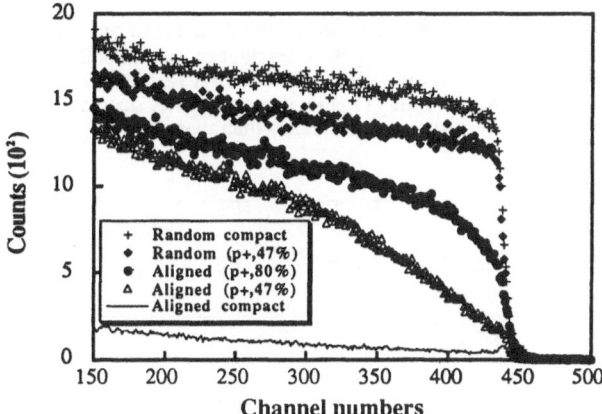

Fig. 19: Spectra in random and channeling geometry for porous (47 and 80 % porosity) and monocrystalline silicon.

Concerning the p+ porous layer of 80% porosity, χ_{min} = 52.5% which seems to indicate that the silicon part of the psl is more disoriented with regard to the substrate than the porous layer of medium porosity. TEM experiments showed that the morphology is similar to that of the layers of medium porosity but there is some differences between the two layers : the mean size of the pores is higher, about 30 nm, and the silicon walls are discontinuous showing many connexions between adjacent pores. Moreover, the curvature of the diffraction spots confirms that the psl of 80% porosity is more disoriented than that of 47%.

In the case of porous layers, the yield in channeling geometry is not easy to interpretate since it is due to two phenomena : 1)- if the porosity is p, a part (1-p) of the incident particles which enter the silicon are dechanneled similarly to compact silicon 2)- a part p of the incident particles enters the pores and have greater probability to be backscattered by silicon atoms on the pore surface. This could explain why the yield of the aligned spectra increases faster with depth in the case of the p+ porous layer than in the case of compact Si.

D - Thermal oxidation and nitridation of porous silicon

We began this work [35, 36] to study 1)- the oxidation and nitridation kinetic of porous silicon in relation to that of compact Si, 2)- the influence of these thermal processes on the photoluminescence and on the defects such as Pb centers, oxygen vacancies etc...

The silicon wafers were p+ material, (100) oriented, boron doped. The porous silicon layers were 1 μm thick with 66 % porosity. We performed the classical preoxidation step at 300°C in a dry atmosphere of $^{16}O_2$ at 12 mbar during 1 hr to prevent the thin Si structure from coalescing during further treatments.

a)- *Thermal oxidation* : The second step of oxidation has been performed in dry $^{18}O_2$ gas enriched to 97% (12 mbar, 1000°C) in order to distinguish the two steps of oxidation. Ion beam analysis allows the distinction between the two oxygen isotopes. For ^{16}O measurements we used $^{16}O(d, p_1)^{17}O^*$ nulear reaction whereas for ^{18}O we used $^{18}O(p, \alpha)^{15}N$ nuclear reaction. Numbers of ^{18}O and ^{16}O atoms/cm^2 have been deduced from a direct comparison with calibrated reference of tantalum anodic oxide formed in $H_2^{18}O$ and in $H_2^{16}O$ solutions respectively.

Fig. 20 shows the ^{18}O content as a function of the oxidation time and the O/Si ratio (the ^{16}O content of the initial oxide formed during the preoxidation step is 600 $10^{15}/cm^2$). The point at 90 min corresponds to a total oxidation : O/Si ≈ 2. The first results [37] we have obtained showed that the PL intensity increases and the (111) Pb centers density decreases when the oxidation time increases but that the PL intensity remains weaker than that of an as prepared psl of high porosity (80 %).

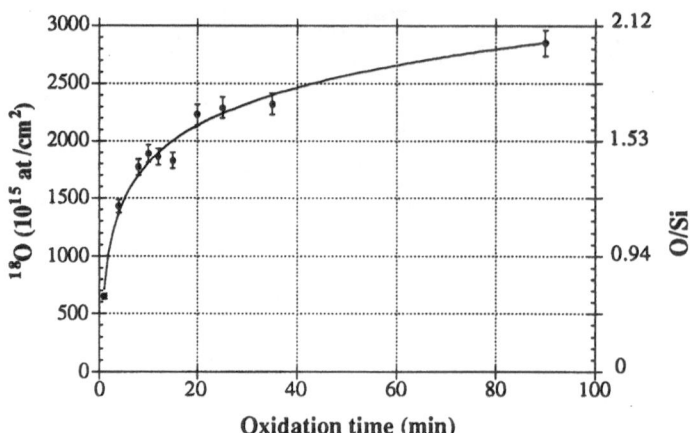

Fig. 20: ^{18}O content as a function of oxidation time and corresponding $(^{18}O+^{16}O)/Si$ ratios.

b)- *Thermal nitridation* : The nitridation has been performed at 1000°C in $^{14}NH_3$, 12 mbar. The ^{14}N content was measured using the $^{14}N(d, \alpha_0)^{12}C$ at Ed ≈ 1500 keV where the cross section is stationnary. As shown in fig. 21, the α_0 peak is well separated from the other particles coming from O, Si, C , ^{14}N etc... Fig 22 shows the N content as a function of time. As the internal surface area is about 200 cm^2/cm^2 for 1 μm thick porous layer the thickness of the layer formed after 90 min corresponds to 5 10^{15} N atoms $/cm^2$.

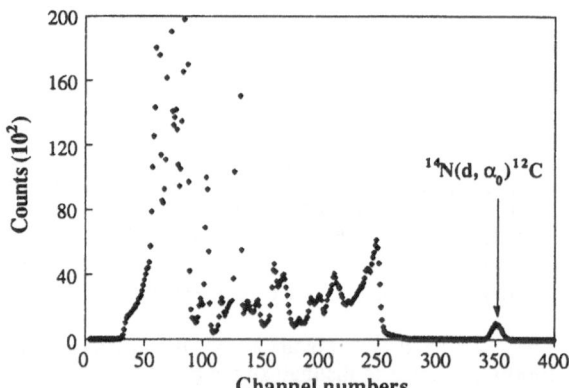

Fig. 20: Typical energy spectrum from deuteron bombardment of 1450 keV on porous silicon nitride (1 μm, 66%). The α_0 peak is well isolated from the α and proton peaks from nuclear reactions on Si, O, C, N.

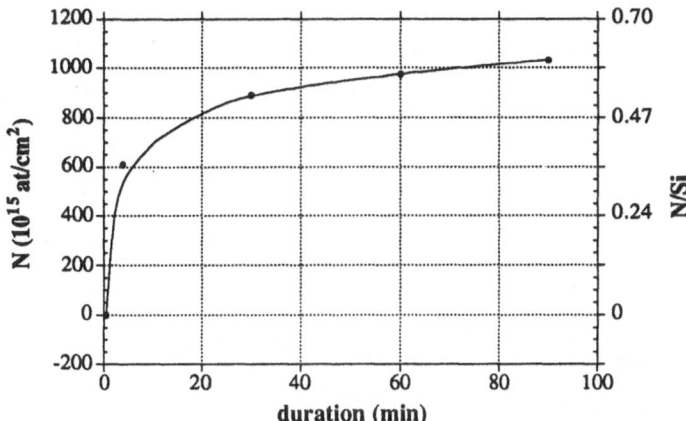

Fig. 22: [14]N content as a function of nitridation time and corresponding ^{14}N/Si ratios.

E - Anodic oxidation of porous silicon

In order to check the validity of the existing quantum models or to improve them, it is important to control the size of the Si wires or dots. With this aim in view, anodic oxidation seems to be a well appropriate method as this technique applied to p-type porous silicon permits to obtain porous layers which exhibits better mechanical properties and more stable and efficient luminescence than the photoluminescent porous layers directly obtained by electrochemical dissolution.
(100) p+ porous layers of 66% porosity were anodically oxidized [38] at constant current in 0.1 M KNO_3-H_2O solution until predetermined potential difference between the sample and a platinum cathode.

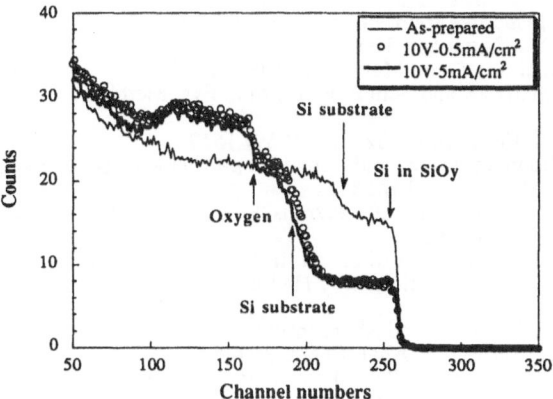

Fig. 23: [4]He RBS spectra from as-prepared and anodic oxidized porous silicon.

C. Ortega et al.

In fig. 23, we have represented the RBS spectra from a p+ psl of 1 μm thick of 66% porosity and from the same layer which was anodically oxidized at 5 mA/cm^2 and 0·5 mA/cm^2 until 10 V. Due to oxidation stage, the stopping power largly increases in the oxidized porous layer so that 1/ the plateau corresponding to the Si atoms is lower and 2/ the front of the Si substrate appears at lower energy than that of the as-prepared psl. Hence, this plateau is flat which indicates that the whole porous layer is uniformly oxidized. Similar result, which was confirmed by SIMS oxygen profils, have been obtained at different current densities which indicates that, the whole p+ porous layer is oxidized (independently of the current density value). The O/Si ratio is equal to 1.6 at 10V and no more increases with the voltage.
This indicates that the break of the electrical contact between the substrate and the porous part occurs around 10 V. Whereas it occurs at 1-2 V for a p-type psl.
An increase of the photoluminescence intensity with the Si electrode potential and a slight shift of the luminescence peak towards higher energies are observed which could be attributed, in a Si quantum confinement model, to the increasing of the number of Si crystallites with lateral sizes low enough to be able to emit visible light [38]. However, the PL intensity we obtained after anodic oxidation is lower than that from an as-prepared p+ porous layer of 80% porosity. This is probably due to a bad passivation of the Si/SiO$_2$ interface as shown by EPR measurements which indicate that, contrary to the as-prepared or thermal oxidized porous layers, anodic oxidized layers contain oxygen vacancies (E' centers) and hence, the trigonal (111) Pb-center is not the dominating paramagnetic defect [37].

REFERENCES

[1]- Feldman L. C. and Mayer J. W., Fundamentals of surface and thin films (Elsevier Science Publishing, Amsterdam, ISBN 0-444-00989-2, 1986).
[2]- Chu W. K., Mayer J. W. and Nicollet M. A., Backscattering Spectrometry (Academic Press, New York, ISBN 0-12-173850-7, 1978).
[3]- Valentin L., L'Univers Mécanique (Herman, Paris, ISBN 2-7056-5956-0, 1983).
[4]- L'Ecuyer J., Davies J. A. and Matsunami N., Nucl. Instr. and Meth. 160 (1979) 337.
[5]- Cameron J. R., Phys. Rev. 90 (1953) 839.
[6]- Blanpain B., Revesz P., Doolittle L. R., Purser K. H. and Mayer J. W., Nucl. Instr. and Meth. in Phys. Res. B 34 (1988) 459.
[7]- Doolittle L. R., Nucl. Instr. and Meth. in Phys. Res. B 9 (1985) 334.
[8]- Kotai A., Nucl. Instr. and Meth. in Phys. Res. B 85 (1994) 588.
[9]- Fano U., Ann. Rev. Nucl. Sci. 13 (1963) 1.
[10]- Lindhard J., Nucl. Instr. and Meth. 132 (1985) 1.
[11]- Lindhard J., Scharff M. and Schiott H. E., Mat. Fys. Medd. Dan. Vid. Selks No 14 (1963) 33.
[12]- Firsov O.B., Zh. Eksp. Teor. Fiz. 36 (1959) 1517.
[13]- Ziegler J. F., Stopping Powers and Ranges in all Elements (Pergamon, Oxford, 1977).
[14]- Bohr N., Phil. Mag. 30 (1915) 581.
[15]- Feldman L. C., Mayer J. W. and Picraux S. T., Materials Analysis by Ion Channeling, (Academic Press, New York, 1982).
[16]- Quillet V., Abel F. and Schott M., Nucl. Instr. and Meth. in Phys. Res. B 83 (1993) 47.
[17]- Szilagyi E., Paszti F., Manuaba A., Hadju C. and Kotai E., Nucl. Instr. and Meth. in Phys. Res. B 43 (1989) 502.
[18]- Vizkelethy G., Nucl. Instr. and Meth. in Phys. Res. B 45 (1990) 1.
[19]- Amsel G. and Samuel D., J. Anal. Chem. 39 (1967) 1689.
[20]- Lennard W. N., Mssoumi G. R., Alkemade P. F. A., Mitchell I. V. and Tong S. Y., Nucl. Instr. and Meth. in Phys. Res. B61 (1991) 1.
[21]- Maurel B., Thesis, Paris 1981.
[22]- Abel F., Amsel G., D'Artemare E., Ortega C., Siejka J. and Vizkelethy G., Nucl. Instr. and Meth. in Phys. Res. B 45 (1990) 100.

[23]- Maurel B. and Amsel G., Nucl. Instr. and Meth. in Phys. Res. 218 (1983) 159.
[24]- Amsel G., Cohen C. and Maurel B., Nucl. Instr. and Meth. in Phys. Res. 218 (1983) 159.
[25]- Ligeon E. and Bontemps A., J. Radioanaly. Chem. 12 (1972) 335.
[26]- Bosseboeuf A., Bouchier D. and Rigo S., J. Electrochem. Soc. 133 (1986) 810.
[27]- Dieumegard D., Maurel B. and Amsel G., Nucl. Instr. and Meth. 168 (1980) 93.
[28]- Amsel G., Nadai J. P., D'Artemare E., David D., Girard E. and Moulin J., Nucl. Instr. and Meth. 92 (1971) 481.
[29]- Ortega C., Siejka J. and Vizkelethy G., Nucl. Instr. and Meth. in Phys. Res. B 45 (1990) 622.
[30]- Grosman A., Ortega C., Siejka J. and Chamarro M., J. Appl. Phys. 74 (1993) 1992.
[31]- Earwaker L. G., Farr J. P. G., Grzeszczyk P. E. and Sturland I., Nucl. Instr. and Meth in Phys. Res. B 9 (1985) 317.
[32]- Steiner P, Weidhass J. and Lang W., "Luminescent Porous Silicon Investigated by accelerator analytics", Mat. Res. Symp. Proc. 281 (1993) 531.
[33]- Sabet-Dariani R. and Haneman D., J. Appl. Phys. 73 (1993) 5.
[34]- Dubin V. M. , Surf. Sci., 274 (1992) 82-92.
[35]- Morazzani V., Chamarro M., Grosman A., Ortega C., Rigo S., Siejka J. and von Bardeleben H. J., J. Lum. 57 (1993) 45.
[36]- Morazzani V., Grosman A., Ortega C., Rigo S. and Siejka J., Nucl. Instr. and Meth. in Phys. Res. B 85 (1994) 287.
[37]- von Bardeleben H. J., Ortega C., Grosman A., Morazzani V., Siejka J. and Stievenard D., J. Lum. 57 (1993) 301.
[38]- Grosman A., Chamarro M., Morazzani V., Ortega C., Rigo S., Siejka J. and von Bardeleben H. J., J. Lum. 57 (1993) 13.

LECTURE 11

IR spectroscopy of porous silicon

W. Theiss

*I. Phys. Inst., Aachen University of Technology, RWTH,
Sommerfeldstr. 28, 52056 Aachen, Germany*

1   INTRODUCTION

IR spectroscopy is a non-destructive analytical tool widely used in material research (Ferraro, 1990). In semiconductor technology it is a standard technique to determine the concentration of dopants and impurities. In addition it can be used to obtain information on the properties of free charge carriers and thicknesses of epitaxial layers.

Apart from these more or less industrial applications optical spectroscopy in the mid and far infrared spectral region plays an important role in basic research work, too. Especially the study of electronic transitions in low dimensional systems such as 2-, 1- or 0-dimensional electron gases with excitation energies much smaller than 1 eV has made use of IR spectroscopy to a large extent.

The investigation of porous silicon - both the more technological and the basic research - can benefit from this technique as well. IR spectra can give information about impurities, free charge carriers, porosities, layer thicknesses and the quality of interfaces. Here we present a short introduction which can serve as a guideline through the given literature.

There are quite different concepts on how to obtain information about a sample using optical spectroscopy. So-called chemometric methods are purely statistical tools: spectra measured on reference samples with known parameters are analysed and the relations between these parameters and spectral features are determined. This can be done by neural networks or other mathematical techniques such as Principal Component Analysis (PCA) or Partial Least Squares (PLS) (Martens, 1989). Applying these methods only little must be known about spectroscopy and there are almost no restrictions on the kind of measurement - provided there are many reference samples and the relation between spectral features and wanted parameters are not too non-linear. Analytical chemistry usually can satisfy the latter conditions easily and hence chemometric methods are by now the dominant tools in IR spectroscopy of organic materials.

In solid state physics the situation is reverse in most cases: there are only a few good samples whose parameters (that determine the observed spectra in a quite non-linear way) are not known very precisely. Therefor the successful analysis of optical spectra measured on solid state samples is much harder and requires a detailed physical understanding of the problem under investigation. Fortunately semiconductor systems of technological relevance are in a good shape on the average, i.e. geometries are simple (e.g. one often has stacks of plane-parallel layers) and interfaces almost perfect in most cases. This leads to the lucky situation that one can generate quite easily theoretical IR spectra on the basis of simple models and compare them quantitatively to measured ones. Provided the used models are chosen carefully a parameter fit resulting in a convincing agreement between experiment and theory then gives a reliable set of numbers characterising the sample under investigation. This kind of doing spectroscopy is described in this article.

First we sketch the necessary theoretical background for the calculations, afterwards experimental techniques are discussed and finally some examples demonstrate what (and how) IR spectroscopy can tell you about porous silicon.

2   THEORY

Doing optical spectroscopy means to expose samples to electromagnetic radiation and analyse the emerging radiation. For the infrared spectral region there are mainly two mechanisms which influence the propagation of light waves: reflections at the boundaries between two (optically) different materials and absorption processes occurring while a wave is travelling through a medium. Both effects are well described by using the concept of the dielectric function which is discussed in the first part of this theory section. Then we make some notes on wave propagation and discuss an algorithm to add all partial waves that combine to the total reflected and transmitted light intensity for a stack of plane-parallel layers (which is the usual geometry in experiments). These are the quantities that can be compared with measured reflectance or transmittance spectra.

Finally we touch the problem of taking into account the inhomogeneities in porous media which can be done in the infrared region applying an appropriate so-called effective-medium theory.

2.1   Excitations by electric fields: dielectric functions

In most cases it is the electric field of the probing light wave that interacts with the sample and hence only excitations can be observed that are going along with a polarisation of the matter. The polarisation $\vec{P}$ induced by an externally applied electric field $\vec{E}$ in a homogeneous material is given by the electric susceptibility χ:

$$\vec{P} = \varepsilon_0 \chi \vec{E} \ . \tag{1}$$

The dielectric function ε which connects the dielectric displacement and the electric field vector is closely related to the susceptibility:

$$\vec{D} = \varepsilon_0 \varepsilon \vec{E}, \qquad \varepsilon = 1 + \chi \ . \tag{2}$$

The frequency dependence of the susceptibility is very characteristic for a material since it incorporates vibrations of the electronic system and the atomic cores. In the following typical features are discussed shortly.

2.1.1   Dielectric background

Fig.1 shows the dielectric function of bulk silicon in the near infrared ('NIR'), visible ('Vis') and ultraviolet (UV) spectral region. The very strong structures in the UV can be assigned to electronic interband transitions (critical points in the joint density of states) (Adachi, 1988, Forouhi and Bloomer, 1990). In the infrared region one is far away from the interband transitions which can be noticed only by an almost constant and real contribution to the dielectric function - the so-called dielectric background. Therefor the optical properties of undoped silicon in the infrared can be described quite well by just a constant and real dielectric function (sometimes called ε_∞ where ∞ stands for 'very high frequency' from the infrared viewpoint).

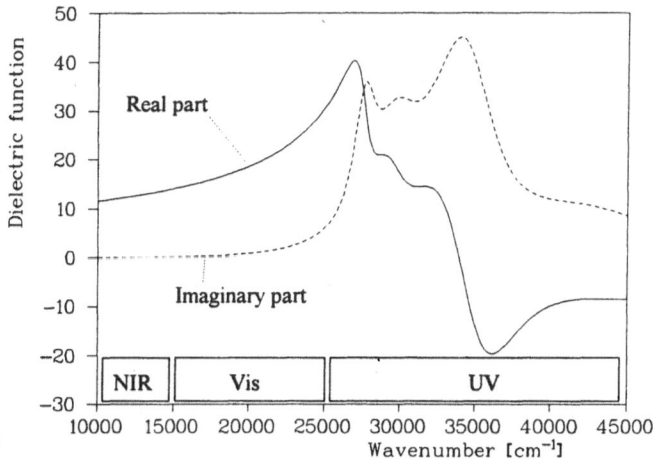

Fig.1: dielectric function of bulk silicon in the near infrared, visible and UV. The strong structures can be related to electronic interband transitions (Kittel, 1976, Adachi, 1988, Forouhi and Bloomer, 1990).

2.1.2 Free charge carriers

In the case of doped silicon the charge carriers set free by the donors or acceptors can be accelerated by very little energies and hence do respond to applied electric fields with frequencies in the infrared region. A simple expression for the susceptibility of free carriers is given by the so-called Drude model where the carrier concentration and a damping constant enters:

$$\chi_{\text{Drude}} = -\frac{\Omega_P^2}{\tilde{v}^2 + i\tilde{v}\Omega_\tau} \quad \text{with} \quad \Omega_P^2 = \frac{ne^2}{\varepsilon_0 m} \quad . \tag{3}$$

Here n is the volume density, e the charge and m the effective mass of the charge carriers. Ω_P is called plasma frequency (Kittel, 1976). A typical example for highly p-doped silicon is shown in fig.2. The formula

$$\varepsilon = \varepsilon_\infty + \chi_{\text{Drude}} = \varepsilon_\infty - \frac{\Omega_P^2}{\tilde{v}^2 + i\tilde{v}\Omega_\tau} \tag{4}$$

has been used with $\varepsilon_\infty = 11.7, \Omega_P = 1800 \text{ cm}^{-1}$ and $\Omega_\tau = 500 \text{ cm}^{-1}$.

2.1.3 Vibration of atomic cores: oscillator model

Microscopic vibrations involving the motion of the atomic nuclei (which are much heavier than the electrons) usually have their resonance frequencies in the infrared region. These characteristic frequencies depend on the oscillating masses and the strength of the bonding between them and hence can be used for material identification (which is the reason for the dominant role IR spectroscopy plays in analytical chemistry). In the case of silicon important impurities like carbon and oxygen are detected by their characteristic vibrational modes. Susceptibilities describing microscopic vibrations can be modelled by harmonic oscillator terms:

$$\chi_{\text{Harmonic oscillator}} = \frac{\Omega_P^2}{\Omega_{TO}^2 - \vec{v}^2 - i\vec{v}\Omega_\tau} \qquad . \tag{5}$$

Ω_P^2 gives the oscillator strength, Ω_τ the damping and Ω_{TO} the resonance position. The index TO is used here since with the same oscillator model transverse optical phonons (lattice vibrations, TO-modes) are described.

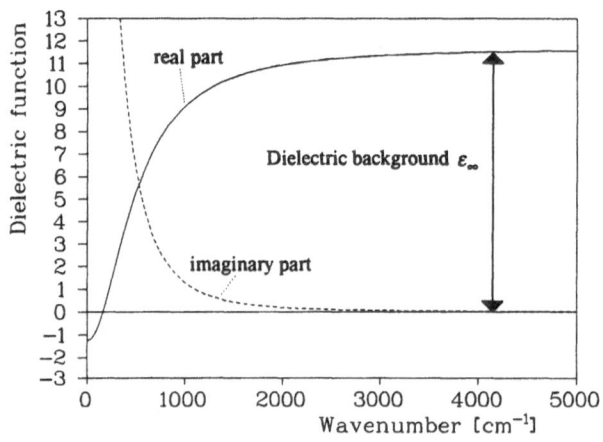

Fig.2: dielectric function of highly p-doped silicon in the infrared spectral range. For large wavenumbers the influence of the free charge carriers is negligible and only the dielectric background (see also fig.1) contributes to the dielectric function. For small wavenumbers - on the other hand - the Drude susceptibility dominates.

2.2   Wave propagation
Now we can insert the response of matter to electric fields (described by the dielectric function) into Maxwell's equations and look for plane-wave solutions (that will represent light waves) of the form

$$\vec{E} = \vec{E}_0 \exp\left(i\vec{k} \cdot \vec{r} - i\omega t\right) \qquad . \tag{6}$$

In non-magnetic media solutions for transverse plane waves exist under the condition

$$\vec{k} \cdot \vec{k} = \frac{\omega^2}{c_0^2} \varepsilon(\omega) = \left(\frac{2\pi}{\lambda_0}\right)^2 \varepsilon(\omega) = k_0^2 \, \varepsilon(\omega) \tag{7}$$

which is called dispersion relation (c_0 is the speed of light in vacuum and λ_0 the vacuum wavelength of the wave) (Jackson, 1975).

2.2.1   Complex refractive index
Inserting the wave vector into (6) one gets

$$\vec{E} = \vec{E}_0 \exp\left(i\vec{k} \cdot \vec{r} - i\omega t\right) = \vec{E}_0 \exp\left(i\sqrt{\varepsilon(\omega)}\,\vec{k}_0 \cdot \vec{r} - i\omega t\right) = \vec{E}_0 \exp\left(i\,n\,\vec{k}_0 \cdot \vec{r} - i\omega t\right)\exp\left(-\kappa\vec{k}_0 \cdot \vec{r}\right) \tag{8}$$

where the complex refractive index $n + i\kappa = \sqrt{\varepsilon(\omega)}$ has been introduced. (8) describes a plane wave exponentially decaying with an absorption constant $K = \kappa k_0 = 2\pi\kappa\tilde{\nu}$ where we have returned to wavenumbers which is used in IR spectroscopy more often than frequencies.

2.2.2 Boundaries

At the interface between two media with different refractive indices an incident wave (incidence angle α) will be partially reflected and transmitted, as sketched in fig.2. This follows from an evaluation of the boundary conditions for electric and magnetic fields. The amplitude reflection and transmission coefficients depend on the polarisation of the wave and the refractive indices of the two adjacent media, explicit expressions can be found in the book of Jackson (1975).

Waves at the interface between two layers:

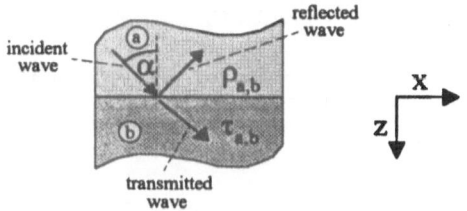

$\rho_{a,b}$: amplitude reflection coefficient

$\tau_{a,b}$: amplitude transmission coefficient

Fig.3: light waves at an interface between two media with different dielectric functions are partially reflected and transmitted, respectively.

For the following treatment of layer stacks it is useful to look at the wave vectors: if the medium 'a' (see fig.2) has a real refractive index n_a then the tangential component of the wave vector

$$k_x = \frac{\omega}{c_0} n_a \sin\alpha \tag{9}$$

is real and will be the same in layer 'b'. The normal component of the wave vector will be different in the two media:

$$k_z = \frac{\omega}{c_0} \sqrt{\varepsilon(\omega) - n_a^2 \sin^2\alpha} = \frac{\omega}{c_0} \tilde{N} \tag{10}$$

Here the generalised complex refractive index $\tilde{N} = \sqrt{\varepsilon(\omega) - n_a^2 \sin^2\alpha}$ has been introduced.

2.3 Optical properties of layer stacks

Knowing the wave propagation characteristics inside homogeneous materials and the reflection and transmission coefficients at the boundaries between them the reflectance and transmittance of a stack of homogeneous layers - which is a very typical geometry in semiconductor physics -

can be calculated straightforward. Here we just sketch the basic ideas without going into details.

Taking into account multiple reflections:

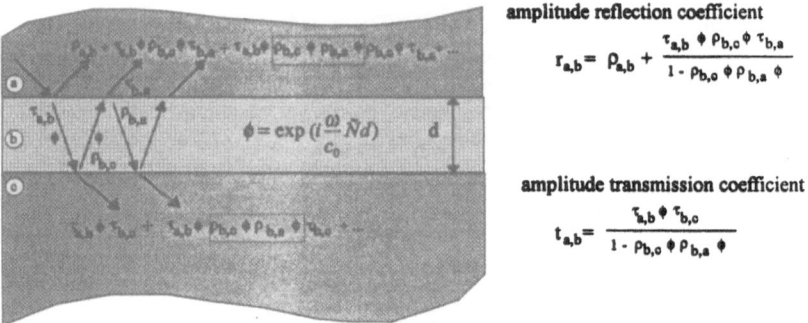

amplitude reflection coefficient

$$r_{a,b} = \rho_{a,b} + \frac{\tau_{a,b} \, \phi \, \rho_{b,c} \, \phi \, \tau_{b,a}}{1 - \rho_{b,c} \, \phi \, \rho_{b,a} \, \phi}$$

amplitude transmission coefficient

$$t_{a,b} = \frac{\tau_{a,b} \, \phi \, \tau_{b,c}}{1 - \rho_{b,c} \, \phi \, \rho_{b,a} \, \phi}$$

Fig.4: amplitude reflection and transmission of a layer obtained by adding up all partial waves as shown: note that the multiply reflected contributions contain powers of the terms in the dashed rectangles which can be combined in a geometric series leading to the denominator expression.

Formalism for complex layer stacks:

reduce the stack repeatedly by the last layer

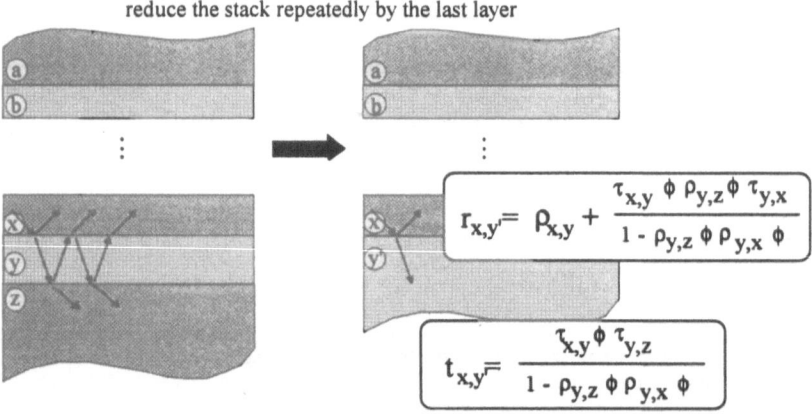

$$r_{x,y} = \rho_{x,y} + \frac{\tau_{x,y} \, \phi \, \rho_{y,z} \, \phi \, \tau_{y,x}}{1 - \rho_{y,z} \, \phi \, \rho_{y,x} \, \phi}$$

$$t_{x,y} = \frac{\tau_{x,y} \, \phi \, \tau_{y,z}}{1 - \rho_{y,z} \, \phi \, \rho_{y,x} \, \phi}$$

Fig.5: treatment of a stack of many layers: after summing up all transmitted and reflected partial waves of the last layer and the bottom halfspace these two are replaced by a new 'artificial' bottom halfspace with amplitude reflection and transmission coefficients as indicated. These reduction of the layer stack is repeated until all layers have been processed. The finally obtained coefficients are then squared to yield the intensity reflection and transmission coefficients which then can be compared to measurements.

The stack is a pile of plane-parallel layers between two halfspaces - the top and the bottom halfspace. There is a light wave incident from the top halfspace onto the first layer with a certain angle of incidence with respect to the surface normal and a certain polarisation. For the top halfspace (which will be vacuum in most cases) we assume a real refractive index. In this case (9) and (10) hold for all layers. The wave is reflected and transmitted at the first interface according to Fresnel's equations. The wave transmitted inside the first layer is now again split at the next interface and so on. The problem to be solved is now to sum up correctly all partial waves to the total reflected and transmitted radiation. This can be done most easily the following way: first consider fig.4 and see how multiple reflections in just one layer can be combined to new reflection and transmission coefficients for the combination of the layer and the bottom halfspace. Then - starting at the back side of the layer stack - the last layer and the bottom halfspace are combined to a new 'quasi'-halfspace (see fig.5) which then in turn is combined with the next layer and so on. Finally the amplitudes of the reflected and transmitted waves are squared to give the corresponding intensities.

2.4   Porous systems: effective dielectric functions

Up to now we have established a description of stacks of homogeneous layers. Nevertheless porous silicon is not homogeneous but - to a first approximation - a two-phase composite consisting of a solid state phase (the former silicon) and the empty pores. Fortunately the characteristic size of the pores is much smaller than typical IR light wavelength and hence the radiation cannot resolve details of the micro-geometry - in particular no light scattering occurs and the system can almost be described like a homogeneous one. The only difference is that we now have to use a so-called effective dielectric function which is obtained by averaging the dielectric functions of the two phases. As has been shown this averaging is non-trivial but involves details of the microgeometry (Theiß, 1994). Some simple 'averaging recipes' like the Maxwell Garnett (Maxwell Garnett, 1904) or the Bruggeman (Bruggeman, 1935) formulas may fail severely in certain cases. Instead models based on the more flexible Bergman representation (Bergman, 1978) are recommended. This is especially important in the Vis and UV where effective dielectric functions can also be used and depend very strongly on the correct choice of the 'averaging rule' (Theiß, 1994, Theiß, 1993).

3   INSTRUMENTATION

3.1   Fourier transform spectroscopy

Today doing IR spectroscopy means to work with Fourier transform spectrometers which are used almost exclusively. They have replaced grating instruments mainly because of their much more efficient use of the infrared radiation (multiplex and throughput advantage) (Bell, 1972, Griffiths, 1986, Geick, 1975).

The principle of Fourier transform spectroscopy is sketched in fig.6. The broad-band light source creates radiation by many spontaneous emission processes which can be considered classically as a random emission of more or less damped wave packets. A beam splitter divides the wave packets into two (ideally equal) parts which are then reflected by a fixed and a movable mirror, respectively. Finally, after passing the beam splitter again, the two copies of one and the same wave packet are superimposed at the location of the detector which responds to the time-averaged square of the total electric field. Besides a constant background this quantity is the autocorrelation function of the stochastic time-dependent electric field where the time delay between the two copies of the electric field is realised by different travelling times to the fixed and movable mirrors. After recording the autocorrelation function $I(x)$ over a sufficient range of distances x the desired frequency spectrum is obtained by Fourier transformation according to the Wiener-Khintchine theorem (Wiener, 1930, Khintchine, 1934).

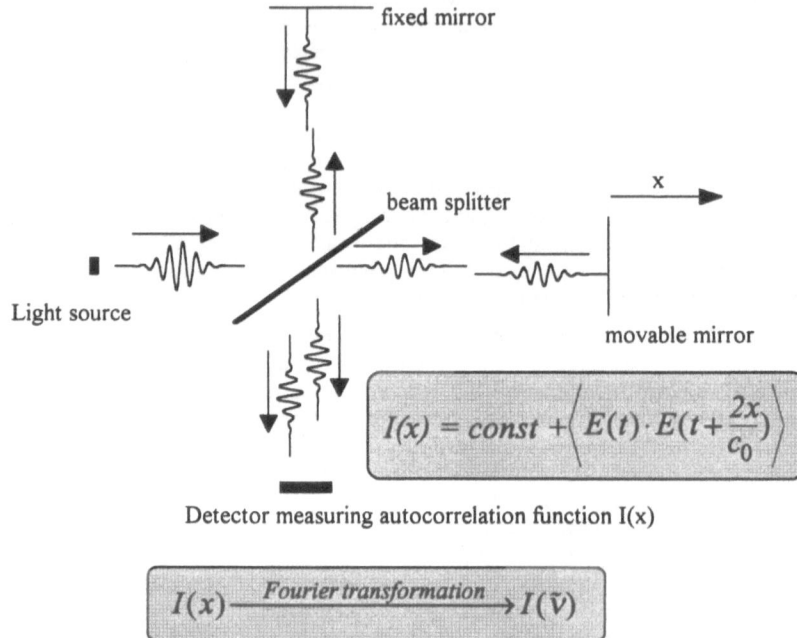

Fig.6: schematic sketch of a Fourier transform spectrometer. See text for details.

3.2   A typical spectrometer

A realisation of the just discussed type of spectrometer is shown in fig.7. The thermal light source (SiC lamp) provides broad band IR radiation which is - after passing the actual Michelson-interferometer and the sample chamber - detected by a pyroelectric detector. Samples can be investigated in transmission and reflection, respectively.

The interferogram of the IR radiation is taken while the movable mirror moves continuously. Its position is determined using the interferogram of a He-Ne laser which is measured by the photodiode shown. Since the frequency spectrum of the monochromatic laser light is a sharp line the corresponding interferogram is a (cosine-like) periodic function of the mirror position x which allows - simply by counting peaks of the photodiode signal - a control of the mirror motion. This kind of measuring the interferogram with continuous motion of the mirror is called 'rapid-scan' technique. Usually many interferograms are co-added to achieve a sufficient signal-to-noise ratio, where one 'scan' (i.e. one interferogram) takes about a second. For certain purposes the so-called 'step can' mode is preferred where the movable mirror stops at the desired positions x and the detector signal is integrated over a long period of time. This is especially useful when the property to be measured can be modulated in time and Lock-In amplifiers can be used to separate very efficiently that part of the radiation that is affected by the modulation from the rest (modulation spectroscopy). The disadvantage of this method is the rather 'bumpy' motion of the mirror which is accelerated and stopped frequently.

Modern spectrometers (above the low-cost price level) use quite sophisticated techniques to control the mirror's motion which is the crucial part of Fourier transform spectroscopy. An example is the use of piezo controllers for the fixed mirror which can handle different tasks at a time: automatic mirror alignment can be achieved (here additional laser beams have to be used), a very slow continuous motion of the moving mirror can be compensated by the piezo-driven 'fixed' mirror to obtain a quasi-step-scan operation (without the acceleration problems) and in addition a modulation of the radiation can be superimposed by the piezo-drives, too.

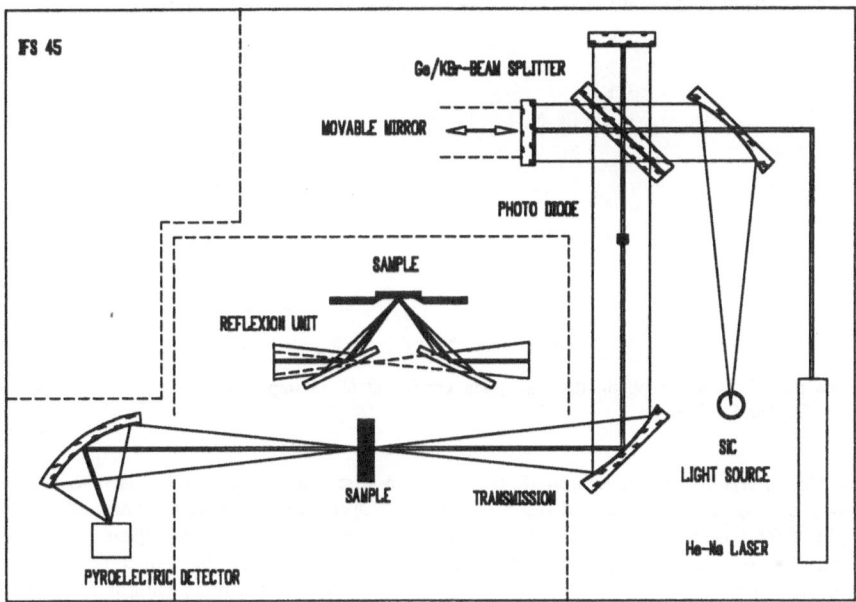

Fig.7: a typical infrared spectrometer (Bruker IFS 45) used to characterise semiconductor layers. See text for details.

3.3 Comparison of various techniques

In an IR experiment usually the intensity reflectance or transmittance of a sample is measured. Fig.7 shows a standard transmission arrangement and also an optional reflection unit which can be placed into the spectrometer's sample chamber alternatively. Besides these standard techniques we will comment in the following also on some other methods to obtain IR spectra.

3.3.1 Transmission

Certainly transmission spectra are recorded most frequently. The reason for this is that the common problem of impurity concentration determination can be solved conveniently with a transmission experiment under certain conditions. If - in a simple one-layer setup as shown in fig.8 -

- the product of sample thickness and absorption coefficient is such that multiply reflected partial waves can be neglected (i.e. no interference structures are observed) and
- the reflection coefficient does not depend on the presence of the substance under investigation, i.e. sample and reference have almost the same reflectance spectrum and

• the transmittance is still significant different from zero

then the so-called absorbance A is given by

$$A(\tilde{v}) = -\ln\left(\frac{T_{\text{Ref+Imp}}}{T_{\text{Ref}}}\right) = -\ln\left(\frac{(1-R(\tilde{v}))^2 \exp\left(-(K_{\text{Ref}} + K_{\text{Imp}})d\right)}{(1-R(\tilde{v}))^2 \exp\left(-K_{\text{Ref}}d\right)}\right) = K_{\text{Imp}}d \qquad (11)$$

and is directly proportional to the impurity absorption coefficient which - in turn - in most cases is proportional to the impurity concentration. $T_{\text{Ref+Imp}}$ is the sample's transmission spectrum whereas T_{Ref} is the reference spectrum without the impurity. Prominent examples are the determination of interstitial oxygen and carbon concentrations in silicon by transmission spectroscopy.

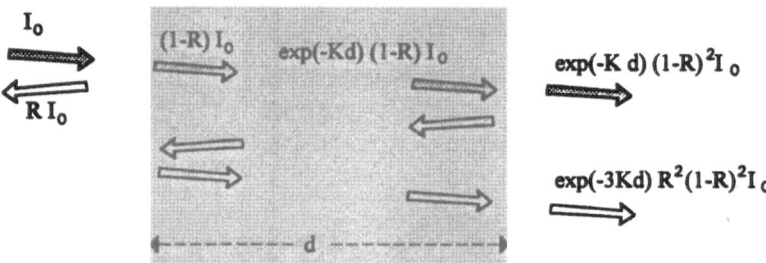

Fig.8: on the conditions to obtain the absorption coefficient of an impurity directly from a transmission spectrum. See text for details.

For characterising porous silicon IR transmission spectroscopy is usually used to show the presence of e.g. oxygen and hydrogen by the corresponding absorption bands. The silicon wafer in which the porous layer has been etched serves as a reference most often. Nevertheless a quantitative treatment on the basis of eq.(11) is not justified since the reflectance spectra of sample and reference are not the same and multiple reflection effects cannot be neglected usually. Of course transmission spectroscopy cannot be performed with highly doped, opaque substrates and metal contacts on the wafer backside (used for etching) have to be removed.

3.3.2 Reflection

Recording reflection spectra (see fig.7) is a very direct way of looking at porous silicon samples. No sample preparation is necessary and no transparent substrate is required. If the thickness of the porous layer is several μm then multiple reflections as sketched in fig.4 must be taken into account. A mixture of both interferences and absorption bands which affect the partial waves passing the porous layer can be found in reflectance spectra and model calculations as explained in the theory section must be used for a reliable interpretation (see examples, e.g. fig.11).

3.3.3 Attenuated Total Reflection (ATR)

The method of Attenuated Total Reflection (ATR) is sketched in fig.9. It makes use of the total reflection of a light wave at an interface to a material with a lower index of refraction. In this case only a rapidly decaying so-called evanescent field penetrates the sample (with penetration depths in the micrometer and submicrometer range). No travelling wave is excited in the sample. Weak absorption processes in the evanescent field cause a significant drop in the 'total' reflection - hence ATR is a good method for thin and ultrathin layers. Using a reflection unit allowing variable angle of incidence of light in combination with an ATR element in the half-cylinder geometry the penetration depths can be varied in a wide range (see fig.9). This can be used to switch on and off the appearance of interferences in the spectra and to distinguish clearly between absorption bands and layer thickness effects.

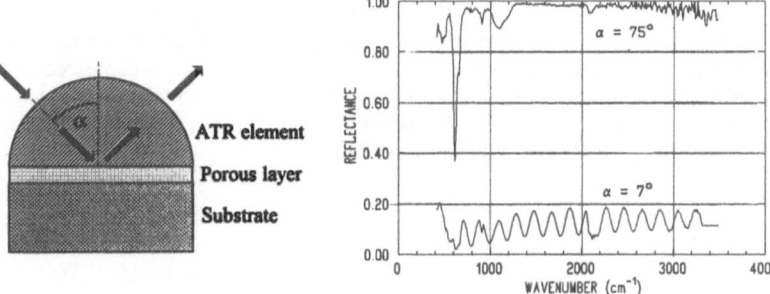

Fig.9: an ATR arrangement (left) using a half cylinder of a material with a high index of refraction (silicon, germanium, KRS5 or KRS6). The incidence angle can be varied in order to realise different penetration depths which is shown by the spectra to the right. The upper curve (incidence angle 75°) was taken well above the critical angle of total reflection: no interferences but only absorption bands are observed. The spectrum below was taken with an incidence angle of 7° and contains information about the layer thickness (17 μm) in the interference pattern. The substrate is p-doped, the porous layer has a porosity of 70%.

3.3.4   IR microscopy

During the last decade IR microscopy has been developed and is now a standard technique: using very sensitive cryogenic MCT detectors (Mercury Cadmium Telluride) cooled by liquid air or nitrogen one can take IR reflectance and transmittance spectra of sample spots with diameters downto 10 μm. IR microscopes usually are add-ons to conventional spectrometers. A visual inspection of the selected sample spot by a conventional 'visible' microscope is provided and in many cases IR microscopes are equipped with a software controlled sample stage for an automatic two-dimensional mapping of sample areas of interest. In the case of porous silicon this can be used for homogeneity investigations or for spectroscopy on small luminescent 'pixels' (Münder, 1993).

3.3.5   Photoacoustic spectroscopy

Finally we would like to mention photoacoustic spectroscopy (PAS) which is a method especially useful for the investigation of porous materials (Hövel, 1992). The principle is sketched in fig.10. IR radiation (which must be intensity modulated) illuminates the sample in a closed gas volume. The absorbed energy causes a periodic heating of the sample (due to the intensity modulation) which results in a temperature oscillation at the sample surface. This in turn causes a pressure variation inside the gas cell which finally is detected by a microphone. Of course, the intensity modulation must be in the 'audio' range. This can be achieved by chopped light or - in the case of a photoacoustic cell placed in the sample position of a Fourier transform spectrometer like the one shown in fig.7 - by the motion of the movable mirror itself: for each frequency there is a cosine-like intensity modulation. Again the frequency contributions can be separated by a Fourier transform of the 'sound interferogram'.

A theoretical simulation of PAS spectra is more complicated as a generation of reflectance spectra. It was shown that for most cases a one-dimensional model is sufficient where material properties change only in one space direction (e.g. the z-axis in fig.2) (Grosse, 1989). The procedure is the following: first solve the optical problem as shown in the theory section of this article. Then calculate the z-dependence of the absorbed light intensity which gives the (oscillating) heat source distribution. For the motion of the generated heat waves the same formalism as for the light waves can be applied yielding reflection and transmission coefficients at boundaries and phase shifts due to travelling times through the layers. Finally the amplitude of the surface temperature modulation is calculated which is proportional to the pressure signal. Calculating the ratio of the sample's temperature modulation and that of a known reference sample a theoretical PAS signal is generated that can be compared to measured values. Of course thermal conductivities strongly influence the PAS signal since they determine the propagation of heat waves in the layers. Hence from a 'fitting' simulation thermal conductivities can be obtained as will be shown below.

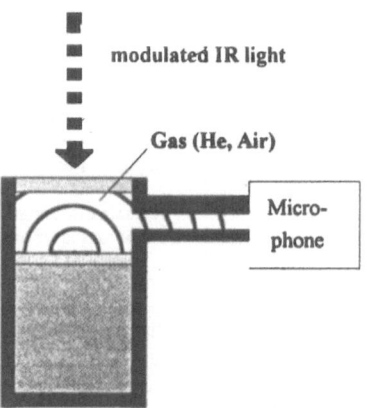

Fig.10: principle of photoacoustic spectroscopy. See text for explanation

4   EXAMPLES

4.1   Single layers

We start this example section with a standard reflection spectrum of a low porosity (36%, non-luminescent) sample on a highly p-doped substrate. With respect to the low porosity it is - to a first approach - a good idea to try the dielectric function of the substrate material for the solid phase of the porous silicon layer which is then averaged using an appropriate Bergman representation (as described e.g. by Hornfeck, 1992) to the resulting effective dielectric function. As fig.11 shows (top curves) the high wavenumber region (dominated by interference structures) is in accordance with experiment but in the low frequency range where the charge carriers are important large deviations occur. To remove these one has to change the dielectric function the following way: the damping of the charge carriers in the Drude model must be very much increased, their concentration a little decreased. The picture is that there are still charge carriers (no complete depletion) which are disturbed in their motion due to collisions with the numerous pore walls. This is known as the classical size effect (Kreibig, 1969). In addition a quite strong absorption band around 640 cm^{-1} is introduced which is probably a Si-H vibration. An alternative interpretation as surface-enhanced multiphonon absorption is still under investigation. The improved 'fit' with the modified dielectric function is shown in fig.11, too (bottom curves).

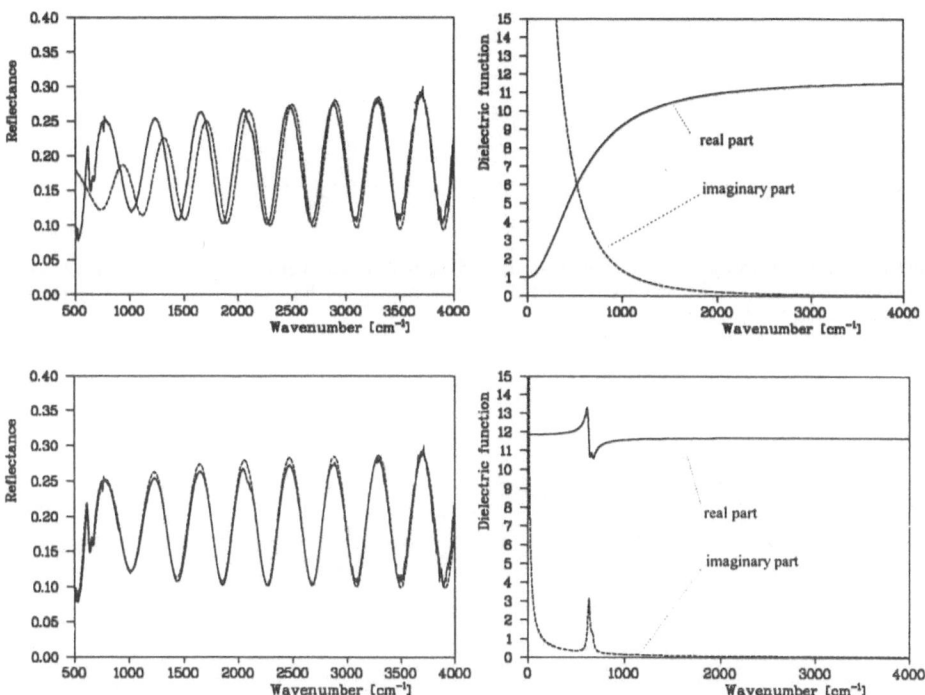

Fig.11: Top: reflectance spectrum of a 36% porosity sample on a highly p-doped substrate (solid line) and a model calculation (dashed) based on the dielectric function of the substrate material shown to the right.
Bottom: the same experimental curve as above, but now the dielectric function of the solid porous silicon component (to the right) has been changed. See text for discussion.

Next we turn to a high porosity (71%, luminescent) sample which requires even stronger modifications of the dielectric function. Charge carriers are not needed any more (complete depletion) but more intense absorption bands show up (Si-H, Si-O). Measurement, fit and the imaginary part of the dielectric function are shown in fig.12.

Now we show how photoacoustic spectroscopy can be used to obtain the thermal conductivity of porous silicon. In a first step a reflectance measurement has been performed and a reliable fit was established as described before (see fig.13, left). Then the only unknown parameter left is the thermal conductivity - which determines the peak heights in the model PAS spectra. Hence it can be adjusted easily until an agreement as good as the one shown in fig.13 (right) is reached. The value obtained for the thermal conductivity is 0.15 W/mK.

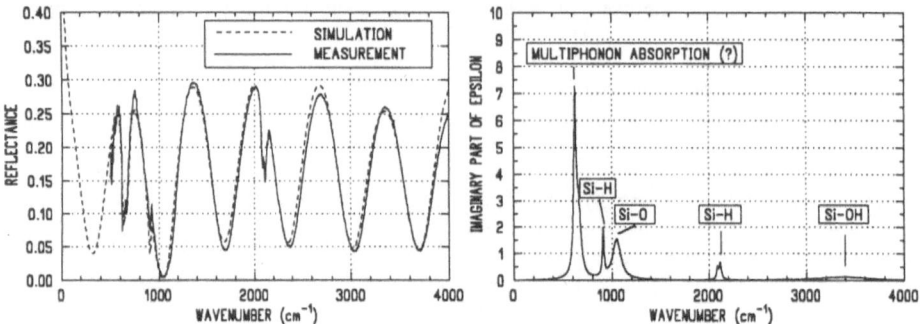

Fig. 12: left: reflectance spectrum of a high porosity (71%) sample (solid line) and the model spectrum (dashed) calculated on the basis of the dielectric function given in the right graph. Note the very strong absorption bands indicating a large hydrogen concentration.

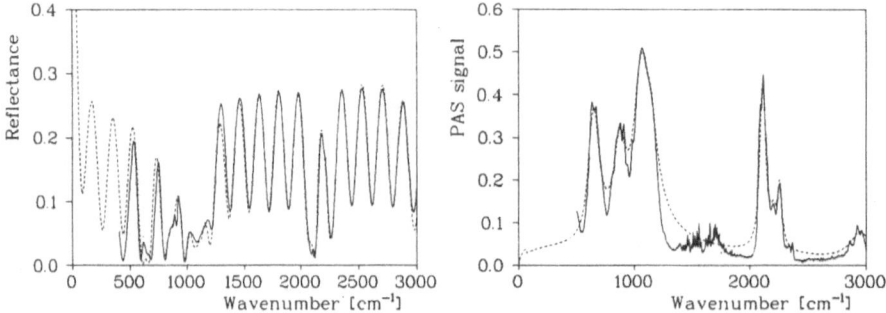

Fig. 13: reflectance and PAS spectra of a 74% porosity layer on a p-doped substrate (measurements: solid line, calculations: dashed lines).

4.2   Investigation of photolithography steps

Of course, in many cases IR spectroscopy can be used without detailed model calculations, but simply by visual inspection of the measured spectra. Here we present as an example the investigation of photolithography processing steps. This standard technique for the production of small silicon devices has been applied to porous silicon as well (Münder, 1993) in order to achieve small light emitting pixels and structured contacts. Fig. 14 shows the basic steps for contact formation and the corresponding IR reflectance spectra.

In fig. 14a) we start with a porous layer on top of the silicon substrate - similar to the one shown in fig. 12. In step b) a photoresist layer (about 3 μm thickness) has been deposited - indicated by additional absorption bands and narrower interferences. After partial illumination the photoresist is developed and removed at the location of the 'future' gold contact. The spectrum at this position (see fig. 14c)) clearly shows that the former porous silicon layer is not recovered unchanged - instead still photoresist bands are found and the reduced interference amplitudes indicate damaged interfaces. The deposition of a gold contact layer is confirmed by the metallic reflection in step d) and finally the remaining photoresist (that was not exposed to light) is lifted off. Here the porous layer is almost the same as before (see fig.14e)) leading to the conclusion that the removal of illuminated photoresist is a critical step in porous silicon photolithography which must be overcome in the future.

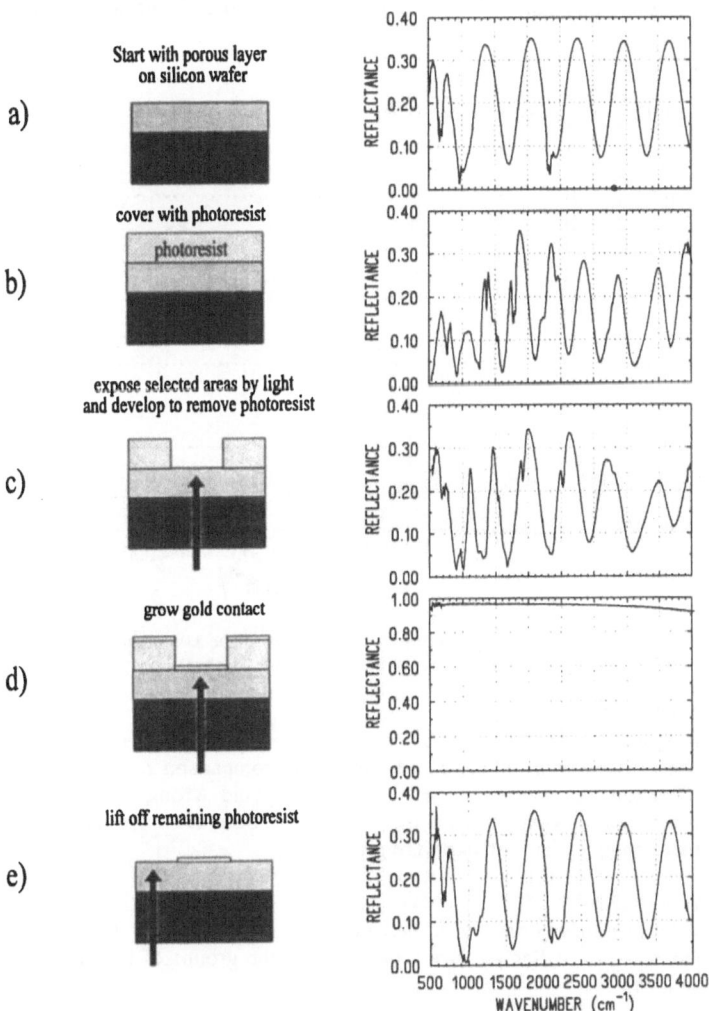

Fig.14: photolithography steps investigated by IR reflectance spectroscopy. Discussion see text

4.3   Porosity superlattices
The realisation of almost arbitrary porosity gradients by a current variation during etching has been verified (Berger, 1994) and, for example, used to achieve porosity superlattices by

switching the current source between two values. Layer stacks with many repetitions of a porosity double layer show remarkable optical properties in the infrared as well as in the visible (see fig.15). In the infrared values of the reflectance even larger than one can be obtained, i.e. the multiply reflected contributions of the many superlattice interfaces do pile up to reflectances larger than that of a gold mirror which is usually used as a reference in the experiment.

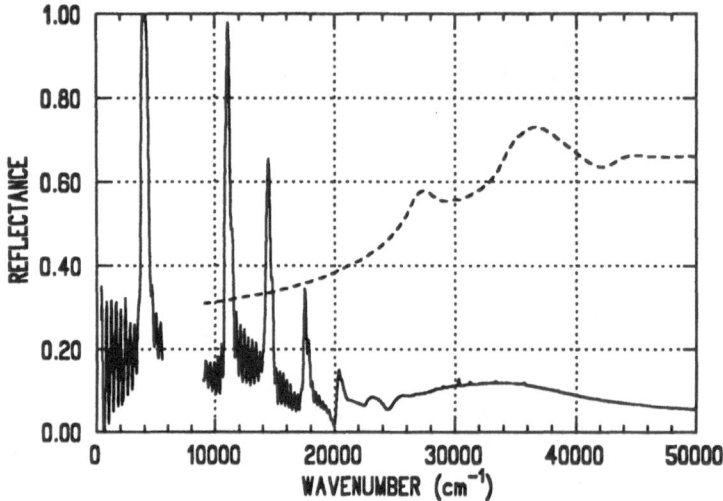

Fig.15: reflectance of a porosity superlattice from the infrared to the UV (solid line). For comparison the spectrum of the silicon substrate without porous layer is shown, too (dashed).

5   SUMMARY

The experimental and theoretical aspects of IR spectroscopy of porous silicon layers have been discussed. An interpretation of measured spectra by comparison to model calculations with well defined parameters was suggested in order to avoid wrong assignments of observed spectral features. Various experimental techniques have been discussed and examples of single layers and more complex layer stacks were given.

ACKNOWLEDGEMENTS

Some of the measurements have been performed by my co-workers M.Arntzen, T.Eickhoff, S.Facsko and M.Wernke. Samples were provided by the groups of R.Herino (Grenoble) and H.Münder (Jülich).

REFERENCES

Adachi S., Phys. Rev. B 38 (1988), 12966
Bell R.J., Introduction to Fourier Transform Spectroscopy (Academic Press, New York 1972)
Berger M.et al., J. Phys. D 27 (1994), 1333
Bergman D., Phys. Rep. C 43 (1978), 377
Bruggeman D.A.G., Ann. Phys. 24 (1935), 636

Ferraro J.R., Krishnan K.(Ed.), Practical Fourier Transform Infrared Spectroscopy (Academic Press, San Diego 1990)

Forouhi A.R., Bloomer I., in: Handbook of opt. const. II (ed. by D. Palik, New York, Academic, 1991), 151

Geick R., Topics in Current Chemistry **58** (Springer, Berlin 1975), 73

Griffiths P.R., de Haseth J.A., Fourier Transform Infrared Spectrometry (John Wiley & Sons, New York 1986)

Grosse P., Wynands R., Appl. Phys. **B 48** (1989), 59

Hövel H., Grosse P., Theiß W., Journal of Noncryst. Solids **145** (1992), 159

Hornfeck M., Clasen R., Theiß W., Journal of Noncryst. Solids **145** (1992), 154

Jackson J.D., Classical Electrodynamics (John Wiley & Sons, New York 1975)

Khintchine A., Math. Ann. **109** (1934) , 604

Kittel C., Introduction to Solid State Physics, 5th Edition (John Wiley & Sons, New York 1976)

Kreibig U., v.Fragstein C., Z.Physik **224** (1969), 307

Martens H., Naes T., Multivariate Calibration, (John Wiley & Sons, New York 1989)

Maxwell Garnett J.C., Philos. Trans. R. Soc. London **203** (1904), 385

Münder H. et al., in 'Optical Properties of Low Dimensional Silicon Structures', ed. by Bensahel D.C., Canham L.T., Ossicini S., (Kluwer Academic Publ., Dordrecht, Boston, London 1993), 75

Theiß W., in Festkörperprobleme/Advances in Solid State Physics **33**, ed. by R.Helbig (Vieweg, Braunschweig, Wiesbaden 1994), 149

Theiß W. et al., Mat. Res. Soc. Symp. Proc. **238** (1993), 215

Wiener N., Act. Math. Stockholm **55** (1930), 117

Nano characterization of porous silicon by transmission electron microscopy

I. Berbezier

*CRMC2, CNRS, Campus de Luminy, Case 913,
13288 Marseille cedex 9, France*

INTRODUCTION

Although the exact origin of luminescence is still controversial, the proposed mechanisms prevalent in the literature are centred around the quantum size effects in very small silicon crystallites which constitute the porous silicon skeleton. To date Luminescent Porous Silicon (LPS) is known to be a very disordered material containing both crystalline and non crystalline domains and networs of pores. Preparation conditions (in particular the porosity level) greatly influence the proportion of crystalline / non crystalline phase ratio. In all cases crystallite with size at the nanometer scale (< 100 Å) are observed. Although no direct correlation has been established so far between the microstructure and the luminescence properties of LPS it is generally accepted that the crystallites features (dimensions, densities ...) play a vital role in luminescence and should be characterized at the nanometer scale.

A wide range of nanocharacterization techniques can be performed using Transmission Electron Microscopy (TEM). Most of them can be devoted to the study of LPS. Electron diffraction provides structural information, whereas morphological aspects are observed by conventional imaging, atom positions are recorded by high resolution imaging and electronic energy levels are measured by spectroscopy (Energy Dispersive Spectrometry : EDS, Electron Energy Loss Spectrometry : EELS or cathodoluminescence). Each source of information will be detailed here and compared to data.

Due to the disordered and unstable structure of LPS peculiar difficulties have been encountered. The main properties studied by TEM are the morphology of the pore network, the crystallite dimension and the local order in the "disordered phase".

The analysis of the very small-angle region of the scattering data should yield statistical information about the network of pores (in particular the dimensios and the morphology of the pores). Indeed some studies have shown that the anisotropy of columnar structure can be detected as an anisotropy in diffraction patterns. However, since the local fluctuations induced by the pores are too low diffraction data failed to reveal anything unless it could be from selected areas sufficiently small to yield some anisotropic texture or other azimuthal structure in the diffuse rings.

The small dimension of the crystallites present in LPS (< 100 Å) induced some difficulties in a quantitative size estimation. Indeed from a mathematical standpoint of diffraction theory the finite boundary of a crystal should be considered as a type of disorder. Moreover the distribution of atoms in and near the surface of a crystal will differ from that within the bulk of a crystal. Therefore as soon as the diffracting crystal dimensions are less than about 100 Å, due to both effects of broadening and displacement of the Bragg peaks, the interpretation of the Transmission Electron Diffraction (TED) patterns becomes difficult. In this field the classical diffraction methods reach their limit of resolution and local order techniques such as extended x-ray absorption fine structure (EXAFS) should be employed. These lead to convenient statistical description which is often adequate to determine first-neighbor numbers and distances. However, the most significant achievement of electron microscopy so far has been to supplement this high-resolution statistical data with information about the large real space structural fluctuations that are frequently visible on the medium range scale (> 20 Å) on micrographs of disordered materials.

Considering both the TED limits (listed above) and the microstructure of LPS (disordered material containing a very small fraction of crystalline phase) TEM images can be much more informative than the conventional wide-beam diffraction patterns. The morphological characterization of LPS should be mainly obtained in operating at a lower-resolution mode, particularly since much lower and therefore less damaging flux densities can be used. High Resolution TEM (HRTEM) images allowed to determine the size, number and shape of the crystalline clusters. However the problem of extracting reliable structural information from high-resolution micrographs was greatly complicated by noise originating from purely random atomic alignments.

Two main difficulties are not solved today (i) the observed structure depends mostly on the specimen preparation and history. Although different preparation modes were used competetively to avoid artefacts the degree of perturbation induced by the extraction of LPS from the Si substrate is still unknown. Furthermore, comparison between materials produced in different laboratories and investigated in different ways, seems to be very difficult. Morphological characterization was performed on samples prepared by argon ion milling and structural characterization on carefully scraped fragments deposited on carbon holder with holes. (ii) The second difficulty concerns the need for a detailed model of the proposed structure for which image predictions can be made and tested in detail against observations. It is a necessary test to ascertain whether the proposed three-dimensional structure is fully consistent with the two-dimensional projected images observed. So far no sufficient detailed model existing can be used for the high resolution simulation. Even if in first approximation it can be supposed that the first nearest-neighbor distance and numbers are very similar to the silicon crystalline phase (which is true in many cases), the distribution of second and further neighbors is however broadened and altered. The second neighbor in the tetrahedrally bonded amorphous semiconductors can for instance be analysed in terms of the variations in the bond angle β but peaks in $\rho(r)$ at higher distances are controlled in a more complex way by factors such as the distribution of the dihedral angle ψ.

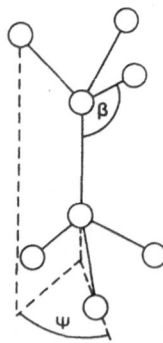

Figure 1 : Schematization of the bond angles in amorphous semiconductors.

A wide variety of models can be considered with at one end polycrystalline or microcrystalline models, and at the other end models such as the random-network structures for covalently bonded amorphous semiconductors. For detailed calculations, any of these models should ideally consist of several hundred or even several thousand atoms arranged with specified coordinates in a quasi-spherical cluster. They can be generated in the computer, starting from some small (non) crystalline seed. These structural models are currently used in High

Resolution Electron Microscopy (HREM) to provide comparison between computed and observed scattering data.

Three LPS specimens were mainly studied and compared here. An abbreviated notation is adopted all along this paper: (85% - p⁻) corresponds to LPS with a porosity of 85% obtained from lightly doped subtrate (p⁻), (65 % - p⁻) to LPS with a porosity of 65% obtained from a lightly doped substrate and (85% - p⁺) to LPS with a porosity of 85% obtained from a highly doped substrate.

Different preparation techniques of TEM samples were tested. Chemical thinning was found to transform the porous silicon material in a microcrystalline one. Take off from the substrate by an intensive electrical pulse at the end of the electrochemical treatment creates a double porous silicon layer including a microcrystalline interfacial layer. Ion milling induces a complete amorphisation of the porous material, excluding this preparation mode for microstructural studies. Nevertheless, morphological information can be gained by this mean. Scraping carefully LPS fragments from the substrate seems to be the less destructive preparation. It allows to obtain thin particles deposited on a carbon grid along different crystallographic directions. Thanks to its easy-making the latter preparation was the most used during both morphological and LPS study.

I. INTRODUCTION TO ELECTRON DIFFRACTION

In electron microscopy the assumptions at the base of the kinematical theory generally are not true. Nevertheless, the kinematical approach is still satisfactory for a mere description of diffraction patterns. It will be briefly related in the first part of this chapter. In contrast it is necessary to use the more realistic dynamical theory of electron diffraction to interpret the details of most images obtained with electron diffraction. It will be the subject of the part III.

I.1 Kinematical approach to electron diffraction from a crystalline specimen

When an electron beam is incident on the top of a thin crystalline electron microscope specimen, specific diffracted beams arise at the bottom exit surface. The strong diffraction beams arise because scattered wavelets are in phase for specific directions in the crystal, leading to a path difference equal to an integral number of the wavelength. Thus a strong reflection is described by the Bragg law:

$$2d_{(hkl)} \sin\theta = n\,\lambda$$

Each individual atom of a crystal scatters the incident electron beam and acts as a source for a spherical wave propagating at an angle 2θ relative to the incident wave direction. The efficiency of the atom in scattering waves is described by the atomic scattering factor:

$$f_\theta = \frac{me^2}{2h^2} \cdot \frac{\lambda}{\sin\theta}^2 (Z-f_x)$$

with θ the scattering angle; Z the atomic number and f_x the atomic scattering factor for x-ray. The structure factor (F) describes the contribution of the entire cell to the diffracted intensity considering both the position of atoms in the reflecting planes and the atomic identity. F is defined as :

$$F = \frac{\text{amplitude of the wave scattered by all the atoms in the unit cell}}{\text{amplitude of the wave scattered by an electron}}$$

$$F = \Sigma f_n \exp(ik\Phi_n) = \Sigma f_n\exp[2\pi i(hx_n + ky_n + lz_n)]$$

with $k = 2\pi/\lambda$ and $\Phi = k \cdot \lambda (hx_n + ky_n + lz_n)$, the resulting phase difference.

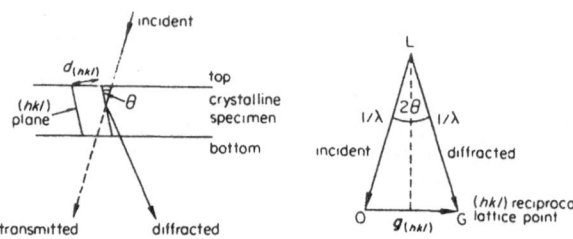

Figure 2: Diffraction by a (hkl) plane in the real space and in the reciprocal space.

The presence or absence of reflections in the crystal structure can be obtained mathematically from the following equation of the diffracted intensity :

$$I = |F|^2 = [\Sigma f_i \cos\{2\pi(hx_i + ky_i + lz_i)\}]^2 + [\Sigma f_i \sin\{2\pi(hx_i + ky_i + lz_i)\}]^2$$

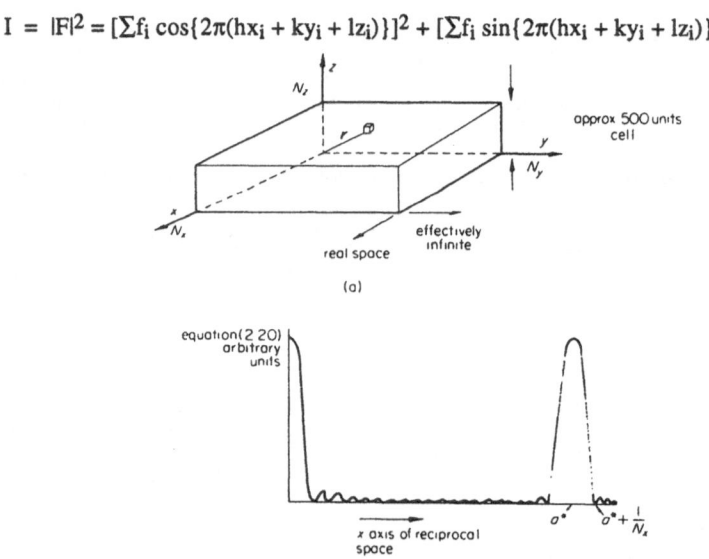

Figure 3 : Representation of the nth unit cell in the crystal and the variation of the intensity along the x axis of the reciprocal space.

Let us now consider the total diffracted intensity (A) from a large array of unit cells that form the diffracting specimen. A is the sum of all the phase differences along the x, y and z axes for N_x, N_y and N_z unit cells:

$$A = |F|^2 \times \frac{\sin^2(1/2\ N_x ka.P)}{\sin^2(1/2\ ka.P)} \times \frac{\sin^2(1/2\ N_y kb.P)}{\sin^2(1/2\ kb.P)} \times \frac{\sin^2(1/2 N_z kc.P)}{\sin^2(1/2 kc.P)}$$

the P diffracted vector is maximum when $P/\lambda = g$.

Thus A is maximum when each term is maximum i.e. when $a\lambda\ (h'a* + k'b* + l'c*) = h\lambda$

and A falls to zero when $h\lambda/N_x$ i.e. $a.h' = 1/N_x$

This shows that the maximum of diffracted intensity is centred on the reciprocal lattice points and falls rapidly to zero on moving a small distance $1/N_x$ from the reciprocal lattice points.

Consequently, when crystals are of the order of 10 unit cells in one (or more) direction the scattering becomes appreciable over an extended range of the angle θ. Considering a small deviation from the exact Bragg position $(g + s) = g'$, this diffracted vector can be easily converted to an angle $\Delta\theta$ with :

$$(2d/\lambda)\sin(\theta_n + \Delta\theta)\Delta\theta = \lambda/Nd\ \cos\theta \text{ where } Nd = L \text{ is the crystal size}$$

Thus the angular breadth (B) measured in radians on the 2θ scale is given by :

$$B = \Delta(2\theta) = \lambda/L\ \cos\theta$$

A more detailed analysis involving the shape of the ring profile as expressed by equation $I = F^2 \sin^2(N\Phi)/\sin^2(\Phi)$ gives the well-known Scherrer equation:

$$L = K\lambda/B\ \cos\theta$$

where K is a constant near unity which is in general taken at 0.91. However the angular breadth is always affected by the design of the diffractometer and a correction for instrumental effects should be done by :

$$B_{(obs.)} = B + b$$

$B_{(obs.)}$breadth experimentally measured, B is the corrected value and b breadth arising from the instrumental effects (that should be estimated using a polycrystalline power).

A good approximation of the diffracting cluster dimension can be obtained by applying this formulae to the case of $(85\% - p^-)$ LPS. The TED micrographs were obtained from a Jeol 200 CX ($\lambda = 2.51 \times 10^{-2}$ Å). The angular breadth measured on the (111) silicon planes is $\Delta(2\theta) = 1.4 \times 10^{-3}$ rd corresponding to a crystal size of ≈ 18 Å. However, considering the poor experimental precision and to avoid some possible artefacts this result should be correlated to other quantitative results.

I.2 Electron diffraction in the TEM

The two diffraction modes currently employed in TEM are characterized by the two main experimental parameters of the electron diffraction (excluding the camera length) (i) the angular aperture of the beam and (ii) the diameter of the irradiated area. Different information can be gained by these two modes.

I.2.1 Selected Area Diffraction (SAD)

SAD consists in virtually selecting a part of the specimen (diameter d_s) by a diaphragm (diameter d'_s) located at the focus plane of the objective. The conjugate of this diaphragm through the objective is in the plane of the specimen. Considering the objective aberrations the diameter of the area effectively selected is :

$$d_s = d's/M + 2C_s(2\theta)^3$$

with C_s spherical aberration and M magnification of the objective.

The smallest diameter of a diaphragm is about 20µm which corresponds to a selected area on the sample of $\approx 0,2$ µm with M $\approx 10^2$. An example of the smallest selected area of LPS is visualised in figure 5 where both LPS fragment and diaphragm aperture are superimposed. If we recall that the diffracting crystals in LPS are embedded in a disordered phase and have dimensions below 50 Å, we can easily imagine the weak intensity in the TED pattern obtained with this mode.

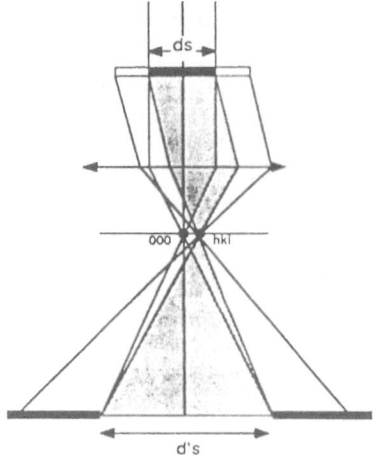

Figure 4: Schematization of the selected area diffraction mode in the microscope.

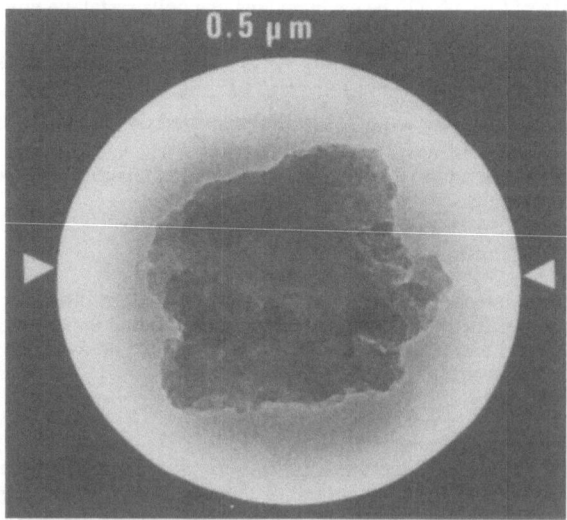

Figure 5 : microscope Jeol 2000 FX , d'$_s$ = 50 µm and d$_s$ = 0.5 µm.

Nevertheless a direct comparison of the LPS microstructure in relation with the porosity level (from 65 % to 85 %) and with the substrate type (p⁻ and p⁺) is performed by SAD. The overall electron diffraction analysis reveals monocrystalline material in the case of (85 % - p⁺) (figure

6a) and of (65 % - p⁻) (figure 6b), and amorphous material in the case of (85 % - p⁻) (figure 6c). Nevertheless, local SAD performed on very small areas on the edges of the particles reveal different interesting features. Indeed they allow to image some structural defects in (65 % - p⁻) (figure 6d) and (85 % - p⁺) (figure 6e) such as (i) small crystallographic planes disorientations from each other (revealed by arc shaped streaking of the monocrystalline spots) and (ii) quite polycrystalline feature in some areas (revealed by well defined superimposed rings). On the contrary different very thin areas of the (85 % - p) sample (figure 6f) present a crystalline structure pointed out by diffracted points on the TED allowing to conclude on the presence of small crystallites.

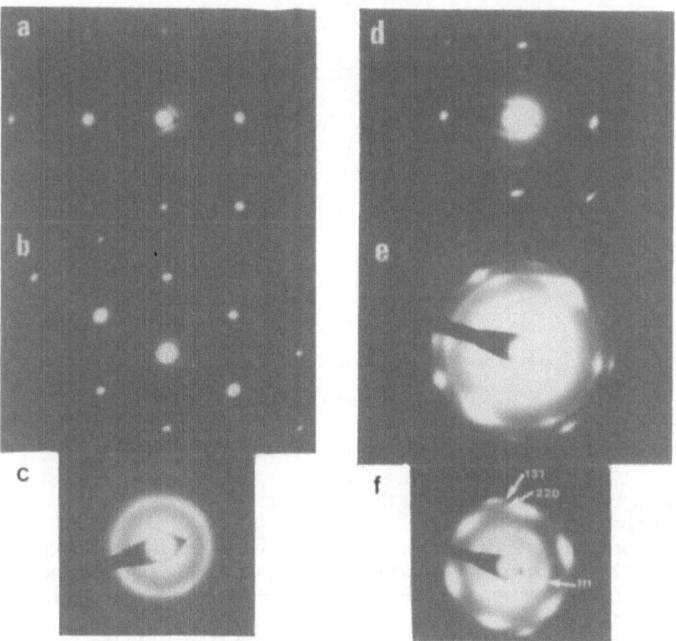

Figure 6 : SAD obtained with d_s = 10 μm for a), b), c) and with ds = 0.2 μm for d), e), f).The samples analysed are (85% - p+) for a and d), (65% - p) for b and e and (85% - p) for c and f.

I.2.2 Convergent beam diffraction (CBD)

CBD consists in focussing the incident beam on the specimen (some nm). The convergence of the beam is made with a convergent-objective lens allowing an angular aperture of $\alpha = 10^{-2}$ rad (instead of 10^{-5} rad in SAD).

In this optical configuration the incident beam can be decomposed in a series of plane waves disoriented one from each other and distributed in the convergent cone of half top angle a. Each incident wave group converges to a point in the image focus plane of the objective at a distance :

$r_i = f.\alpha_i$ of the objective axis

α_i is the angle between the ith wave group and the objective axis and f the focus distance of the objective (figure 7).

The diffracted beams are constituted by a series of parallel wave group each corresponding to one of the incident wave group and inclined of a_i from the mean direction of diffraction. The

intensity of one diffracted wave group is registered in a point inside each diffraction disk. Thus the diffraction disks represent the map of the intensity transmitted or diffracted in accordance with the orientation of the plane waves superimposed in the incident beam.
The diameter of the diffraction disks can be deduced from :

$$D = 2.M.\alpha.f$$

where M is the magnification of the primary diagram by the projective lenses (the camera length L = M.f).

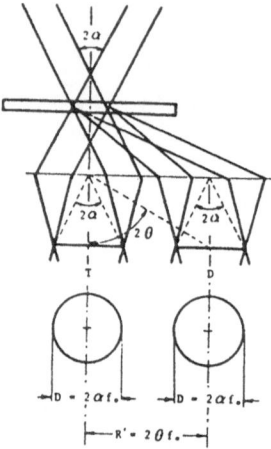

Figure 7 : Schematization of the convergent beam diffraction mode.

The main advantages of this technique are the following:
- high spatial resolution (some nm)
- thickness measurement : the fine structure inside the diffraction disks correspond to the intensity variation in accordance to the deviation from the Bragg angle (for a specimen with a constant thickness). The position and the number of the fringes observed is highly sensitive to the sample thickness.
- lattice parameter measurement : the displacement of the Holz lines inside the central disk is correlated to the variation of the lattice parameter. The relative displacement of these lines (proportinal to $\Delta\theta$) is related to the variation of the interplanar distance by the Bragg law :

$$\Delta\theta \,/\, \theta = \Delta g \,/\, g$$

$\Delta\theta$ is as much as large as the Bragg angle is higher. Moreover the line breadth decreases when g increases giving a higher precision in the measurements.
Practically it has been shown that a relative variation of the lattice parameter of 2 for 10^{-4} should be measured by this technique.
However the determination of the absolute value of the lattice parameter with this technique gives a precision similar to the other techniques.

II- IMAGE ANALYSIS BY CONVENTIONAL T.E.M.

In the microscope the images of the objects produced in the image plane of the objective (direct space) are created by the partial recombination of the scattered and /or diffracted waves at the

exit face of a specimen. The image is observable if it presents some contrast difference between different points of the observation plane. The contrast arises from two reasons :
- the electron scattering by the atom of the object. Scattering effects are highly preponderant over the absorption differences (that can be neglected for very thin samples).
- the limited aperture of the objective diaphragm which creates a diffraction contrast by subtracting in the image the diffracted beams under an angle higher than a .

II.1 Bright-field imaging (BF)

The objective diaphragm located in the focal plane of the objective is centred on the (000) transmitted beam. All the diffracted and / or the scattered beams are stopped by the diaphragm. In this configuration there is an amplitude contrast and the optical resolution is of 5-10 Å (limited by the diameter of the objective diaphragm). The image is formed only with the direct transmitted beam. The diffracting object appears in dark on a bright field with a contrast directly related to the number of crystalline planes in the Bragg conditions (the higher contrast is obtained along a main zone axis).

In order to compare the morphological aspects of various LPS (from lightly doped p⁻ substrates) BF transversal observations were performed on samples prepared by ion milling. Although, this preparation mode induces amorphisation, this can be neglected during a morphological study. Obviously, any structural study can be performed in these conditions.
Due to the overlapping phenomena, the pore networks are not clearly visualized when the porosity is less than 70 %. At higher porosity, some individual pores with mean diameter of 3 nm can be distinguished (figure 8). At this level of porosity, we can also remark that the electrochemical process employed to fabricate the LPS induces an electropolishing of the silicon substrate / porous silicon interface. The results show that this electropolishing phenomenon increases with the porosity, since the roughness of the interface is reduced by a factor two (from 6 to 3 nm) when the porosity increased from 65 to 85 %.

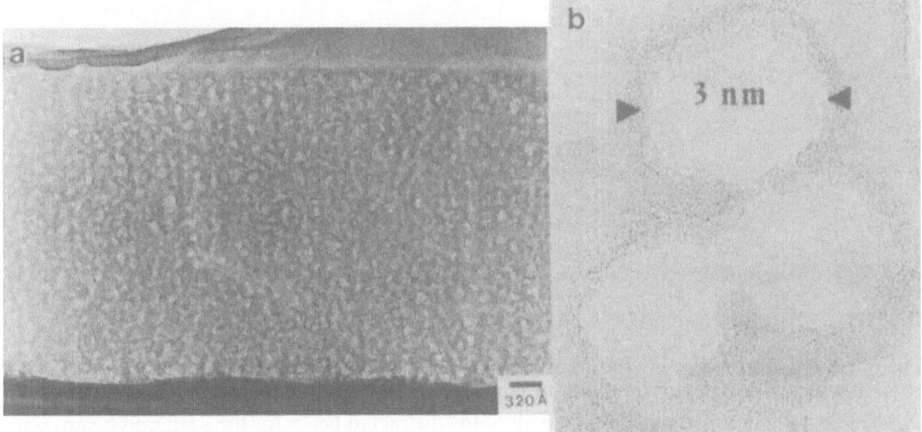

Figure 8 : (a) Morphological aspect of (85 % - p⁻) porous silicon on silicon substrate; (b) visualisation of pore coalescence.

Plane view observations on scraped (85 % - p⁻) LPS fragments reveal the isotropic morphology of the material. As we shall see below, the small spherical components with diameter of 3 nm emerging on the LPS edges (figure 9) do not correspond to crystalline particles.
On the contrary, the anisotropic nature of the (85 % - p⁺) together with branching greatly appears, revealing a network of large pores (approximatly 50 nm) separated by silicon rods (figure 10).

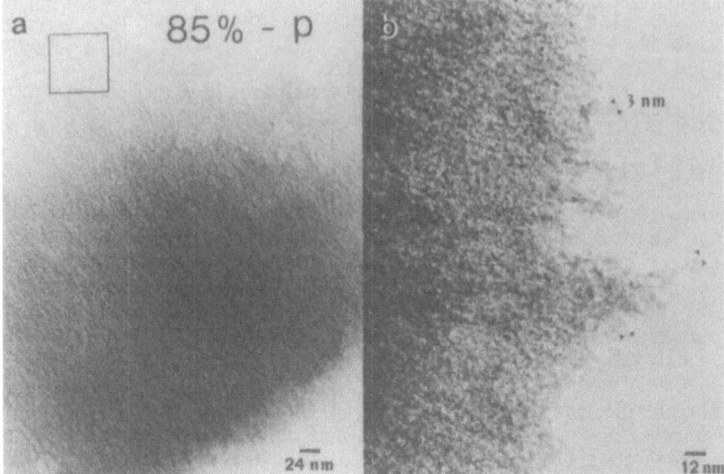

Figure 9 : (a) Overall observation of a (85 % - p⁻) scraped fragment; (b) magnification of the square area.

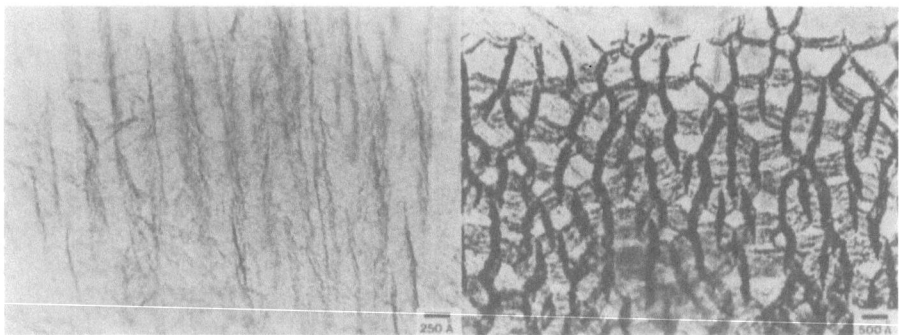

Figure 10 : (a) Overall observation of a (85 % - p⁺) LPS scraped fragment and (b) tilted sample.

II.2 Dark field imaging (DF)

The objective diaphragm is now centred on a diffracted beam (hkl). In order to obtain this configuration without a considerable loss of resolution (in particular due to astigmatism) the

incident beam is inclined of 2θ (hkl) to align the diffracted beam (or the scattered beam) with the microscope axis. Only the diffracting units appear in bright on a dark field. This imaging mode is very well adapted to visualize small crystallites and to determine the level of disorder in crystals. The image obtained presents a random granulometry with statistical properties due to the incoherent response of the optical system. Each image presenting differences in comparison to this statistic contains structural information. The random feature of the image can be tested by changing either the response of the microscope or the projection mode of the sample.

The DF / BF comparison of the three LPS specimens is presented below. White particles representing the crystallites in Bragg position emerge from the rest of the image with accurate dimensions and densities.

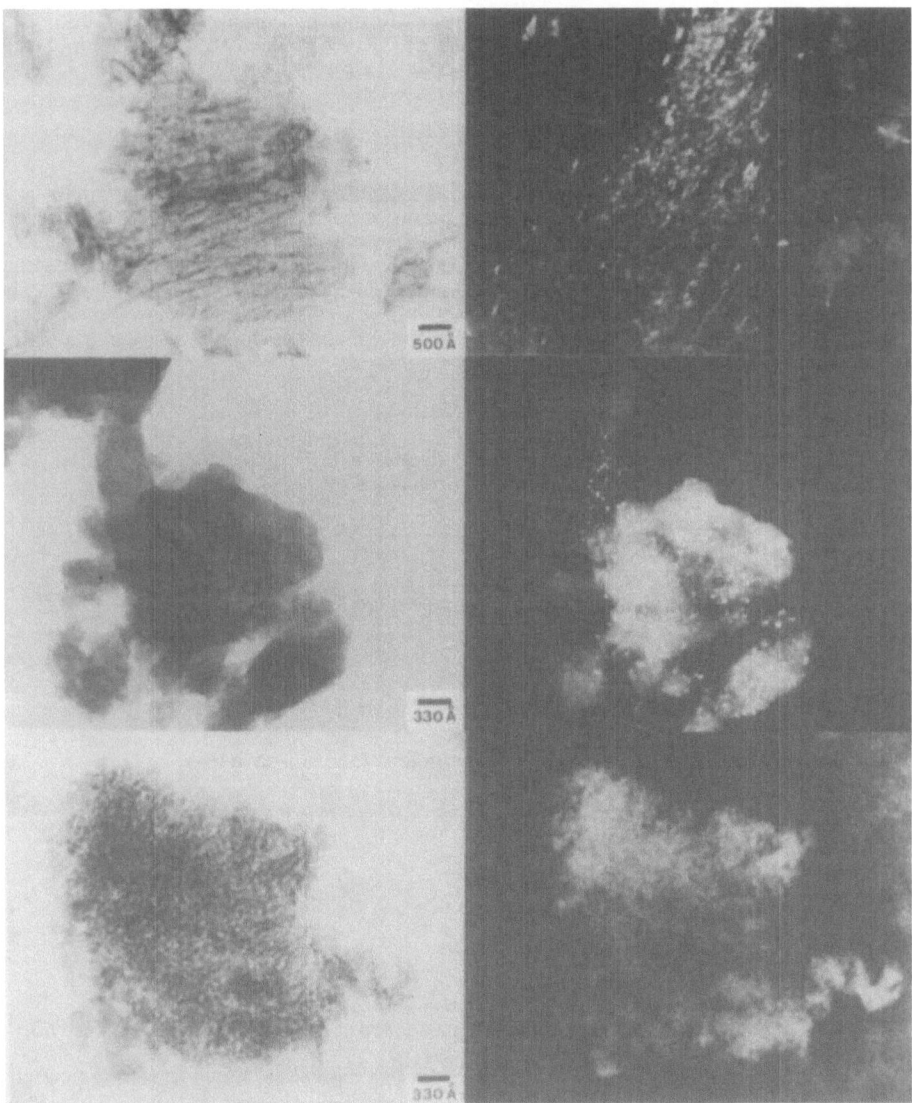

Figure 11 : (a), (b), (c) bright field images and (d), (e), (f) the corresponding dark field images. The corresponding porosities are indicated on the pictures.

Different qualitative information arise from the images presented in figure 11: (i) the larger monocrystalline domains ($\approx$ 400 Å) are observed in the (85% - p^+); (ii) (65% - p^-) LPS is mainly microcrystalline with diffracting crystals of 60 Å mean dimension; (iii) the crystalline phase present in the (85% - p^-) can not be resolved by the DF spatial resolution since the image only present a random granulometry representative of the statistical response of the microscope. This result proves the very small size of the crystalline clusters and shows the limitation of conventional microscopy in the study of highly disordered material.

Thus these observations confirm the preceeding classification concerning the ratio of the crystal phase on the disordered one in relation with the LPS type. Indeed the disorder proportion increases from (85% - p$^+$) to (65% - p$^-$) and (85% - p$^-$) LPS.

III- HIGH RESOLUTION IMAGE ANALYSIS

Because of the strong multiple scattering and the additional effects due to the microscope aberrations and instabilities, the resultant Bragg beams cannot be directly related to the structure factors. In order to solve the structure for a given high resolution lattice image it is imperative to compute the image for a trial structure and to compare with the experimental results. Electron diffraction can be dealt with either as i) the solution of the Schrödinger equation for a fast electron in a periodic potential (Bloch-Wave method), or ii) the solution of the variation in the intensity of diffracted beams as the electron wavefront propagates through a crystal (multislice approach).

III.1 Dynamical theory of electron diffraction

Due to the optical system aberrations and the electron energy dispersion, the electron wave function is modified inside the microscope and is obtained by :

$$\vec{\Psi}(r) = \vec{\Psi}_0(r) \cdot \vec{T}(g, \Delta z) \cdot \vec{D}(r)$$

- $\vec{T}(g, \Delta z)$ is the transfer function of the microscope given by :

$$\vec{T}(g, \Delta z) = \exp i \, \vec{\phi}(g, \Delta z) \times \tau(\alpha^*, \delta)$$

• $\vec{\phi}(g, z) = -2 \pi / \lambda \, [C_s \lambda^4 g^4 / 4 + \Delta z \lambda^2 g^2 / 2]$

λ :wavelength ; Δz : defocus ; C_s : spherical aberration of the objective lens.

• $\tau(\alpha^*, \delta) = \exp [-1/2 \pi^2 \alpha^{*2} (g^{*3} + \Delta g^*)^2] \exp [-1/2 \pi^2 \delta^2 \lambda^2 g^{*4}]$

$g^* = (C_s \lambda^3)^{1/4} g$ (g : transmitted frequency)

$\alpha^* = (C_s / \lambda)^{1/4} \alpha$ (α : source aperture)

$\delta = C_s (\Delta E / E)$ (δ : electron energy dispersion)

$\Delta z^* = -\Delta z (C_s \lambda)^{-1/2}$ (Δz : defocus)

- $\vec{D}(r)$ is the the objective diaphragm function (positioned at the focal plane of the objective).

The wavefunction of fast electrons accelerated through a potential E and travelling through a material with an inner potential $\Phi(r)$ is described by the Schrödinger equation :

$$\frac{h^2}{2m\gamma} \Delta\vec{\Psi}(r) + e \vec{\Phi(r)} \vec{\Psi}(r) = E \vec{\Psi}(r)$$

Considering a periodic potential $\Phi(r)$ (the real crystal potential) the solution $\Psi(r)$ of the Schrödinger equation is the linear sum of N Bloch waves, each with a different wavevector k$^{(j)}$

$$\vec{\Psi^{(j)}}(x) = \Sigma C_g^{(j)} \exp (i k_g^{(j)} \cdot r) \cdot \exp 2\pi i \, \vec{g} \cdot \vec{r}$$

C_g are known as the Bloch wave coefficients

The total wavefunction of the fast electrons within the crystal is therefore given by

$$\vec{\Psi_0}(r) = \sum_{j=0}^{N} \sum_{g=0}^{N} C_0^{(j)} C_g^{(j)} \exp i \, \vec{k^{(j)}}.\vec{r} \exp i \, 2\pi \, \vec{g} . \vec{r}$$

$C_0^{(j)}$ is the amplitude (often called the excitation amplitude) of the jth Bloch wave $\Psi^j(r)$; N corresponds to the number of g accepted in the diaphragm aperture.

Considering the boundary conditions at the exit face and the absorption (which introduces an imaginary eigenvalue $\gamma^{(j)} + iq^{(j)}$ leading to different attenuations of the Bloch waves):

$$\vec{\Psi_0}(r) = \sum_{j=0}^{N} \sum_{g=0}^{N} C_0^{(j)} C_g^{(j)} \exp 2\pi i \, \gamma^{(j)} z . \exp -q^{(j)} z . \exp i \, (\vec{k^{(j)}} + 2\pi \, \vec{g}) \, \vec{r}$$

The Bloch wave excitations ($C_0^{(j)}$, $C_g^{(j)}$) and the $k^{(j)}$ are the eigenvalue and eigenvector of a diffraction matrix. They are easily calculated with a computer and they give the amplitude and the phase of the wave at the exit face.
The intensity in the image is given by:

$$\vec{I}(r) = \vec{\Psi}(r) . \vec{\Psi^*}(r)$$

It depends on the orientation of the beam, the thickness of the specimen, the defocus and the objective diaphragm aperture. All those parameters can be changed by the experimentalist.

Physically the crystal acts as an interferometer. An incident electron of fixed energy and wavevector is partitionned by the crystal into a set of Bloch waves of different wavevectors k. As each Bloch wave propagates it becomes out of phase with those diffracted by neighbouring atoms (due to its different wavevector). Hence interference occurs.
The two beam approximation is generally a valid first approximation for 100 kV incident electrons. However it breaks down even as a first approximation for higher energy incident electrons. Hence for all accurate quantitative work, a many beam dynamical theory is necessary and the many beam equations require a computer.
For example during the LPS study the images of the silicon crystal are often obtained with the [110] silicon axis parallel to the electron beam (in the Laüe symmetrical conditions). Only the <111> silicon planes are visualised. Hence four diffracted beams are admitted in the objective diaphragm and the five beam equations should be calculated.

The intensity of the image presents different terms :
- terms participating to the background,
- terms corresponding to the fondamental frequencies (cos $2\pi g.x$)
- terms corresponding to the double frequencies (cos $4\pi \, g.x$) or intermodulation frequencies.

Only the fondamental terms are affected by the function:

$$T(\vec{k}, \Phi) = \cos (\vec{k_0} - \vec{k_g} - \phi_g)$$

When $T = \pm 1$ the frequencies pass through maxima of contrast; the atoms are either white (T = +1) or black (T = -1) mainly depending on the defocus and the thickness. When T = 0 only the frequencies of intermodulation pass in the image and there is no more direct relation between the image and the crystal structure.

III.2 Multislice method

The multislice method is the most probably widely used approach to the simulation of many-beam high resolution lattice images. This is a physical optics approach, based on a theory developped by Cowley and Moodie (1957). The basic scheme is that we imagine the crystal divided up into number of slices parallel to the surface of the film.

The approximation is made that the scattering from any individual slice occurs on a single plane and may be described by a "transmission function". A "propagation function" is used to describe the transfer of the wave function to the next plane, through a uniform medium. The process of transmission followed by propagation is repeated for each slice sequentially until the bottom surface is reached, yielding the exit wavefunction of the electron.

$\Psi(\vec{x}, z)$ propagating along the z direction can be deduced from the $\Psi(\vec{x}, 0)$ wave by convolution.

Putting $\Psi(\vec{x}, z) = \exp ikz \cdot \tau(\vec{x}, z)$,

the propagation of Ψ is given by the equation :

$$\tau(\vec{x}, z) = -i/\lambda\, z\, (i\pi\, \vec{x^2} / \lambda z) * \tau(\vec{x}, 0) = P(\vec{x}, z) * \tau(\vec{x}, 0)$$

with $P(\vec{x}, z)$ propagation function

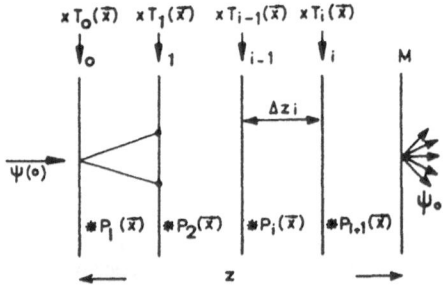

Figure 12 : Basic schema of the multislice principle.

When a plane wave passes through an object both its amplitude and its phase may be altered, giving a transmission function of the form :

$$T_i(\vec{x}) = \exp [i \chi_i(\vec{x}) - i\mu(\vec{r})]$$

where $\chi_i(\vec{x}) = -\sigma\, \phi_i P(\vec{x})\, \Delta z_i$ describes the local phase changes and $\mu(\vec{r})$ the local absorption. This term is generally neglected (for very thin slabs of crystal)

σ is the interaction constant ($= 2\pi\, me\, \lambda/ h^2$). $\phi_i P(\vec{x})$ is the projected optical potential averaged on the thickness Δz_i.

At the exit of a very thin slice i the wave is :

$$\Psi_i(\vec{x}) = T_i(\vec{x}) \cdot [\Psi_{i-1}(\vec{x}) * P_i(\vec{x}, \Delta z_i)]$$

$$\Psi_i(\vec{x}) = \exp -i\sigma\, \phi_i P(\vec{x})\, \Delta z_i\, [\Psi_{i-1}(\vec{x}) * -i / \lambda\Delta z_i \exp (i\pi\, \vec{x^2} / \lambda\Delta z_i)$$

The wave at the exit face of the crystal is obtained by the successive application of these recurrence formulae on the M slices (M Δz_i = z).

HR analysis of the (65% - p⁻) reveals local disordered areas of the silicon lattice. Different types of disorder are observed : i) small disorientations of the crystal planes from a crystalline area to another and ii) variations of the crystal plane distance evidenced by the undulated shape of the (111) silicon planes. Thus partial degradations of the monocrystalline lattice are pointed out, revealing the starting point for the crystallinity loss. At this level of porosity, the structureless phase already mentioned is not identified.

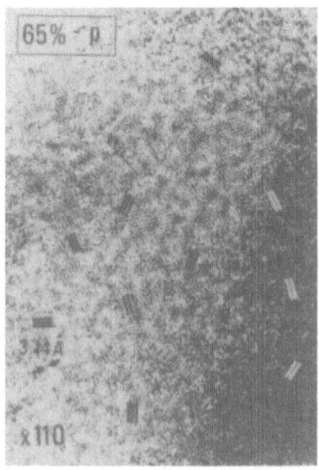

Figure 13 : HREM image of (65% - p⁻) fragment with the (1-10) plane deposited on the carbon grid allowing the visualisation of the <111> planes (d = 3.14 Å).

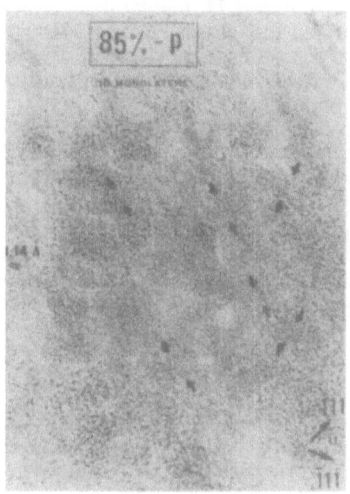

Figure 14 : HREM image of (85% - p⁻) along the [1-10] crystallographic direction.

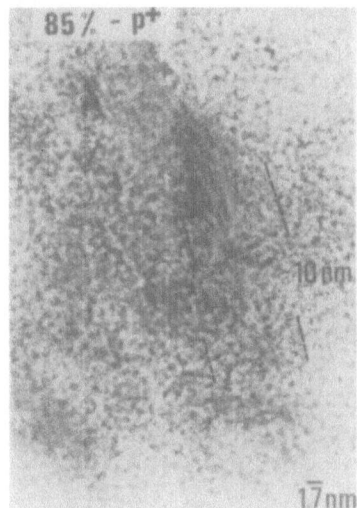

Figure 15 : HREM image of (85% - p⁺) LPS along the [1-10] crystallographic direction.

On the contrary, (85% - p⁻) LPS presents two phases (figure 14) : i) one phase composed of clusters with dimensions in the range 20 - 30 Å and ii) a second phase presenting a statistical random contrast. Different tilt angles were used to check the random feature of this contrast and confirmed that any coherent information arised from these areas. Considering that coherent diffraction arises from a cube of three cells edge, we conclude that the contrast observed results from a disordered phase either nanocrystalline with crystallite dimension less than 15 Å or amorphous.

From numerous HRTEM images, the mean concentration of clusters was estimated to about 10 clusters / cm with mean dimension of 2 nm. This result is in agreement with the electron diffraction study (part I.1).

Larger and anisotropic crystalline areas are visualised in the (85% - p^+). The two main components emerging consist of i) long monocrystalline rods with diameter of 10 nm or more and ii) small rotated silicon crystallites with size ranging from 2 to 6 nm (figure 15). Due to the large variety of dimensions of the crystalline areas, the quantitative estimation of the clusters density and size is rather difficult to obtain.

IV- MICROANALYSIS

• EDS is the most currently used microanalytical technique. This allows a precise chemical analysis of the material. For example it should allow to ascertain the absence of oxygen in the LPS. The first informations obtained are in this sense.

• EELS spectrometry should be used to provide both the number of and the distances to the nearest-neighbor atoms about the silicon atoms. The comparison between the local order in the disordered phase (using EXELFS) and in the crystallites is promising. The first experimental data obtained need further investigation to bring an accurate interpretation.

• A very promising technique is the cathodoluminescence (CL) in scanning TEM (STEM) since it is 1) sensitive to the very low concentration of impurities that may be electrically important (usually present in concentrations too low to be detected by EDS, EELS or HREM), 2) capable of sufficient spatial resolution to isolate individual cristallites and 3) able to provide sufficiently high spectral-energy resolution to study the electronic states of interest.

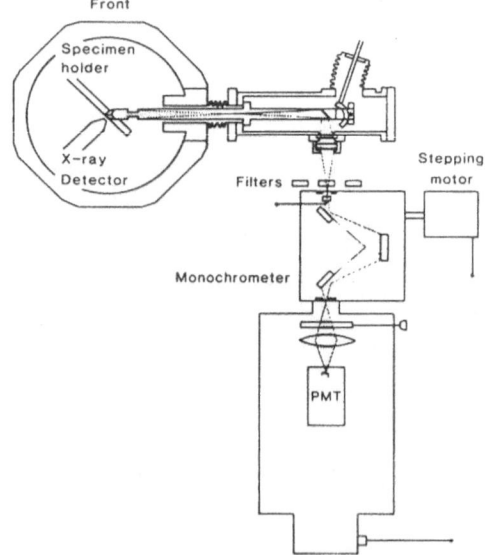

Figure 16 : Cathodoluminescence system mounted in a 100 kV microscope (Philips EM 400).

Here the optical emission excited by the electron beam in passing through a thin sample is collected by a small mirror and passed to a convenientional optical spectrometer for analysis (figure 16).
By forming a small electron probe and plotting the CL intensity within a small spectral range as a function of electron-probe position, one may also obtain a scanning - monochromatic CL image, giving a spatial map of the impurity or electronic state of interest.
The usefulness of the STEM-CL technique for the study of defects in GaAlAs / GaAs heterostructures was first demonstrated by the pionneering work of Petroff and others (1980).

The spatial resolution appears to greatly depend on the specimen thickness for thin samples. In general it is expected to be given by :

$d_r = (d_P^2 + d_G^2 + d_D^2)^{1/2}$
where d_P is the electron probe diameter, d_G is the electron - hole pair generation volume and d_D is the carrier-diffusion length. For thin samples, d_P and d_G are small compared with d_D which is dominated by surface recombination. As a rough approximation, this diffusion length is expected to give (Brown, 1984) :

$$d_D = \frac{t}{6}$$

where t is the specimen thickness.
For such very thin specimens, the emission intensity is very small requiring efficient light - collection optics. The spatial resolution of a monochromatic - scanning image is not influenced by the optical - resolution limit of the light collection mirror or lens; however this lens does affect the amount of stray light that is collected and hence the signal to noise ratio in the image.
From such a small volume a spectral resolution of perhaps 10 Å may be obtained from very low concentrations of impurities and this spectral information can be collected together with the corresponding HREM image.
As an example of STEM - CL work in the infrared range, Graham et al. (1986) have recorded spectra in the 0.8 to 1.0 eV range from groups of straight dislocations in silicon.

Acknowledgements

We are grateful to J. Derrien for interesting discussions and manuscript review. The author thanks also F. Quintric for helpful assistance during the photographic work.

References

[1]   G. W. Brindley and G. Brown Ed., "Crystal structures of clay minerals and their x-ray identification".
[2]   J.W. Edington Ed., "Practical electron microscopy in materials science".
[3]   A. Baronnet, "Techniques d'imageries et de microanalyses par microscopie électronique" private communication.
[4]   B. Jouffrey, A. Bourret and C. Colliex, "Microscopie électronique en sciences des matériaux", Bombannes, 1981.
[5]   P. Buseck, J. Cowley, Le Roy Eyring, "High-Resolution Transmission Electron Microscopy and Associated Techniques", Oxford Science Publications, 1992.
[6]   C. Meneau - d'Anterroches, Thesis, Université de Grenoble, 1982.
[7]   Petroff P., Logan R.A. and Savage A., Phys. Rev. Lett., 44, 287 (1980)
[8]   Graham R.J., Spence J.C.H. and Alexander A., 82, 235 in "Characterization of defects in materials", ed. by A. Siegel, R. Sinclair and J. Weertman, North Holland, New York, 1987.
[9]   J.M. Cowley, "Electron Diffraction Techniques" Volume 1, Ed by , International union of Crystallography.
[10]   I. Berbezier and A. Halimaoui, J. Appl. Phys., 74, 7, 1993.

LECTURE 13

Electron paramagnetic resonance spectroscopy: Defect and structural analysis of solids

H.J. von Bardeleben and M. Schoisswohl

Groupe de Physique des Solides, Universités Paris 6 & 7,
URA 17 du CNRS, 2 place Jussieu, 75005 Paris, France

1. INTRODUCTION

Electron paramagnetic resonance (EPR) spectroscopy is one of the most powerful techniques for the structural analysis of point defects in solids. It has been widely applied in the last thirty years to the study of defects in all relevant semiconductor materials and a large amount of information on defects related with impurity dopings but also about intrinsic defects is now available for semiconductor materials such as Si, Ge, III-V's and II-VI's. Defect studies in new type of materials related to special growth processes or preparation conditions such as low temperature MBE grown GaAs or porous Si are expected to be related to these previous studies not precluding new developments.

The structural analysis of point defects is one of the problems, which can be tackled by the EPR technique; it includes the determination of the site of a particular defect, its point symmetry, its chemical nature, its electronic groundstate configuration as well as possible perturbations induced by its presence on the nearest neighbour configurations. But once a defect is identified, we can use also the variation of its properties to study the material properties such as local strains, misorientations, Fermi level positions. This will particularly apply to interface defects, whose properties are expected to react strongly to local environment changes.

Since its first experimental realization some 50 years ago EPR spectroscopy has found a variety of further developments, which range from its classical form of detection via microwave absorption at resonance, to optical and electrical detection schemes, pulsed EPR spectroscopy and multiple magnetic resonance spectroscopy, combining different techniques such as optically detected ENDOR.

As often, useful application of one technique (EPR) requires complementary measurements by other experimental techniques such as optical absorption, local vibrational mode studies and transport measurements. In the case of complicated materials like porous Si additional structural analyses by techniques such as electron microscopy, Raman spectroscopy or global impurity analysis by nuclear reaction techniques are certainly expected to be of great help.

2. BASIC PRINCIPLES OF EPR SPECTROSCOPY

2.1. The Spin Hamiltonian

EPR spectroscopy [1-6] is a technique based on the resonant absorption of electromagnetic radiation by a spin system, for which the degeneracy of its electronic groundstate has been lifted by an external magnetic field. Whereas the perfect semiconductor material is diamagnetic, defects at the surface or in the bulk both of intrinsic or extrinsic nature as well as free carriers can be paramagnetic and can thus be studied by the EPR technique.

The Hamiltonian H for such a defect, incorporated in a crystal, can generally be decomposed in different terms of decreasing importance allowing a perturbation theory approach:

$$H = H_{cin} + H_{coul} + H_{per}$$
$$\text{and } H_{per} = H_{s\text{-}o} + H_{ss} + H_{si} + H_{li} + H_{eZ} + H_{nZ} + H_{CF}$$

with H_{cin} being the kinetic energy, H_{coul} the Coulomb energy term, $H_{s\text{-}o}$ the spin-orbit interaction between the electron spin and its orbital momentum, H_{ss} the spin-spin interaction, H_{si} and H_{li} the hyperfine interaction , H_{eZ} and H_{nZ} the electronic and nuclear Zeeman interactions, and H_{CF} the crystal field terms. All terms are small as compared to the first two one's ($\approx$eV): $H_{s\text{-}o} = 10^{-2}$eV, $H_{ss} = 10^{-4}$eV, $H_{si} = H_{li} \approx 10^{-6}$eV, $H_{eZ} \approx 10^{-4}$eV, $H_{nZ} \approx 10^{-7}$eV and $H_{CF} \approx 10^{-2}$eV and can thus be treated as a perturbation on the first two terms. For defects incorporated in solids the orbital momentum is normally quenched and the Zeeman interaction part reduces to its spin part:

$$H_{eZ} = g\beta \mathbf{BS}$$

However the terms involving L contribute due to the spin-orbit interaction in second order perturbation theory, which leads to a deviation of the g-factor from the free electron value.

The experimental results are analyzed with the help of a phenomenological Spin Hamiltonian, in which the degeneracy of the groundstate is described by an effective spin S. This Hamiltonian is only composed of spin operator terms. The concept of the Spin Hamiltonian as a means to describe the energies of an electronic groundstate term , whose degeneracy is described by an effective spin S, goes back to Abragam and Bleaney [4]; it is based on perturbation theory within the crystal field approach. The terms, which have to be considered, depend on the spin of the groundstate as well as on the point symmetry of the defect. The Hamiltonian of a defect of effective electron spin S and nuclear spin I can be written as

$$H = \beta \; \mathbf{B} \; g \; \mathbf{S} + H_{Fine \; St.} + \mathbf{S} \; \mathbf{A} \; \mathbf{I} + \mathbf{I} \; \mathbf{Q} \; \mathbf{I} - g_N \beta_N \mathbf{B} \mathbf{I}$$

where $H_{Fine \; St.}$ represents second and higher order terms in electron spin components. The parameters of the spin Hamiltonian are the meeting point between experiment and theory. Analysis of the experimental spectra allows the determination of the different parameters: g-tensor, A-tensor, fine structure coefficients etc. The numerical values must then be interpreted within a particular model.

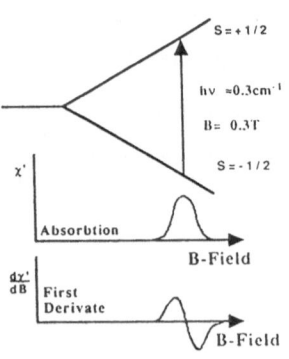

Fig.1 Energy levels for a S=1/2 system

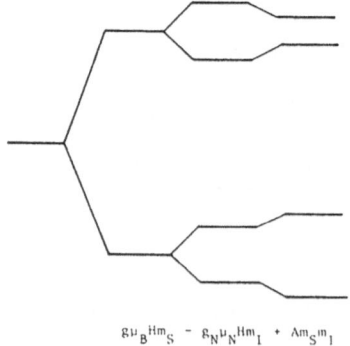

Fig.2 Energy levels for a S=1/2,I=3/2 system

To be more specific let us consider the case of a defect with spin S=1/2, I=0 and an isotropic g-factor. In this simple case, the spin Hamiltonian reduces to the electronic Zeeman term as

$$H = g\beta S \; B$$

the energies of which are $E_\pm = \pm 1/2 \; g\beta B$ (fig.1). If the nucleus of the central atom has also a nuclear spin $\neq 0$, the nuclear Zeeman interaction and the hyperfine interaction between the electron spin and the nuclear spin will further split and shift the Zeeman levels (fig.2):

$$H = \beta BgS + SAI - g_N\beta_N BI$$

The solution of this spin Hamiltonian in the case of isotropic interactions, choosing the four spin functions $/m_S, m_I>$ as base functions and applying the magnetic field along the z direction, are given by

$$E = g\beta B m_S - g_N\beta_N B m_I + A \; m_S m_I$$

When we submit this system to a linearly polarized oscillating magnetic field, oriented perpendicularly to the static one, we can induce magnetic dipole-transitions (selection rules $\Delta m_S = \pm 1$ and $\Delta m_I = 0$) between the Zeeman sublevels, if the energy of the oscillating field corresponds to the splitting between the levels E_1 and E_3 and between E_2 and E_4.

We see, that we have a correspondance at resonance between the magnetic field and the frequency of the oscillating magnetic field; magnetic resonance might thus be observed by application of a fixed static magnetic field and an oscillating field of variable frequency or -what is generally done due to practical reasons - EPR spectroscopy is performed at a fixed frequency ν_e and a varying magnetic field. Typical experimental conditions correspond to fixed microwave frequencies at X,K or Q band, leading to magnetic resonance fields in the 3 to 12kG range for a g value of 2.

Generally we will not work on single spins, but in macroscopic samples of mm sizes we will measure the properties of spin systems composed of typically 10^{10} to 10^{15} spins. At thermal equilibrium the population N_i of the different Zeeman sublevels i will be given by the Boltzman factor:

$$\frac{N_j}{N_i} = \exp(-\frac{\Delta E}{kT}) = \exp\frac{-g\beta B_0}{kT}$$

If no relaxation processes were present, to assure that the spin system stays in equilibrium even in the presence of the oscillating magnetic field, the latter would rapidly equalize the populations N_i and no resonance would be observed.

The rate of absorption of the microwave energy P -the quantity, which is detected in an EPR experiment- is proportional to the microwave transition probability W and the population difference between two adjacent Zeeman levels $\Delta N = N_j - N_i$:

$$P = h\nu W \Delta N$$

The microwave field induced transition probabilities $W_{i->j}$, $W_{j->i}$ after Fermi's golden rule are identical and equal to $\frac{2\pi}{h}|< \Psi|H'|\Psi >|^2 \partial(E_j - E_i - h\nu)$; the thermally induced transition probabilities are $W^{th}_{i->j}$ and $W^{th}_{j->i}$. Combining the two processes we find the following rate equation:

$$\frac{d\Delta N}{dt} = -2W\Delta N + \frac{\Delta N_0 - \Delta N}{T_1} \quad \text{with the steady state solution} \quad \Delta N = \frac{\Delta N_0}{1 + 2WT_1} \; .$$

In the absence of saturation ($2WT_1 << 1$) the absorbed power P becomes

$$P = \frac{\pi\omega^2}{4kT}|< \Psi|\mu|\Psi >|^2 h_0^2 N f(\omega) \; .$$

2.2. EPR spectroscopy

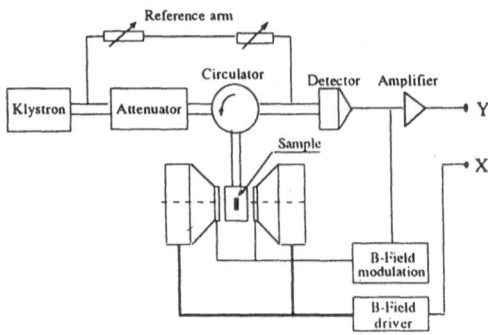

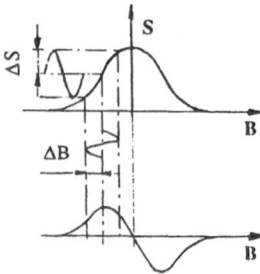

Fig.3 A typical EPR spectrometer   Fig.4 Magnetic field modulation scheme

A typical EPR spectrometer is composed of a microwave source (klystron or Gunn diode), the microwave bridge containing the signal arm directed to the cavity and the reference arm, as well as the microwave detector. In order to observe the EPR absorption the sample is placed into a resonant cavity, which is part of the critically coupled microwave bridge (fig.3); the impedance of the cavity, characterized by the resistance and inductance of an equivalent coil is given by $Z = R_0 + iL_0\omega(1 + 4\pi\chi)$, χ being the complex susceptibility, which in a classical description (Bloch equations) is given by

$$\chi = \chi' - i\chi'' = -\chi_0\omega_0 T_2 \frac{(\omega_0 - \omega)T_2}{1 + (\omega_0 - \omega)^2 T_2^2} - \chi_0\omega_0 T_2 \frac{i}{1 + (\omega - \omega_0)^2 T_2^2}$$

The microwave bridge can be tuned to detect the dispersive part (χ') or the absorbance part (χ'') of the susceptibility. To improve the signal to noise ratio the static magnetic field is modulated with a small amplitude of ≈ 0.1 to 10 G at a frequency of some 10^1 to 10^4 Hz to allow phase sensitive detection with a lock-in amplifier. As a consequence first derivative curves are detected instead of the normal absorption curves (fig.4).

We had seen before, that the hyperfine interaction leads to a splitting of the Zeeman sublevels and a corresponding splitting of the EPR lines in multiplets; this splitting gives important information on the wavefunction extension of the paramagnetic electron and allows at the same time the identification of the chemical nature of the paramagnetic defect. From the magnitude of the hyperfine interaction we can distinguish the interaction with the central nucleus, which is often directly resolved, and the interaction with the surrounding ligand nuclei, the resolution of which requires normally the use of the electron nuclear double resonance technique (ENDOR).

An important aspect of EPR spectroscopy is its quantitative character; by use of spin standard samples the number of defects in a given sample can be determined quantitatively. The sensitivity limitations of modern EPR spectrometers are of $\approx 10^{10}$ spins/Gauss. As typical sample sizes for bulk samples are of 10mm^2 in X band spectrometers, defect concentrations of 10^{12} cm^{-3} are detectable, if the linewidth is ≈ 1G. On the other hand considering interface defects with typical concentrations of 10^{12} cm^{-2}, the minimum sample areas needed to allow detection are ≈ 10 mm^2. Larger linewidths and long relaxation times can increase this sensitivity limit rapidly to higher values.

The detection of an EPR resonance is not only possible via the detection of the associated microwave absorption, which might be called the classical EPR experiment, but its detection is also possible via the perturbations induced on optical emission and absorption processes (ODMR,MCDA) or via the change in lifetime of free carriers (electrically detected EPR). For more details on the EPR technique excellent introductory books are available [1-6].

3. SOME INSTRUCTIONAL EXAMPLES

The EPR spectroscopy can be applied to various types of materials and used to resolve many different problems. Its range of application goes from the study of organic materials, insulating anorganic compounds to semiconductors and metals; materials both in liquid and solid phases can be studied. In semiconductor physics, the field which interests us in connection with porous silicon, the studied defects can be classified in four different groups, the modelling of which requires different theoretical approaches: effective mass like defects, transition metal defects, rare earth defects and deep defects such as intrinsic defects. To give some instructional examples of recent applications in the field of semiconductor physics we would like to present you the following three cases: first, EPR results on intrinsic, bulk defects in GaAs, which show you how a microscopic model of the defects can be developed with the help of tight binding model calculations. As a second example we present you the results obtained in the study of interface defects and we have chosen the Si/SiO$_2$ interface, for which many results are available and which is directly relevant for the understanding of the results obtained in porous silicon. Besides bulk and interface defects, defects in amorphous materials can equally be studied by the EPR technique; to illustrate this, we will rapidly overview the results obtained for the case of amorphous silicon and once again, as we will show you later , a connection with porous materials exists.

3.1 Intrinsic Defects in GaAs [8,9]

Intrinsic defects, such as vacancies, antisites and interstitials are the simplest point defects from a structural point of view; they can exist as isolated defects but they can also interact with themselves or impurity related defects and form associates. These defects are a 'natural' consequence of the growth conditions and are thus generally found as native defects; they can also be introduced artificially by particle bombardment, plastic deformation or thermal annealings. The higher the crystal purity, the more these defect are expected to dominate the extrinsic electrical and optical properties of real crystals. Undoped meltgrown GaAs is a instructive example for this situation with a predominance of anionic antisite defects.

Arsenic antisite (AA) defects [10] are deep defects and thus their properties can be well modelled in a tight binding approximation [11]: AA defects are important native defects in LEC, Bridgman and MBE grown crystals, where they introduce two deep donor states (0/+),(+/2+). In the following we will show that a simple molecular description of this defect allows to predict one of its basic properties, the central hyperfine interaction constant, which can be used for its microscopic identification.

In a molecular approach the wavefunctions of the defect atom in a tetrahedral semiconductor, such as GaAs, can be viewed as hybridized sp^3 orbitals centered on the four first nearest neighbour atoms and s and p orbitals on the central AA atom; to take advantage of the point symmetry T$_D$ of the defect one can construct a basis of symmetrized functions :

$v=1/2\{\sum X_i\}$, $t_1=1/2(X_1+X_2-X_3-X_4)$; $t_2=1/2(X_2+X_3-X_1-X_4)$; $t_3=1/2(X_1+X_3-X_2-X_4)$

v being the basis set function for the irreducible representation A$_1$ and t$_i$'s are the set for T$_2$. The interaction of the AA s function s$_i$ and v, being of the same symmetry , gives rise to the formation of two A$_1$ levels, A$_1$(a) and A$_1$(b), corresponding to the antibonding and the bonding orbitals, associated with the s$_i$-v hybridization respectively:

$$|A_1(b)\rangle = \alpha|s_i\rangle + (1-\alpha^2)^{\frac{1}{2}}|v\rangle \text{ and } |A_1(a)\rangle = (1-\alpha^2)^{\frac{1}{2}}|s_i\rangle + \alpha|v\rangle \text{ with}$$

$$\alpha^2 = \frac{1}{2}\left(1+\frac{\partial}{(\partial^2+V^2)^{1/2}}\right) \text{ and } \partial = \frac{1}{2}(<v|H|v> - <s_i|H|s_i>) \equiv \frac{3}{4}\frac{E_p-E_s}{2} \text{ .}$$

In the same way the T$_2$ gap states t$_i$ of the Ga vacancy interact with the p$_i$ states of the AA defect and form two T$_2$ levels, T$_2$(a) and T$_2$(b).In the neutral charge state the A$_1$(b),T$_2$(b) and A$_1$(a) states are all occupied, the T$_2$(a) being empty and resonant with the conduction band. In the paramagnetic charge state 1+ the AA defect contains one unpaired electron in the A$_1$(a) state, whose properties will determine the associated EPR spectrum. The central hyperfine interaction of this electron with the arsenic nucleus, which has a spin I=3/2 and an isotopic abundance of 100%, is proportional to the s density at r=0: $\Phi_{s^2}(0) = \langle A_1(a)|\partial(r)|A_1(a)\rangle$;

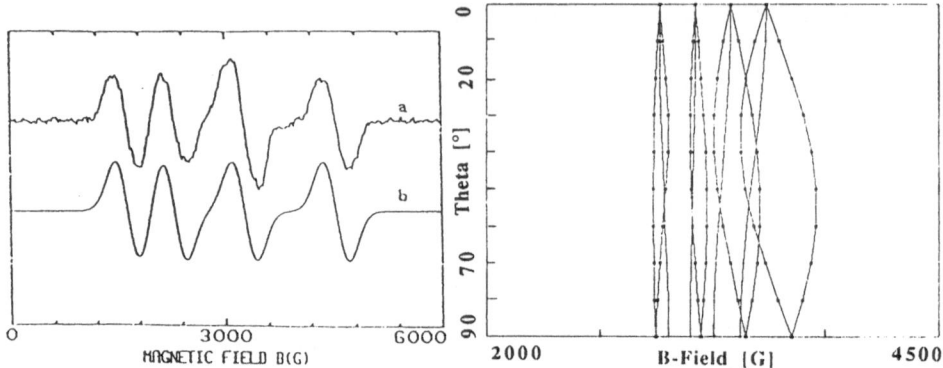

Fig.5 EPR spectrum of As$_{Ga}$ in GaAs   Fig.6 Angular variation of the EPR spectrum of V$_{Ga}$ in
a)experimental; b) simulated   GaAs for the rotation of B in the (110) plane

from this the hyperfine interaction constant A can be calculated:

$$A = \frac{8\pi}{3} g\mu_B g_N \mu_N \left|\Phi_s(0)\right|^2 = A^0\left(1 - \alpha^2\right)$$

With E_s= -17.33eV and E_p= -7.91 eV we find α^2 = 0.84 and thus A = 780x10^{-4}cm^{-1}.
From this result we can simulate the expected EPR spectra for this defect and compare it with
those experimentally found in GaAs (fig.5) . For the simulation of the EPR spectrum we use
the following parameters: S=1/2, I=3/2 (100%), A$_{iso}$= 780x10^{-4}cm^{-1} and g$_{iso}$ = 2.00. The
appropriate Spin Hamiltonian is H = gβ **B S** + A **I S** ; its eigenvalues can be expressed in a
closed form. The selection rule for the magnetic dipole transitions Δm_S=±1, Δm_I=0 gives rise
to a characteristic quadruplet spectrum; spectra very similar to this 'theoretical' one have been
indeed detected in GaAs and have been attributed to the arsenic antisite defect. The real
situation is nevertheless more complex, as different arsenic antisite EPR spectra , for which
the A constant varies up to 25%, have been found experimentally; in this case additional
(ENDOR) measurements are necessary, to establish the corresponding microscopic models.
A slightly more difficult case of a non isotropic defect is the Gallium vacancy defect[12], for
which the point symmetry is lowered to C$_{3v}$ and for which the splitting due to the central
hyperfine interaction becomes comparable to the individual linewidth. In this case once again
S=1/2 and I=3/2; but due to the nonisotropic g-tensors and hyperfine tensors simulations of
the spectra are required to allow the determination of the Spin Hamiltonian parameters from the
experimental spectrum (fig.6); these parameters have equally been theoretically predicted,
requiring however much more elaborate theoretical treatments[12].

3.2 Si/SiO$_2$ Interface Defects

The defects at the interfaces of Si with its oxide have been the object of many studies,
including EPR studies, due to the technological importance of these interfaces in silicon based
microelectronics[13,14]. Whereas electrical techniques are very sensitive for the analysis of
interface defects, they are in general unable to reveal the microscopic structure of the defects
involved; once again EPR has been shown to be a key technique for such studies. As might be
naively expected the nature and concentration of interface defects depend strongly on the
crystallographic orientation of the interface and the oxidation conditions. Detailed studies have
only been obtained for the (111) and (100) Si/SiO$_2$ interfaces[15,16]. The results can be
resumed as follows: high temperature oxidation on well prepared surfaces leads to the
formation of atomically flat interfaces of some 10^{2}Å extension, separated by steps. Typical
defect concentrations, reflecting the mismatch between the Si crystal structure and the
amorphous SiO$_2$, are in the 10^{12}cm^{-2} range. Among the possible intrinsic defects the isolated

silicon dangling bond defects (DB) have turned out to be the dominant interface defects. The symmetry and properties of these defects are different for (111) and (100) interfaces. As the dangling bond defect at the interface is a two electron system, paramagnetic in the neutral charge state, it is suitable for EPR studies. More detailed EPR studies have revealed in addition further properties of the interface, such as strain and strain distribution and their relation with the oxide quality, the existence of a transition region with substoichiometric oxides, the sensitivity of interface defects to electrical fields, their possible passivation by hydrogen, silicon precipitations in the interface region etc. The EPR parameters of the DB center are sensitive to these interface properties and allows thus a characterization of the surface.

A first approach to the electronic structure of the dangling bond defect, which has a very localized electronic wavefunction, can be obtained in a molecular model in the tight binding approximation[17].

The only treated case is that of the DB at the (111) surface: let us consider a silicon atom at this surface, backbonded to three tetrahedrally oriented first nearest neighbour silicon atoms (fig.7).The properties of the bulk material are dominated by the coupling of pairs of sp^3 hybrides forming nearest neighbour bonds, which leads to the formation of bonding and antibonding states; these are further broadened by weaker interbond interactions leading to the formation of the valence and conduction bands respectively. The rupture of one bond, such as for a (111) surface atom, leaves one uncoupled sp^3 orbital, whose energy is midway between the bonding and antibonding states. Interband coupling will modify the pure sp^3 character of this DB and lead to some delocalization on the next nearest neighbours.

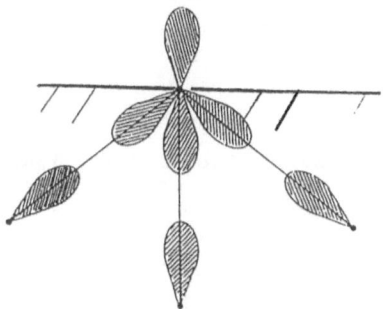

Fig.7 The isolated DB defect

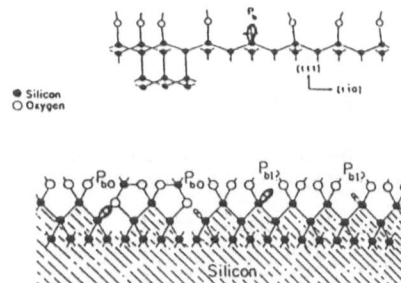

Fig.8. DB defects at (111) and (100) surfaces

Within this model, the g-tensor and the central hyperfine tensor of this defect in its paramagnetic neutral charge state can be calculated.

Let $|0\rangle$ be the sp^3 dangling bond state occupied by one electron and $|n\rangle$ the bonding and antibonding states belonging to the backbond k with energies E_A and E_B :

$$|n_{A,B}\rangle = \frac{|k\rangle \pm |k'\rangle}{\sqrt{2(1 \pm S)}}$$

$$E_{A,B} = E_0 \pm \frac{|\beta|}{1 \pm S}$$

ß being the matrix element $ß = \langle k|H|k' \rangle$; S the overlap $S = \langle k|k' \rangle$; E_0 the sp^3 energy.
The deviation of the g-tensor from the free electron value due to the spin orbit interaction is in first order:

$$(\Delta g_0)_{ij} = -2\,\mathrm{Re}\sum_{n=0}\frac{\left\langle 0|L_i|n \right\rangle \left\langle n\left|\sum_N \lambda(r_N)(L_N)_i\right|0\right\rangle}{E_n - E_0}$$

$$= \frac{\sum_k \left\langle 0|L_i|\Psi_{B,k}\right\rangle\left\langle\Psi_{B,k}|L_i|0\right\rangle}{E_B} - \frac{\sum_k\left\langle 0|L_i|\Psi_{A,k}\right\rangle\left\langle\Psi_{A,k}|L_i|0\right\rangle}{E_A}$$

$$= \frac{3\lambda S}{|ß|}\begin{bmatrix}0 & 0 & 0 \\ 0 & 1 & 0 \\ 0 & 0 & 1\end{bmatrix}$$

The first conclusion is that Δg will have only two components ($\Delta g_{/\!/}$, $\Delta g_\perp$, with $/\!/$ and $\perp$ referring to the <111> axis of the DB center); one of them is zero ($/\!/$) and the other one is positive ($\perp$). With $|ß| = 3\,\mathrm{eV}$, $\lambda = 0.02\,\mathrm{eV}$ and assuming a reduction factor of 60% due to delocalization we can estimate the numerical values of the g tensor to :

$$\Delta g_{/\!/} = 0 \quad \text{and} \quad \Delta g_\perp = 0.0085 \quad \text{i.e.}$$

$$g_{/\!/} = 2.0023 \quad \text{and} \quad g_\perp = 2.0108$$

Within the same model we can also calculate the hyperfine interaction tensor A, which will also have axial symmetry with two components $A_{/\!/}$ and $A_\perp$:

$$A_{/\!/} = a + 2b \quad \text{and} \quad A_\perp = a - b$$

with $\quad a = \dfrac{16\pi}{3}g_n ß_n ß|\Psi_{3S}(0)|^2 \quad$ and $\quad b = \dfrac{4}{5}g_n ß_n ß < \dfrac{1}{r^3} >_{3p}$

Assuming the same reduction factor of 60% we obtain the numerical values:
$a = 170\times10^{-4}\,\mathrm{cm}^{-1}$ and $b = 13\times10^{-4}\,\mathrm{cm}^{-1}$. How do these values compare to the experimental results?

Experimental results:

The electronic structure of the DB center at the Si(111)/SiO$_2$ interface, also called P$_b$ center, has been actively studied by EPR spectroscopy. Its first observation has been reported in 1971 by Nishi[18], who has observed the corresponding EPR spectrum and shown its localization within some 400Å of the interface. The microscopic structure of this defect was not clear at that time. One key experimental result for the identification of its microscopic structure was the observation of the central hyperfine interaction with one Si nucleus by Brower [19]. He was able to show that the main P$_b$ center line is accompanied by a low intensity anisotropic doublet structure, the intensity of which was not incompatible with the Si^{29} isotope abundance of 4.7%. The observation of the P$_b$ center line and a fortiori of its 50 times lower intensity hyperfine spectrum is difficult due to sensitivity reasons: the P$_b$ center concentrations of typically $10^{12}\,\mathrm{cm}^{-2}$ require the use of large sample surfaces of $\approx 10\,\mathrm{cm}^2$, which can only be incorporated in an EPR X-band cavity by using stacks of samples, difficult to handle. An important experimental result is the trigonal symmetry of the P$_b$ defect; this property can be used to test the alignment of the interface with the crystallographic orientations of the substrate.

Some representative theoretical and experimental results are summarised in the following table :

g//	g⊥	a(10^{-4}cm^{-1})	b(10^{-4}cm^{-1})	Ref.
2.0023	2.0108	170	13	Lannoo et al [17]
--------	---------	120	27	Cook et al [20]
--------	--------	133	17	Edwards [21]
2.0020	2.0070	-------------	-------------	Nishi [18]
2.0012	2.0086	--------	-------------	Caplan et al [22]
2.0013	2.0086	----------	------------	Poindexter [16]
2.0016	2.0090	105	20	Brower [19]
2.0013	2.0090	----------	-----------	Stesmans [53]
2.0011	2.0080	99	21	Carlos [24]

Table I: Comparison of the predicted and experimental g-values and hyperfine constants.

Comparing the simple theoretical predictions with experiment, we see that the main features of the model are confirmed by the experiment. However, we have a scatter of the numerical values, which is beyond the experimental error, which -when care is taken -can be below 0.0001; the values of Nishi should be taken apart , as in this study the angular dependence of this defect was not measured with sufficient precision. A possible explanation for this scatter is the sensitivity of the EPR parameters of the P_b center to the local configuration ; if the interfaces are of different quality as concerns the stress, this will lead to variations of the perpendicular component of the g-tensor and the isotropic part of the hyperfine interaction 'a'.

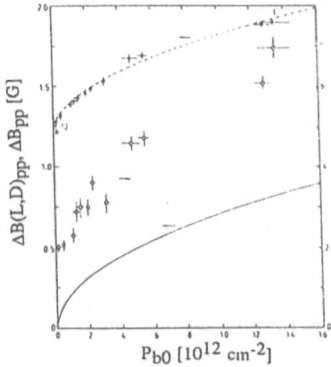

Fig.9 Dipolar broadening of P_b (ref.26)

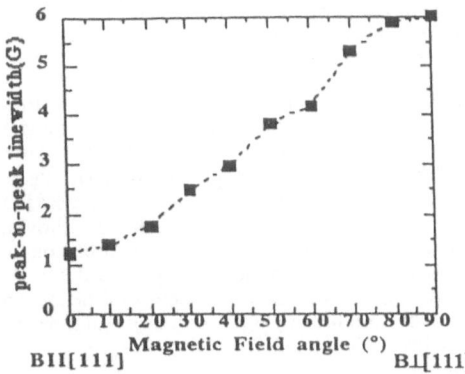

Fig.10 K-band Linewidth vs orientation for P_b after (27)

In fact, more refined calculations[20,21] have supported this idea: Cook and White [20] have published cluster calculations, which show that the value of the 'a' parameter is very sensitive to displacements of the DB atom normal to the (111)surface: a relaxation of Δz by 0.01Å reduces its value by 15%. Due to this strong dependence the parameter 'a' is an excellent probe for the presence of interface stress and we can anticipate, that its value will depend on the type of the oxide grown.

Many more details of the P_b center have been studied since its first EPR observation: to mention just some: the dependence of its concentration on the oxidation conditions, its evolution with post growth treatments, its possible passivation by hydrogen , its relation with degradation of the interface, the position of the associated levels $P_b^{0/+}$ and $P_b^{-/0}$ in the gap , the correlation of P_b with the total interface state density...We would like to discuss in more detail two further aspects of the P_b centers, which are instructive for the discussion of the results in porous silicon: (i) the influence of the dipolar interaction on the lineshape [25,26] and (ii) the manifestation of strain distributions at the interface as deduced from the EPR [27]spectra.

(i): the dipolar interaction between P_b centers depends on their relative distance and is only expected to be important for close neighbour pairs. Splittings of the EPR spectrum can be expected in the case of very close pairs (n<6) or line broadening effects in the case of more distant pairs (6<n<10)[25]. If the P_b centers are statistically distributed, the linewidth can then be related to the average defect concentration. Typical concentrations for high temperature thermal oxides are $10^{12} cm^{-2}$ with a variation by a factor of 10 to 0.1 depending on the oxidation conditions. On the other hand, if the concentration of the P_b defect is known, the question, whether the P_b centers are distributed statistically over the entire surface or whether local clusters might exist, can be addressed. Experimentally close pairs, giving rise to line splittings, have not been observed up to now; but linewidth variation due to high concentrations of the P_b center have been observed; they have been interpreted by distant dipolar interaction [26]. Figure 9 shows the results obtained on one and the same sample, which has been submitted to hydrogenation and depassivation treatments in order to modify the concentration of the paramagnetic P_b centers for a given interface [26]. The concentration varied between 1×10^{11} and $1.5 \times 10^{13} cm^{-2}$. These results seem to indicate a universal relationship between the EPR linewidth (for $B // [111]$) and the P_b concentration. Extrapolation to zero concentration gives a natural linewidth in the absence of dipolar interaction of 1.25G. Given, that the EPR linewidth for this particular magnetic field orientation is not sensitive to strain, the same quantitative relationship should hold for all $Si(111)/SiO_2$ interfaces.

In cases, where the surface area is difficult to assess (porous silicon for example), the simultaneous determination of the P_b center linewidth and the number of the centers might then be used to deduce a lower limit for the surface area . Dipolar interactions are equally revealed by the change in lineshape, which develops broad shoulders for concentrations higher than $\approx 5\%$ [26].

(ii)Whereas in general the linewidth of the EPR spectrum of a point defect in silicon is not dependent on the orientation of the static magnetic field relative to the crystal axes, the P_b center shows strong angular variations. An example is given in figure 10 published by Brower for 1000°C oxidized p⁻-type sample[27]. We see that the total peak-to-peak linewidth as measured in K-band varies from 1.25G for $B // [111]$ to up to 6G for $B // [110]$. Measurements at different microwave frequencies have shown, that the broadening is proportional to the microwave frequency; from this it can be concluded that its origin is a distribution of $g_\perp$ values. Such a distribution is expected if the P_b centers experience different strains at the interface. Using the EPR parameters of the P_b center at a given interface we can thus evaluate the strain from the average value of $g_\perp$ and the strain distribution from the angular dependence of the linewidth.

3.3 Intrinsic Defects in Amorphous Silicon

In the previous two cases we have shown examples of EPR studies of intrinsic defects in the bulk and at surface/interfaces of monocrystalline silicon. EPR spectroscopy has equally turned out to be efficient for the analysis of defects in amorphous or disordered silicon. As this is an issue, which is still under discussion for the properties of porous silicon, we will recall very briefly the basic results; different review articles exist, which give detailed information on this subject, see for example [28].

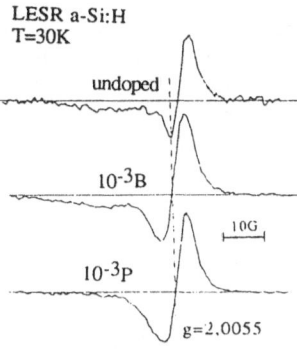

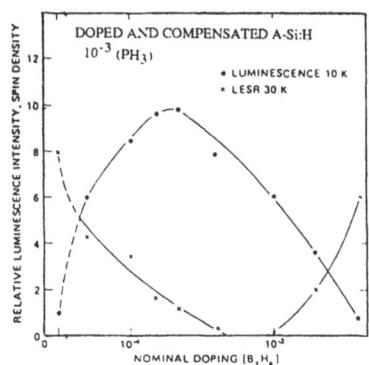

Fig.11 EPR spectra in a-Si (ref.30) showing DB and light induced signals

Fig.12 Correlation between PL & DB density (ref.31)

Amorphous silicon (a-Si) which can be prepared by evaporation or sputtering at low temperatures is characterized by the fact that whereas the nearest neighbour configuration is almost identical to that of monocrystalline material the radial distribution function shows strong bond angle and dihedral angle variations for neighbours further away than 5A. As a result, amorphous material is highly defective with defect densities of $\approx 10^{20} cm^{-3}$. Three paramagnetic defects have been observed by EPR spectroscopy (fig.11): in undoped material a single line spectrum of linewidth ≈ 5 to 8 G and a g factor of 2.0055 was observed[29]. This spectrum has been attributed to the neutral dangling bond defect of a tri-coordinated Si atom. It can be considered as a sort of equivalent of the P_b center in a volume version. However as the preferred orientation is lost in amorphous material the spectrum is isotropic. In doped samples light induced EPR has shown the existence of two additional paramagnetic defects, which have been attributed to band tail states associated with the conduction or valence bands[30]. Their respective g-factors are 2.004 and 2.01.

In hydrogenated a-Si the apparent defect density of the DB center is typically reduced by some two orders of magnitude as compared to non-hydrogenated material. The reduction is mainly due to the passivation of the DB defects by hydrogen bonding. Exodiffusion of the hydrogen by annealing in the 450°C range allows to reveal the total DB density.

DB centers do equally intervene in the recombination processes of free carriers. The photoluminescence efficiency of a-Si-H is strongly influenced by the DB centers, which are recombination centers. As shown in fig.12 an increase in the DB density from 10^{16} to 10^{18} cm^{-3} decreases the Pl efficiency by three orders of magnitude[31]. Recombination via the neutral DB center is also weakly radiative with an emission centered at 0.9eV.

The occurrence of the g=2.0055 EPR spectrum can also be taken as evidence for the existance of disordered or amorphous silicon material.

4. APPLICATION OF EPR TO THE CASE OF POROUS SILICON

4.1 Defect Analysis

As explained in the introductory part on EPR spectroscopy the characterisation of a paramagnetic defect proceeds via the determinantion of its spin Hamiltonian parameters, i.e. its electron spin S, g-tensor, A tensor etc. For a non isotropic center this requires to perform an angular variation study of the EPR spectrum in order to determine the six independent elements of the g and A tensors. In cases of defects with a high point symmetry, like C_{3v}, a rotation in one suitably chosen plane might be sufficient. It is clear, that the measurement of an EPR spectrum for one or two orientations is insufficient for its characterization and identification; it has only a meaning, if the defect has already been identified before. Further, the numerical values of g and A have to be determined with sufficient numerical accuracy, which requires special care in the case of centers such as the P_b center. A statement such as "a narrow line near g=2" contains no real information given the number of different defects, which correspond to such a description.

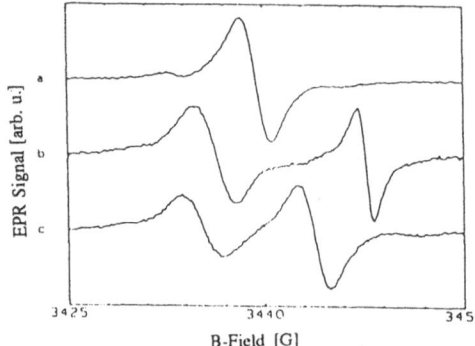

Fig.13 EPR spectra in p^+ porous Si for three orientations
(a):B∥[001], (b): [111] and (c): [110]

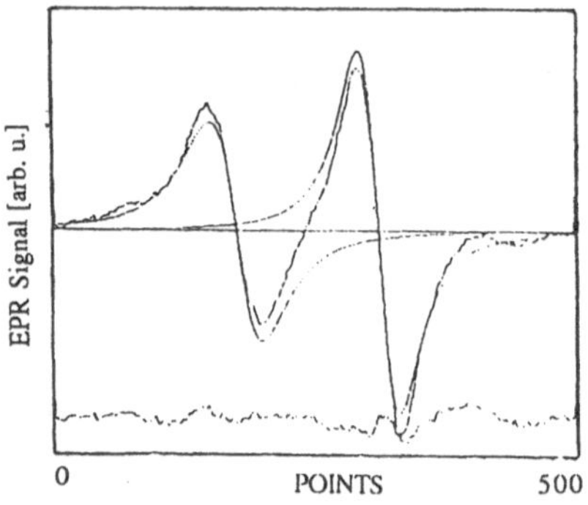

Fig.14 Decomposition of the exp.
spectrum with Lorentzian lines

4.1.1. Defects in as Prepared n^+,p^+,p,p^- Layers

Different groups have studied the defects in as prepared porous silicon by X-band EPR spectroscopy at room temperature[32-36]. We know now, that the term of porous silicon covers a wide variety of different structures depending on the substrate material, its doping level, the electrochemical preparation conditions, the porosity, the layer thickness, the fact whether the layer is free-standing or supported, the controlled or non controlled post preparation conditions etc. Nevertheless, it is interesting to start the discussion of the results with a global approach corresponding to defects in as-prepared porous silicon. Anyway, in most of the published EPR studies only partial information on the material aspects is given.

In spite of a undeniably complex situation the results obtained seem to be rather homogenous at a first glance; they can be resumed as following: n^+, p^-, p and p^+ layers exposed to air show independent on porosity (40% to 85%) at thermal equilibrium conditions at 300K one dominant anisotropic EPR spectrum. It is characterised by a spin S=1/2, trigonal symmetry of the g-tensor as related to the [111] directions of the substrate and absolute values of the g-tensor. The most significant published results based on the analysis of angular dependencies are gathered in table II.

In order to be more specific and allow to appreciate the experimental accuracy with which the defect parameters can be determined we will present in more detail the results for a 30μm thick supported layer prepared from a (100) p^+ substrate with a gravimetric porosity of 60%. For a rotation of the magnetic field in the (110) substrate plane, we observe for an arbitrary orientation a three line spectrum, which for B∥[001] , B∥[111], B∥[110] reduces to a one line and two line spectrum respectively(fig.13).

$g_{/\!/}$	$g_{\perp}$	$A_{/\!/}$	$A_{\perp}$	doping	author	ref.
2.0023	2.0086	-------	-------	p⁻	Mao 93	[32]
2.0022	2.0078	-------	-------	p⁻	Yokomishi 93	[33]
2.0024	2.0080	-------	-------	p⁻	Ushida 93	[34]
2.0020	2.0088	-------	-------	p⁻(111)	Ushida 93	[34]
2.0018	2.0085	146	78	p	Rong 93	[35]
2.0017	2.0091	139	73	p⁺	v.Bardeleben 93	[36]

Table II: g-values and hyperfine constants for different types of porous Si.

The splitting of the lines is of the order of the linewidth. Further, the linewidth is angular dependent. For most of the orientations, Spectrum simulations are necessary in order to decompose the spectrum in its individual lines; then the Spin Hamiltonian (fig.14) parameters can be determined from the line positions. This procedure reveals equally a partially resolved superhyperfine interaction corresponding to a doublet structure associated with each transition. This structure is best seen for $B/\!/[001]$.The complete angular dependence of the three central lines for each orientation are then simultaneously fitted with a Spin Hamiltonian for S=1/2 in C_{3v} symmetry with four equivalent centers. The result of such a fit for both, the central lines (I=0) and the hyperfinelines (I=±1/2) is shown in figure15. The fit gives the principal g-values of $g_{/\!/}$ = 2.0020±0.0001 and $g_{\perp}$ = 2.0086±0.0001 with $/\!/$ and $\perp$ referred to the four equivalent [111] orientations of the substrate (p+ sample with 60% porosity).

If we examine the spectrum at a higher amplification (typically x50) we observe in addition an anisotropic central hyperfine interaction with a nuclear spin I=1/2 for each of the central lines(fig.16). They are fitted with the same procedure and the following principal values of the A tensor, which has equally C_{3v} symmetry, are obtained: $A_{/\!/}$=(139±3) x 10^{-4}cm^{-1} and $A_{\perp}$ =(73±3) x 10^{-4}cm^{-1}. The intensity ratio of the hyperfine lines (I=1/2) and the central lines (I=0) can be used to identify the chemical nature of the atom on which the electron is bound; numerical integration gives a ratio of 0.041. This ratio identifies the central atom as the silicon atom, which has a natural isotopic abundance of [^{29}Si] / [^{28}Si + ^{30}Si] = 4.7/95.3= 0.048.

The symmetry and the absolute values of the spin Hamiltonian parameters are very close to those of the P_b center at the (111) Si/ SiO$_2$ interface and consequently the defect observed in porous silicon has been identified with the (111)P_b center.

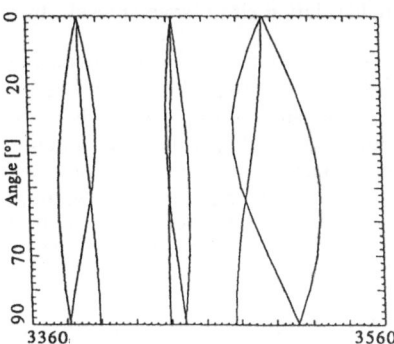

Fig.15 g-factor variation for the P_b defect for a variation of B in a (110)-plane.

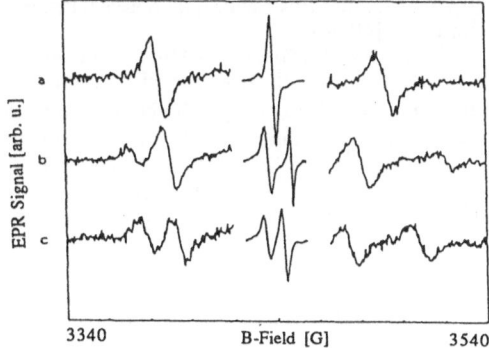

Fig.16 P_b center hyperfine interaction for $B/\!/[001],[111],[110]$

The trigonal symmetry identifies also unambiguously the crystallographic nature of the Si surface at the oxide interface as (111); at (100) Si /SiO$_2$ interfaces, the only other surface studied in detail for bulk silicon, two different P$_b$ centers P$_{b0}$, P$_{b1}$[22] have been observed simultaneously. The dominant defect at the (100) interface, P$_{b1}$, has a lower symmetry and a different g-tensor; P$_{b0}$ is actually a (111) interface defect at a non atomically flat (100) interface (table III).

	(100) interface Pb0	(100) interface Pb1	(111) interface Pb
g0	2.0060	2.0032	2.0013
g1	2.0015 60° off<100>	2.0012 30° off<100>	2.0012 II<111>
g2	2.0080	2.0076	2.0086
g3	2.0087	2.0052	2.0086

Table III: EPR g-values for the different P$_b$ centers in oxidized (111) and (100) silicon wafers.
The value with B$_0$ normal to the wafer surface is denoted g$_0$. For further information see ref.[22].

By comparison with a spin standard sample we have determined the total number of the neutral P$_b$ centers in the various sample types. Typically the n$^+$, p$^-$ and p type layers have an order of magnitude higher spin concentration than the p$^+$ layers. However as the P$_b$ center is an interface defect, it is meaningless to refer its concentration to the sample volume; the relevant information is its surface concentration. Evidently it is not the external surface of the sample, which must be taken as a reference, but the real surface, which has been determined[23] for p$^+$ layers to 200m^2/cm^3 ; this value is independent on the porosity for porosities between 35% and 70%. We will assume in a first approximation, that this specific surface value holds also for the n$^+$ layers, whereas p-type layers have higher specific surfaces of 600m^2/cm^{-3}. Further, assuming that approximately 50% of the surface will be made up by (111) or equivalent planes, we find P$_b$ center concentrations of 5x10^{9}cm^{-2} for the p$^+$ layers and 5x10^{10}cm^{-2} for the n$^+$ and p layers.

Such low defect concentrations are astonishing, as the oxide formed at room temperature is expected to have high ($\geq$10^{12}cm^{-2}) interface state concentrations. It is clear, that the measured P$_b$ center concentrations are just lower limits as we had in a first approximation assumed, that all the internal surfaces consisted of oxidized (111) planes. What we really need to know, is the area of the oxidized (111) oriented specific surface or its fraction relative to the total surface. From an impurity analysis by nuclear reaction techniques, the fraction of oxidized surface in as-prepared, but air exposed samples has been determined to $\approx$ 10% to 20%, but this fraction increases when the sample is left at air for several weeks by a factor of $\approx$5 [40, 45].

Nevertheless, an independent information of the concentration of the neutral P$_b$ centers can be obtained directly from the EPR spectrum: as described before [25] the peak-to-peak width of the EPR line corresponding to the magnetic field orientation B||[111] depends on the defect concentration due to dipole-dipole interaction. For low neutral P$_b$ center concentration (<1x10^{12}cm^{-2}) the linewidth is equal to 1.2G; for higher concentrations it increases continuously up to 2.0G for P$_b$ center concentrations of 1.6x 10^{13}cm^{-2}; the linewidth of 2.0G is a higher limit found in high temperature thermal oxidation studies. In the porous silicon samples studied here the peak to peak linewidth is between 1.2G and 1.6G which implies, that its surface concentration must be between 10^{12} cm^{-2} and 10^{13} cm^{-2}.

4.1.2. EPR of Photoexcited Free Electrons: an Indication for Spatial Separation of Photoexcited Carriers

Photoexcitation with near infrared (1μm) or visible excitation at temperatures below T=50K gives rise in both n$^+$ and p$^+$ material to the same high intensity anisotropic EPR spectrum (fig.17)[37]. The peak to peak linewidth is of the order of 2.0G. The g-factor is below 2.00 and has values of 1.9985 and 1.9988 for B//[001] and B//[110] respectively. The effective g-factor of the EPR spectrum is close to the one of conduction electron spin resonance CESR in n-type Si g=1.9988. In spite of the high intensity of the spectrum no hyperfine structure could be found. From the high concentration, its g-factor, the absence of hyperfine splitting and its observation in both n+ and p+ material we ascribe this spectrum to the spin resonance of optically excited free electrons in quantum confined Si structures, in which the degeneracy of the six X-valleys has been lifted by the uniaxial stress present in this material. The uniaxial stress perpendicular to the (100) plane of the substrate will lead to an anisotropy of the resonance spectra and may further give rise to valley repopulation effects. The anisotropy of the EPR spectrum demonstrates directly, that the spectrum originates from the monocrystalline Si layer and cannot be associated with the amorphous oxide layer present at the surface.

For temperatures below 50K the spectrum can be observed easily due to long lifetimes τ of the photoexcited electrons - τ≈100s at 4K- but for temperatures higher than 50K it is no longer observable due to a strong reduction in the lifetime in the 50±10K temperature range. The attribution of the spectrum to triplet S=1 excitons, a model recently proposed[38] for the interpretation of the visible photoluminescence band at 1.8eV, can be excluded by the very different lifetimes: the lifetime of the proposed triplet excitons has been measured as 10ms, whereas the lifetime of the photocarriers detected in EPR is of the order of 100s. However both lifetimes, that of the free electrons and the supposed triplet groundstate decrease strongly in the same temperature range at ≈50K and with a very similar activation energy(fig.18).Our results indicate, that either free holes become available at this temperature giving rise to recombination of the photoexcited electrons or the free electrons are captured by defects with a thermally activated capture barrier corresponding to 50K.

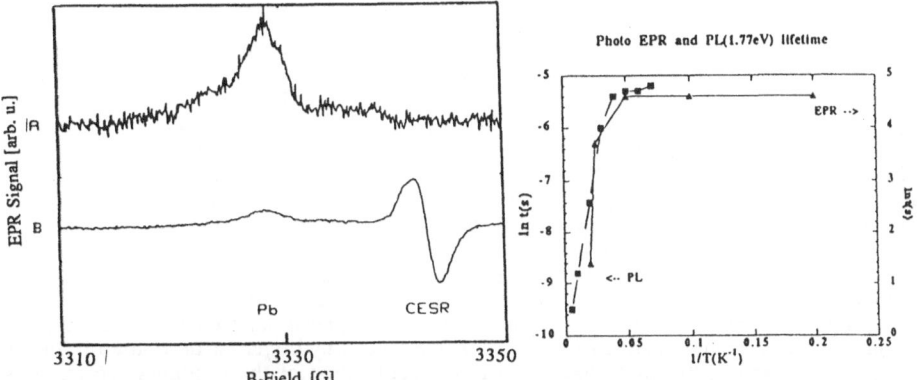

Fig.17 EPR spectrum at T=4K of a p+ sample at thermal equilibrium (A) and under 1μm photoexcitation(B)

Fig.18.temperature dependence of EPR-lifetime and PL-lifetime [38]

4.1.3. EPR of the Dopants B,P

The dopants originally present in the substrate material are expected to remain in the porous Si layers at the same concentration, as no preferential dissolution has been reported. In fact preliminary nuclear reaction analysis results on p+ type layers, which show an unreduced boron concentration in the porous layer as compared to the substrate, have confirmed this view[40]. The presence of the dopants in the n+ and p+ layers can also be tested by EPR and this in two ways: first via the associated electrical conductivity of the layers, which can be appreciated from the cavity loading, and second, directly by the spin resonance of the (paramagnetic) neutral dopants. Surprisingly, the introduction of both, the p+ and n+ free standing samples, in which the effect of the substrate is eliminated, lead even at room temperature to only a small reduction of the cavity Q factor , demonstrating the <u>absence of free</u> <u>carriers</u>; the material behaves as an intrinsic one. A possible explanation is, that the carriers are frozen out on deep defects with thermal ionisation energies higher than ≈0.6 eV. The loading of the cavity is practically independent on temperature down to 4K. This is in strong contrast with bulk silicon ,where carrier freeze out is only observed for temperatures below 40K.

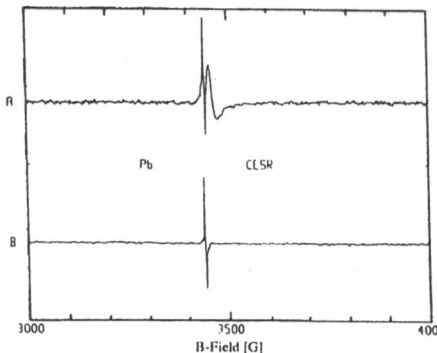

Fig.19 EPR spectrum of a supported (A) and a free standing (B) n⁺ porous layer at 300K

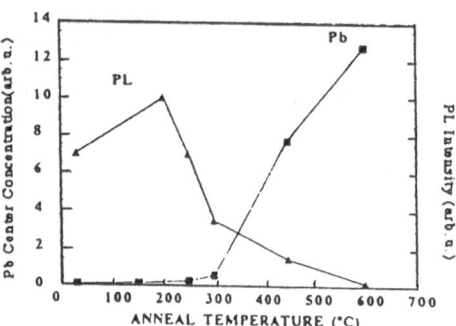

Fig.20 [P_b] variation as function of isochronal annealing in vacuum for a p⁺ layer

Whereas it is difficult to observe the EPR of the neutral B acceptor in bulk silicon due to strain broadening effects [39], this is not the case for P , which can easily be detected with high sensitivity in n-type material for T<40K. Inspite of a careful search in the temperature range of 4K to 300K no donor related EPR spectrum is observed in the n⁺ porous Si samples. Figure 19 illustrates this result for the case of n⁺ degenerate layers: it shows the corresponding EPR spectra at T=300K of a supported (layer + substrate) n⁺ sample (p=80%,thickness t=10μm) (fig19a), which displays a broad line corresponding to the conduction electron spin resonance (CESR) of the bulk substrate and the sharp line of the P_b center in the porous layer. In fig 19b we show the corresponding spectrum of a free standing (no substrate) n⁺ sample (p=80%, t=50μm): the CESR line is no longer observed. This demonstrates the absence of free carriers even at 300K, as has been previously directly observed by electrical measurements. Among possible explanations we have considered the strong hydrogen contamination of the as-prepared layers; hydrogen incorporation in the % concentration range[40] might have lead to the passivation of the P donors. To test this hypothesis we have performed thermal annealings in the 300°C to 400°C range, where depassivation of donors is known to take place in bulk n-type Si[41]. But still no donor resonance is observed in the annealed layers, a strong argument against the hydrogen passivation model. Further, previous results on bulk Si had anyway shown that in n⁺ Si only incomplete P passivation is possible [41]. We conclude from these negative results, that the Fermi level in n⁺ porous Si is no longer pinned by the donor impurity, but must have moved to the midgap position; this conclusion is also supported by the EPR observation of the neutral P_b centers under thermal equilibrium conditions, not expected for n⁺ or p⁺ material.

4.1.4. Influence of Thermal Annealing under Ultra High Vacuum

The effect of vacuum annealing or high temperature annealing under different atmospheres such as N₂, O₂ has been studied by different authors[32,34,36,42,43]. We have performed a series of isochronal (t=30min) thermal annealings in vacuum in the 100 to 600°C temperature range on a series of 40μm thick, 80% porosity n^+ and p^+ samples and determined the variations in the neutral Pb_0 center concentration[36,42,43] (fig.20). Thermal annealings, which are known to degrade the luminescence efficiency, are expected to modify the Pb center concentration by different means: first, already existing but hydrogen passivated Pb centers, which are EPR inactive , will be depassivated at typically 400°C due to hydrogen exodiffusion [44]; second, previously hydrogen passivated surfaces can be depassivated - at 400°C for (111) surfaces and ≈600°C for (100) surfaces due to the different bonding configurations Si-H and Si-H$_2$ respectively- . When the sample is again exposed to ambient condition after the annealing these surfaces will oxidise and additional Pb centers will be created. As shown in fig.20 the neutral Pb center concentration for both n^+ and p^+ type samples is constant up to 300°C. For an annealing at 450°C we have a strong increase by ≈ 10^2 for all types of samples; for still higher annealing temperature of 600°C the Pb center in the p^+ samples increases further by a factor of 2, whereas it decreases for the n^+ samples. We attribute the sharp increase at 450°C mainly to a depassivation of existing Pb_0 centers as we know from Nuclear Reaction Analysis, that the fraction of the oxidized surface is much higher than 1% [40,45]. Further, previous results concerning the depassivation kinetics of Pb centers in this temperature range [44] support this model.

The EPR linewidth for the annealed samples is smaller than for the as-prepared one's. For B‖[111] it is still between 1.2G and 1.5G in these annealed samples, which allows once again an estimation of the oxidized fraction of the surface. Typical sample volumes are 3×10^{-4} cm^3 and the absolute number of Pb centers measured after a 450°C anneal ≈10^{15} spins; assuming a surface concentration of 10^{12}cm^{-2} we can deduce that the oxidized surface must be greater than 10^3cm^2; this number is to be compared to the total internal surface of such a sample, which is - based on an internal surface to volume ratio of 200m^2/cm^3 - equal to 6×10^2cm^2; from this we can conclude, that the total surface must consequently be oxidized. But as the (111) surfaces are only a fraction of the total internal surface, it follows, that the interface defect density at the (111) Si/SiO$_2$ interfaces must exceed 10^{12}cm^{-2}.

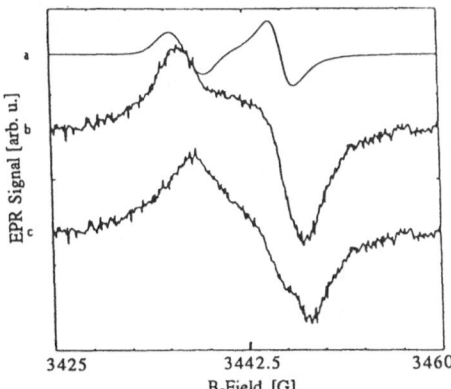

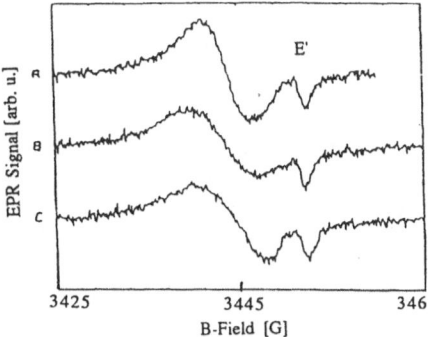

Fig.21 Total EPR spectrum for B‖[110] of a 80% porosity 450°C annealed p^+ sample (b); subtraction of the Pb center spectrum (a) shows the presence of an isotropic line (c) with g=2.0055 and linewidth 8G.

Fig. 22 EPR spectra for (A): B‖[001], (B): B‖[111], (C): B‖[110] of an anodically oxidized p+ layer of 65% porosity, displaying Pb centers from different interfaces and E'.

Whereas the thermal anneal under ultrahigh vacuum leads to an increase of the Pb center concentration for both low and high porosity n^+ and p^+ samples, the high porosity samples show after a 450°C annealing a modified EPR spectrum (fig.21)[36,37] ; this new EPR spectrum (fig.21b) can be decomposed into the P_b center spectrum (fig.21) and an additional isotropic spectrum characterised by a g-factor of 2.0055 and a linewidth of 8G. The decomposition of this spectrum is confirmed by the measurement of the central hyperfine doublet associated with the P_b center spectrum. The new isotropic spectrum can be identified from its g-factor and linewidth with the dangling bond defect in amorphous or disordered silicon. From the EPR measurements alone it is not possible to decide whether the dangling bond defect is a consequence of the thermal annealing or whether this defect existed already in the as-grown porous silicon but in a hydrogen passivated diamagnetic configuration . In this second case once again the thermal anneal would correspond to the depassivation of the dangling bond center by hydrogen exodiffusion. In fact, the annealing conditions (T=450°C) are also typical for the depassivation of hydrogen passivated dangling bonds in hydrogenated amorphous silicon. However Raman studies on the vacuum annealing effect on the structural change in porous Si indicate, that the disordered parts are formed during the annealing process and do not exist in the as-prepared layers[46]. The formation of amorphous inclusions, as manifested by the observation of the dangling bond defect, has also been reported for open furnace annealed p-type porous silicon[47]; in this case the dangling bond defect could be observed already after a 200°C annealing. The dangling bond concentrations observed in our samples referred to the entire sample volume are of the order of 10^{19}cm^{-3} if the porosity of the sample is taken into account. Such volume concentrations are typical for annealed hydrogenated amorphous silicon. This result seems at first contradictory to the conclusions obtained from the P_b center analysis, but the only consistent interpretation seems to be that the disordered regions have in fact much higher dangling bond concentrations.

4.1.5. High Temperature (1000°C) Thermally Oxidized Layers

The analysis of as-prepared porous Si by techniques such as EPR, NRA, IR absorption had shown, that the surface of samples exposed to air is covered to a high fraction by a thin SiO$_2$-layer [40,45]. The quality of the interface is an important parameter in the sense that a high interface defect (P_b) concentration will be detrimental for the radiative emission efficiency; indeed, P_b centers have high capture cross sections for electrons and holes ($10^{-14}...10^{-15}$cm^2), allowing efficient non radiative recombination by phonon emission[48]. It is well known from oxidation studies on bulk Si, that oxides with low interface state densities can be grown at high temperature in controlled oxygen atmospheres. For this reason we have studied a series of p$^+$ samples of 65% initial porosity oxidized at 1000°C for 1 to 10 minutes in dry oxygen at a pressure of 12mbar[37,42,43].

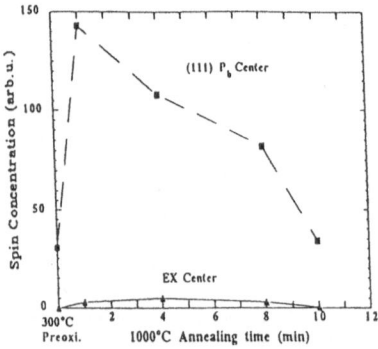

Fig.23 [P_b] and [EX] variation with oxidation time

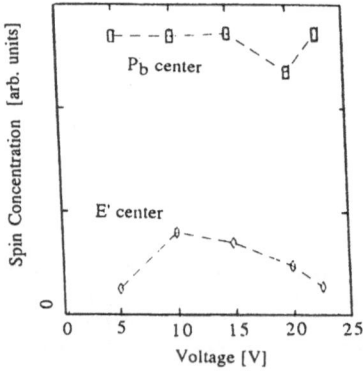

Fig.24 Spin concentrations versus voltage for the P_b and the E' center for anodically oxidation of a p$^+$ layer with a porosity of 65%.

The EPR results obtained for these samples are shown in fig.23. The layers were preoxidized at 300°C for 60min. After 1 min oxidation the samples contain a strongly increased concentration of P_b centers with an absolute concentration estimated to $2 \times 10^{12} cm^{-2}$; this value is once again referred to the total internal surface of the porous Si ($200 m^2/cm^3$) and represents only a lower limit. The P_b center concentration is two orders of magnitude higher than in the non purposely oxidized, aged samples and thus a first conclusion is, that high temperature oxidation - under these conditions- is not adequate for the preparation of layers, in which high radiative recombination efficiencies are searched for. The reason for this 'disappointing' result is that the high temperature oxidation leads to an exodiffusion of hydrogen, which had passivated most of the P_b centers present in as-prepared samples. Combined oxidation /passivation procedures seem to be necessary in order to improve the results. The P_{b1} center, the dominant interface defect for (100) Si/SiO$_2$ interfaces is not observed in these samples even though it can be assumed, that the entire surface of the layers will be oxidized. For longer annealing times the P_b center concentration decreases approximately linearly with the oxidation time. The decrease may be due to an improved interface quality, if the total surface is assumed to stay constant.

The 1000°C oxidized samples present in addition to the P_b center a different EPR spectrum. The spectrum is characterised by an isotropic line with g-factor of 2.0025, and a linewidth of 0.5G. This spectrum seems to be identical to the EX center observed by Stesmans [15]; the exact microscopic structure of this defect has not been identified up to now, but the hyperfine interaction observed indicated an interaction with three equivalent Si atoms. This defect is located in the SiO$_2$ layer, not directly at the interface. The concentration of this defect varies only little with the oxidation time.

Surprisingly no dangling bond defects characteristic for disordered and amorphous silicon were observed. A possible explanation is given by our previous results[36], in which we had shown that the formation of amorphous silicon in porous silicon is only effective in high porosity material ($p \approx 85\%$).

4.1.6. Anodically Oxidized p$^+$ Layers

Similar to high temperature oxidation the process of anodic oxidation can be used to reduce the lateral size of the Si wires in p$^+$ porous Si in order to achieve wires of dimensions below $\approx 40A$, required for visible light emission and to increase the mechanical stability of these structures. Whereas anodic oxidation has been studied for p type porous Si before and shown to improve the visible photoluminescence efficiency, its effect on p$^+$ layers, which have a different pore structure, has not been studied in detail . We report here the results of an EPR study on 10µm thick p$^+$ porous Si of 65% porosity anodically oxidized at a constant current in a 0.1 molar KNO$_3$-H$_2$O solution up to various (5V to 25V) potential differences between the sample and a platinum cathode. For more details on the oxidation process and an oxygen analysis by nuclear reaction see reference [50].

Fig. 22 shows the room temperature EPR spectrum of a sample oxidized up to a potential difference of 5V for three orientations of the magnetic field. It is clear from the spectrum, that the P_b center is no longer the dominant paramagnetic defect in these layers. For BII[001] the total spectrum is composed of a 5G broad line at g=2.0056 and a small intensity, isotropic line with a powder lineshape and g factors of g_{II}=2.0018 and g_I=2.0004, which identify this second spectrum as the E' center[51,52]. The E' center is one of the main irradiation induced intrinsic defect in SiO$_2$ and has been attributed to the oxygen vacancy defect. In the anodically oxidized p$^+$ layers the defect is a native defect. The E' center is a bulk defect, whose concentration will scale with the amount of oxide formed.

Concerning the main spectrum at g=2.0056 for BII[001] an angular variation of the magnetic field shows it to be anisotropic and to be composed of different lines. For BII[111] a line of width of $\approx 1G$ and a g-factor of 2.0017 is emerging from the broad line. This line is the BII[111] line of the P_b center; its intensity allows to estimate the contribution of the P_b center to the broad spectrum as $\approx 10\%$. As a model concerning the origin of the rest of the spectrum at g≈2.005 we can exclude the formation of amorphous regions as in this case we would expect an isotropic EPR spectrum. The most probable model seems to be, that the spectrum results from the superposition of P_b centers from different interface planes (111), (100) etc. Further simulation of the spectrum are done right now but we are faced with the problem that the EPR parameters for dangling bond centers at other interfaces than [111] and [001] are not known.

In fig.24 we have plotted the total spin density of the two defects - the sum of P_b as well as the E' centers- as a function of the oxidation potential. We see, that at a potential of 5V the maximum P_b center concentration is already obtained, which implies that the oxidation of all internal surfaces is already completed. The E' center concentration increases with further oxidation as expected for a volume defect in SiO_2. The decrease of its intensity for potentials higher than 10V must be ascribed to a modification of the stoichiometry of the oxide formed under these conditions.

The proposed model - P_b centers at different interface planes - raises the question why this spectrum is not observed in high temperature thermally oxidized porous Si layers. A possible answer to this question is given by the relaxation properties of the different surfaces (hkl). It is known, that the (100) surfaces reconstruct themselves in a way to minimize the dangling bond centers; this can be more easily achieved for the (100) surfaces than for the (111) surfaces: in the first case we have two singly occupied dangling bonds on the same surface atom, which can transform into a doubly occupied dangling bond level and an empty one. The main difference in the oxidation processes is the temperature at which the oxides are formed - room temperature and high temperatures respectively- allowing the relaxation of the (100) surfaces in the second case but not in the first. In order to test this hypothesis we have annealed the anodically oxidized (20V) layer at 300°C. This annealing leads to a reduction of the spin density of the main spectrum by 80%, which gives good support for the proposed model.

4.2. STRUCTURE ANALYSIS: MONOCRYSTALLINE / DISORDERED SI

The (111)P_b center is equally an excellent probe for the microstructural analysis of the internal surface/interface of porous silicon. The angular dependence of its EPR spectrum displays the axial symmetry of the defect wavefunction around the [111] axes normal to the (111) surfaces as well as their alignment with the underlying substrate; if the (111) interfaces of the porous layer were no longer aligned with the substrate (111) surfaces, this should show up in the angular variation of the EPR spectrum; as this is not observed, it directly demonstrates the monocrystalline character of the porous silicon interface structure and its alignment with the substrate. These results have been obtained for layers of various thicknesses (5μm to 300μm) and porosities (40% to 80%).

Whereas electron microscopy techniques have been used for the study of the pore structure of porous silicon the information on the crystallographic orientations (hkl) of the specific surface is nearly absent. The EPR results seem to indicate, that, at least as concerns the oxidized fraction of the surface, we have a preferential development of (111) interfaces. The at first surprising result of the non observation of interface defects at the Si/SiO$_2$ interface from surfaces with other crystallographic orientations than (111) is ascribed to multiple reasons such as the preferential formation of (111) surfaces in the etching process, the easy relaxation of the (100) surface allowing a minimisation of singly occupied dangling bond centers as well as different oxidation kinetics for different surfaces. However the importance of each mechanism needs to be determined in further studies.

The simultaneous knowledge of the absolute defect number and the defect concentration can be used to estimate the area of the internal (111) surfaces or more precisely of the oxidized fraction of this surface. Even non intentionally oxidized samples, exposed to ambient conditions over periods of weeks, are found to be oxidized to a high fraction. This is important for the analysis of the luminescence experiments, in which the presence of an substoichiometric silicon oxide layer in all types of material has to be considered. This aspect must also be important for the band structure calculations of quantum wires and dots, in which only H passivated surfaces are generally considered; though, this seems not to be a realistic situation in real crystals.

The EPR results published up to now demonstrate clearly the non observation of disordered or amorphous inclusions in as-prepared or room temperature aged porous silicon. However, disordered inclusions have been observed in high porosity thermally annealed layers -for porosities higher than 80%-.The localisation of the amorphous regions in the thermally annealed high porosity samples as well as their possible correlation with the luminescence properties is still an open question.

The observation of long lifetime ($\tau \approx 10^2$s) photocarriers has been interpreted by us in the terms of spatial separation of photocarriers in selected parts of the porous material, which are characterized by the impossibility of radiative or non radiative recombination of the photoelectrons; nanocrystals free of interface defects seem to be good candidates; further correlation studies between the concentration of photoelectrons and PL efficiency should be performed.

The EPR results on the (111) P_b center had shown, that the notion of "the" P_b center is insufficient for the characterisation, as clearly P_b centers with distinct EPR parameters have been found. These variations must reflect the interface quality of the different porous materials, a question which has not yet been addressed at all in the context of porous silicon.

5. REFERENCES

1. J.E.Wertz and J.R.Bolton, Electron Spin Resonance, McGraw Hill, New-York (1986)
2. C.P.Poole,Jr, Electron Spin Resonance, John Wiley, New-York (1983)
3. S.Geschwind,Optical Techniques in EPR in Solids, Plenum, New-York (1972)
4. A.Abragam and B.Bleaney, Electron Paramagnetic Resonance of Transition Ions, Clarendon , Oxford 1970
5. J.H.Pilbrow,Transition Ion Paramagnetic Resonance, Oxford Science Publ., Oxford (1992)
6. J.M.Spaeth,J.R.Niklas and R.H.Bartram, Structural Analysis of Point Defects in Solids, Springer Ser. in Sol.State Sciences 43, Berlin (1992)
7. J.Bourgoin and M.Lannoo, Point Defects in Semiconductors II, Springer Ser. in Sol.State Sciences 35, Berlin (1983)
8. D.Stievenard, X.Boddaert,J.C.Bourgoin and H.J.von Bardeleben, Phys.Rev.B41 ,5271 (1990)
9. H.J.von Bardeleben,C.Delerue,D.Stievenard,Proc.17ICDS,ed.by H.Heinrich and W.Jantsch, Mat.Sc.Forum 143-147,223 (1993)
10. R.J.Wagner,J.T.Krebs,G.H.Stauss and A.M.White, Sol.State Commun.36,15(1980)
11. A.Mauger,H.J.von Bardeleben,J.C.Bourgoin and M.Lannoo, Phys.Rev.B36,5982(1987)
12. Y.Q.Jia, H.J.von Bardeleben,D.Stievenard,C.Delerue, Phys.Rev.B45,1645(1992)
13. for a review see:KL.LBrower, Mat.Science Forum10-12,181 (1986)
14. M.Lannoo and P.Friedel, Atomic and Electronic Structure of Surfaces, Springer Ser. in Surface Science 16, Berlin (1992)
15. A.Stesmans, Phys.Rev.B45,9501(1992)
16. E.H.Poindexter,P.J.Caplan, B.E.Deal and R.R.Razouk, J.Appl.Phys.52,879(1981)
17. M.Lannoo and J.C.Bourgoin, Physica 116B,85,(1983)
18. Y. Nishi,Jpn.J.Appl.Phys.10,52(1971)
19. K.L.Brower,Appl.Phys.Lett.43,1111 (1986)
20. M.Cook and C.T.White, Phys.Rev.Lett.59,1741,(1987)
21. A.Edwards, Phys.Rev.B36,9638 (1987)
22. P.J.Caplan,E.H.Poindexter, B.E.Deal and R.R.Razouk,in The Physics of SiO_2 and its Interfaces, ed. by S.Pantelides(Pergamon, New-York)(1978),p.306
23. G. Bomchil, A. Herino, Appl. Surface Science 41/42,(1989),604
24. W.E.Carlos, Appl;Phys.Lett.50,1450(1987)
25. K.L.Brower and T.J.Headley, Phys.Rev.B34,3610(1986)
26. A.Stesmans and G. van Gorp, Phys.Rev.B42,3765 (1990)
27. K.L.Brower, Phys.Rev.B33, 4471(1986)
28. S.R.Elliot, Philosophical Mag.B38,325(1978)
29. P.A.Thomas,M.H.Brodsky,D.Kaplan and D.Lepine,Phys.Rev.B18,3059(1978)
30. R.A.Street and D.K.Biegelson, J.of Non-Crystalline Solids 35&36,651(1980)
31. D.K.Biegelsen, Mat.Res.Soc.Proc.Vol.14,75(1983)
32. J.C.Mao,Y.Q.Jia,J.S.Fu,E.Wu,B.R.Zhang,L.Z.Zhang and G.G.Qin, Appl.Phys.Lett.62, 1408 (1993)

33.H.Yokomichi,H.Takakarua and M.Kondo, Jap.J.Appl.Phys. 32, L365 (1993)
34.Y.Ushida,N.Koshida, H.Koyama and Y.Yamamoto, Appl.Phys.Lett. 63,961 (1993)
35.F.C.Rong,J.F.Harvey,E.H.Poindexter and G.J.Gerardi,Appl.Phys.Lett.63,920(1993)
36.H.J. von Bardeleben, D.Stievenard,A.Grosman ,C.Ortega and J.Siejka, Phys.Rev.B47, 10899 (1993)
37.H.J. von Bardeleben, C.Ortega,A.Grosman ,V.Morazzani, J.Siejka and D.Stievenard,
J. of Luminescence 57(1993),301
38.P.D.J.Calcott,K.J.Nash,L.T.Canham,M.J.Kane and D.Brumhead, J.Phys.C5, L91 (1993)
39.H. Neubrandt   , Phys.Stat.Sol.b86,269 (1978)
40.A. Grosman, C.Ortega , J.Siejka and M.Chamarro, J.of Appl.Phys.(1993)
41.K.Murakami ,S.Fujita and K.Masuda , Mat.Science Forum 83-87,75(1992)
42.H.J. von Bardeleben, M.Chamarro,A.Grosman ,V.Morazzani, C.Ortega ,J.Siejka and S.Rigo, J. of
Luminescence 57(1993),39
43.H.J. von Bardeleben, A.Grosman ,V.Morazzani, C.Ortega and J.Siejka, Proc. of 17 ICDS, ed by H.Heinrich
et W.Jantsch,Mat.Sci.Forum 143-147,1447(1993)
44.E.H.Poindexter,M.Harmatz,W.L.Warren,E.H.Nicollian,E.H.Edwards,J.Electrochem.
Soc. 138,3765 (1991)
45.C.Ortega,J.Siejka and G.Vizkelethy, Nucl.Instrum. Methods Phys. Res., Sec.B45, 622 (1990)
46.H.Münder,M.G.Berger,S.Frohnhoff,H.Lüth,U.Rossow,U.Frotscher and W.Richter,
Mat.Res.Soc.Symposium Proc.256,Boston 1992,p.?
47.S.M.Prokes.W.E.Carlos and V.M.Bermudez, Appl.Phys.Lett.61,1447(1992)
48.D.Goguenheim and M.Lannoo, Phys.Rev.B44,1724(1991)
49.V.Morazzani,M.Chamarro,A.Grosman,C.Ortega,S.Rigo,J.Siejka and H.J.von Bardeleben, J. of
Luminescence 57(1993),45
50.A.Grosman,M.Chamarro,V.Morazzani,C.Ortega,S.Rigo,J.Siejka and H.J.von Bardeleben,J. of
Luminescence 57(1993),13

51.R.H.Silsbee,J.Appl.Phys.32,1459 (1961)
52.D.L.Griscom,E.J.Friebele and G.H.Sigel, Sol.State Commun.15,479, (1974)
53.A.Stesmans,Appl.Phys.Lett.48,972 (1986)

Raman scattering in silicon nanostructures

B. Champagnon, I. Gregora, Y. Monin, E. Duval, L. Saviot

Laboratoire de Physico-Chimie des Matériaux Luminescents
URA 442 du CNRS, Université Lyon I,
69622 Villeurbanne, France

1 Photon-phonon interactions in bulk materials

Before a description of the Raman scattering in nanostructures it is useful to summarise some well known results in bulk materials. Photon-phonon interaction is easily described from the dispersion curves in figure 1 showing the longitudinal and transverse optical and acoustical modes (LO-TO and LA-TA respectively). Optical and acoustical modes interact with laser light (energy hν) to give an inelastic scattering at the energy hν', respectively Raman and Brillouin scattering. Near K= 0, for silicon these dispersion curves can be approximately described by a parabola $\omega = A - BK^2$ for optical modes and by $\omega = v_s\,K$ for acoustical modes, v_s being the sound velocity.

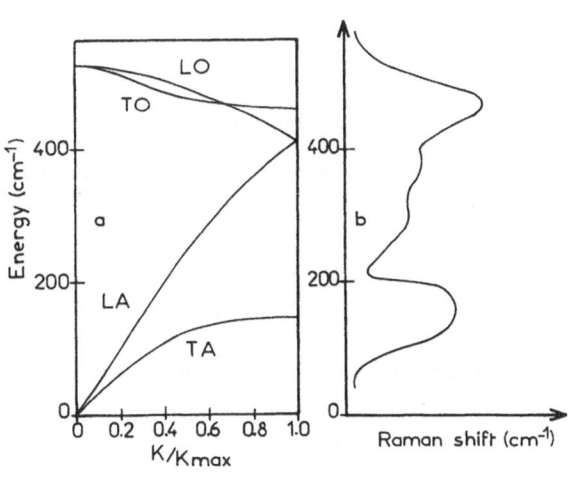

Figure 1 : Dispersion curves of silicon (a) in the direction (001) and Raman spectrum of amorphous silicon (b)

In the bulk material the wave vectors and energy conservation laws between the phonons K and

the incident k, and scattered k', photons ($k = \frac{2\pi}{\lambda}$) are expressed as:

$$k = k' + K$$

$$h\nu = h\nu' \pm h\omega$$

Due to the small values of the **k** and **k'** wavevectors for the visible light compared to the K $_{max}$

wavevector ($|K|_{max} = \frac{\Pi}{2a}$) the photon-phonon interactions are limited to the center of the
Brillouin zone **K**= 0.

Selection rules taking into account the given symmetry of the semiconductor (m3m for silicon) show further that the only first order Raman transition allowed corresponds to the triply degenerate (1 LO+ 2TO) phonon at 520cm^{-1}, at **K**= 0. Second order effects involving pairs of phonons with opposite **K** vectors give rise in the bulk material to bands of lower intensity (Temple and Hathaway, 1973).

2. Size effects in quantum dots

The direct consequence of the confinement effects in quantum dots is due to the lack of long range order . The wavefunction e $^{-iKr}$ of the infinite crystal is now replaced by :

$$\Psi = e^{-iKr} \times W(r,\Lambda)$$

where the limited correlation length Λ corresponds to the limited size of the crystal.

The effect of this spatial damping factor is first to relax the **K**= 0 selection rule (Brodsky, 1983). In the case of the amorphous silicon this relaxation leads to a Raman spectrum which clearly reflects the density of states (figure 1). Amorphous silicon, where only the short range order is retained, can be considered as the limit case of a nanocrystal.

When the size of nanocrystals decreases an increasing part of the dispersion curve is able to interact with light. The detailed behaviour of a Raman line with decreasing has been extensively used these last years in the case of porous silicon with the objective to characterize the nanostructures. From the results of Richter et al.(Richter, Wang, Ley, 1981). and Campbell and Fauchet (1986) it has been shown that it is possible to reproduce the broadened nanocrystal Raman line in porous silicon by a simple model which considers the curvature of the LO dispersion curves. It was shown by the above authors that the Raman intensity of the line is given by:

$$I(\omega) = \int \frac{/C(0,K)/^2 \, d^3K}{(\omega - \omega(K))^2 + \Gamma_0/2} \tag{1}$$

where C(0 ,K) is a Fourier coefficient obtained from the W (r, Λ) function.

For example from transmission electron microscopy observations the spherical shape approximation is valid for p-doped samples with a porosity of 62%.
The simplest case considering spherical nanocrystallites and a Gaussian weighting functions

$$W(r,\Lambda) = e^{-\frac{8\pi^2 r^2}{\Lambda^2}} \tag{2}$$

leads to the better fit for
$\Lambda \approx 4$ nm (figure 2).The
same experimental curve is
compared with a model
calculation for three
different sizes : 3.5 nm,
4 nm, 5 nm. However the
poor fit on the low energy
wing of the line shows that
other contributions have to
be considered, as for
example, amorphous
silica(SiO_2) which is
postulated in several recent
papers as being at the origin
of the efficient visible
luminescence in porous
silicon (Sacilotti et al.,
1993), (Prokes, 1993).

The simple model where
very few parameters are
introduced can also be
refined in different ways
which allow a better fitt of
the experimental curves.

At first, shapes different
from spheres can be
considered.In the paper of
Richter et al.(1981) the
shape corresponding to thin
films and columns are
calculated. Fauchet (1991)
shows recently that for the
same number of atoms the
peak shift is almost
insensitive to the shape
whereas the low energy tail
is a function of this shape:
for example the line is
broader and more
asymmetric for a disk-like
shape than for a sphere
containing the same number
of atoms.

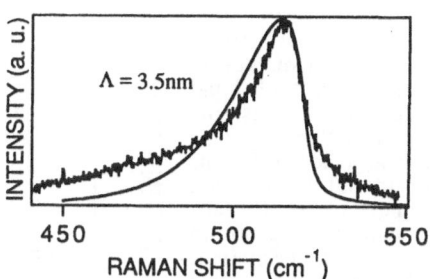

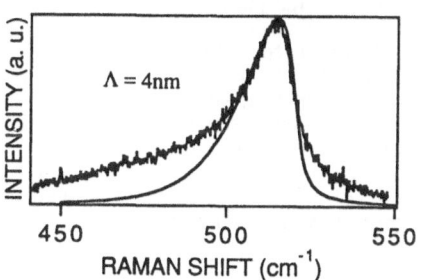

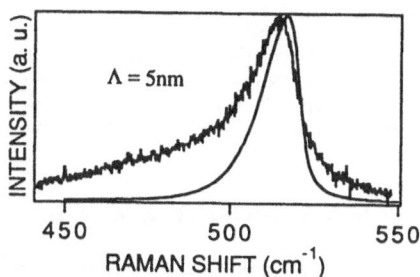

Figure 2 : Fit of the first order Raman line
with formulas (1) and (2) for different sizes Λ
for a p-doped, 62 % porous silicon

The form of the $W(r,\Lambda)$ function is also discussed by Richter et al.(1981), Fauchet and
Young (1991) and Munder et al.(1993) One could use the weighting function appropriate to the
hard sphere model :

$$W(r,\Lambda) = \frac{\sin(2\pi r/\Lambda)}{2\pi r/\Lambda}$$

where the phonon amplitude at the boundary is zero. So far, there is no general agreement on the best choice of the weighting function (Campbell and Fauchet, 1986) (Munder et al.,1993).Several other physical phenomena can modify this line shape: a size distribution, of course occurs and broadens the line. Strains are also considered by Fauchet (1991) et Munder (1993) Surface modes, very clearly observed in GaP microcrystals (Hayashi, Ruppin, 1985) have also been taken into account. We will emphasize (figure 4)that for a same particle size a different Raman line shape is expected if we take into account LO or TO modes. This is due to the different curvatures of the dispersion curves for TO and LO modes (Figure 1); however in view of the width of the lines no splitting between LO and TO was observed experimentally (Tsu, Shen, Dutta, 1992).

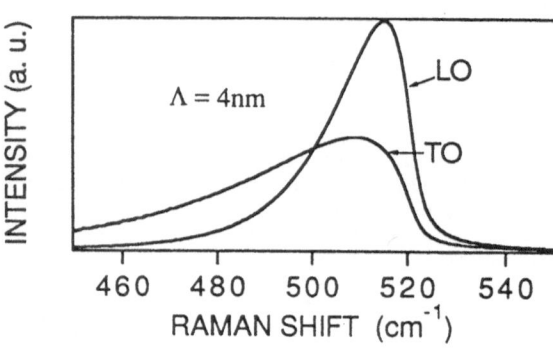

Figure 3 : Calculated shape of the first order Raman line considering LO or TO dispersion curves for Λ = 4mm

To summarise this analysis of the TO-LO Raman line shape one has to emphasize that confinement is mainly responsible for the change in the line shape. The simplest model (Fauchet, 1991) gives good results on the size of the nanostructures with no other parameter than the diameter of the particles. Refinements taking into account size distributions, strains, surface modes , TO modes have also to be considered. Care is, however, needed since the number of the fitting parameters quickly increases : this can lead to good apparent fits but with poor physical meaning.

3. LOw Frequency Inelastic Scattering (LOFIS)

Analysis of the low energy part of the Raman spectrum is also very fruitful as it corresponds to the interaction of light with structures at the nanometer scale.The LOw Frequency Inelastic Scattering (LOFIS) was first studied in the case of glassceramics where nanocrystallites are embedded in a glassy matrix (Duval, Boukenter, Champagnon, 1986) Theoretical description and selection rules were established by Duval (1992) : a particle, assumed spherical in a first step, has vibration eigenmodes which can be calculated from the equation of motion for an elastic sphere.

Among these modes only spheroidal modes, breathing modes (l=0) and ellipsoïdal modes (l=2), can interact with light to induce inelastic scattering (figure 4) . The frequency of these modes can be calculated:

$$\omega = S\frac{v_l}{d} \quad (3)$$

v_l being the logitudinal sound velocity and d the diameter of the particles. The coefficient S depends on the ratio v_l/v_t and on the shape of the particles. For silicon , taking the mean size velocities for longitudinal modes v_l=8780 ms^{-1} and transverse modes v_t=5257 ms^{-1} we obtain v_l/v_t=1.67, this value close to $\sqrt{3}$ corresponds for the breathing modes to S=0.8. The different modes, breathing mode or ellipsoidal mode can be distinguished by their effect on the polarisation of light as it was recently shown in the case of CdSe nanocrystals in an amorphous matrix (Saviot et al.,1994). From an experimental point of view these measurements can only be made with a high rejection rate near the Rayleigh line. In our case this is obtained with a five stages monochromator Dilor Z40 and a single-channel photon counting technique.

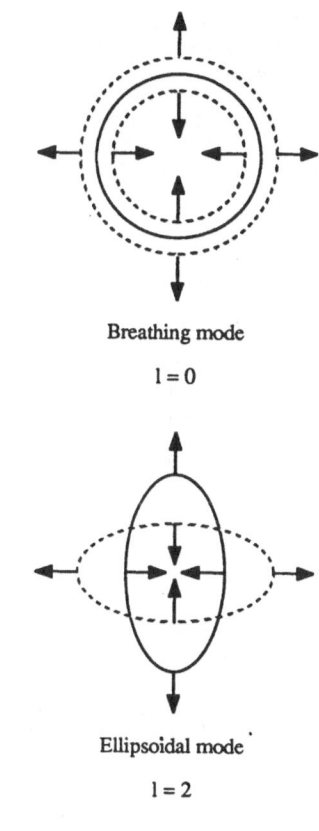

Breathing mode

l = 0

Ellipsoidal mode

l = 2

Spheroidal modes

Figure 4 : Spheroidal modes of a particle corresponding to l = 0 (breathing mode) or l = 2 (ellipsoïdal mode)

For porous silicon the particles considered above are the remaining part of silicon after the chemical etching. From the above formula (3) and considering the shape as determined by transmission electron microscopy the diameter of spheres or columns were determined

4. Luminescence and temperature measurements

The increase of the local temperature of the sample under the laser focused beam is strongly dependent on the porosity of the samples, mainly in microRaman where the exciting light is focused by a microscope objective. To control this effect the comparison of Stokes and antiStokes lines intensities can be done both for the LO line and in the LOFIS region. This is an unique way to measure in situ the true temperatures and was carefully done by Gregora et al. (1993, 1994).

The luminescence can often mask Raman scattering. In the case of porous silicon where the maximum of the luminescence is near 700 nm it is sufficiently far from the usual green or blue lines of the argon laser not to modify significantly the Raman spectra. A very nice advantage of Raman is that the luminescence of porous silicon is generally excited by the laser lines and then it is possible to record both the Raman scattering and the photoluminescence signal from the same spot on the sample. This advantage is very important in the case of inhomogeneous samples (Monin et al., 1994)

5. Conclusion

Raman scattering is a very useful non destructive tool to determine the size of nanostructures in porous silicon. Size determination is possible by two independent ways which lead to similar results in the samples studied until now. Simultaneous measurements of luminescence and Raman scattering are expected to give the quantitative correlation between the shift of the emission and the size of nanocrystals in order to check the quantum size effects. Temperature measurements can be done systematically to avoid spurious effects.

From a fundamental point of view porous silicon stimulates theoretical analysis of line shape and LOFIS spectra which can be extended in different quantum dots.

Aknowledgments

We gratefully aknowledge A.Halimaoui (France-Telecom Meylan) which provide us with various samples, C.Esnouf and A.Thollet (GEMPPM-Insa-Lyon) for TEM and fruitful discussions.

References

Campbell I.H. and Fauchet M.P.*Solid State Comm.*.**58**, (1986) 739

Duval E., Boukenter A.and Champagnon B.*Phys. Rev. Letters* **56**, (1986) 2052

Duval E.*Phys. Rev. B* .**46**, (1992) 5795

Gregora I.,Champagnon B.and Halimaoui A. *J. of Luminescence* **57**, (1993)73

Gregora I., Champagnon B.and Halimaoui A.*J.App. Phys*. Mars 94

Hayashi S.and Ruppin R.*J.Phys. C. Solid State Phys.* **18**, (1985)2583

Monin Y., Saviot L., Champagnon B., Esnouf C., Halimaoui A. *Thin Solid Films* (1994), to be published

Munder H., Andrzejak C., Berger M.G., Klemradt U. and Lüth H.*Thin Solid Films* **221**,(1993)27

Prokes *App. Phys. Letters* **62**, (1993)2676

Richter H., Wang Z.P.and Ley L.*Solid State Comm.* **39**, (1981)625

Sacilotti M., Champagnon B., Abraham P., Monteil Y. and Bouix J. *J. Luminescence* **57**,(1993)33

Saviot L.,Champagnon B., Duval E.,Ekimov A.and Kudriavstev I. to be published

Temple P.A.and Hathaway C.E.*Phys. Rev. B* **7**, (1973)3685

Tsu. R.,Shen H.and Dutta M.*Appl. Phys. Lett.* **60**, (1992)112

Brodsky M.H.,Light scattering in solids (I ed. M.Cardona Topics in Applied Physics Vol.**8**, Springer-Verlag, Berlin, 1983)205.

Fauchet P.M.,Light Scattering in semiconductor Structures and Superlattices (ed. Lockwood and J.F. Young Plenum Press, New York, 1991)229.

Scattering of X-rays

A. Naudon, P. Goudeau and V. Vezin[1]

Laboratoire de Métallurgie Physique, Université de Poitiers,
URA 131 du CNRS, 40 Avenue du Recteur Pineau,
86022 Poitiers cedex, France

1. INTRODUCTION

Small-Angle Scattering (SAS) or scattering at low angles can give information on size, shape, quantity and arrangement of *"scattering objets"* or *"particles"* whose *"characteristic length"* is large by comparison to the interatomic spacing.

(SAS) of X-rays (or neutrons) is a suitable method for examination of periodic and nonperiodic systems. It is indeed different from wide-angle diffraction (crystallography) which gives interatomic distances d. However, in a crystallized material, there is a continuity between the small-angle and the wide-angle techniques and in general the complete diffraction pattern is of interest, because there is also a scattering around Bragg peaks when **"inhomogeneities"** are present in the sample.

The physical principles of scattering of X-rays (SAXS) which are electromagnetic waves are the same for wide-angle diffraction and SAS. The electric field of the incoming wave induces dipole oscillations in the atoms where all electrons are concerned. Secondary waves are generated and add at large distances to the overall scattering amplitude. The intensity is collected in a detector as a function of the scattering angle.

For X-rays, **SAXS** is sensitive to the **electron density** of the investigated sample. For neutrons (particle waves which interact with the nuclei) the situation for scattering is nearly the same, and then small-angle scattering of neutrons (**SANS**) detects the variation of the so-called **scattering length density.**

The angle dependence of the scattering amplitude is related to the electron-density of the scatterers by a Fourier transformation. This holds for both wide-angle diffraction and small-angle X-ray scattering. In the former case, the Fourier transform corresponds with a Fourier series. In the case of SAS, the Fourier transform of a non-periodic limited structure corresponds to a Fourier integral. The consequence is that in SAS we measure a continuous angle-dependent scattering intensity instead of point-like spots in crystallography. That means the sample is treated as a continuum and characterized by an average density and the

---

[1] *Present address:* Department of Metallurgy, Kyoto University, Sakyo-ku, Kyoto 606, Japan

fluctuations around it. Consequently, the scattering is described as arising from "**scattering objets**" of some scattering density embedded in a medium of another density.

As a consequence of the Fourier transform, the SAS pattern is obtained in the reciprocal space. Consequently it is difficult (sometimes impossible) to recover the structure of the real space, because phases are lost and information is averaged over all orientations in space. It is the reason why imaging methods such as transmission electron microscopy (TEM) or atomic force microscopy (AFM) are necessary (when possible) to characterize inhomogeneities of mesoscopic size in the studied materials, and it is interesting to compare these different techniques.

We shall consider here that the SAS region can be isolated from the rest of the scattering pattern. The scattering intensity is usually larger at small angles than at wide angles. In a fine powder of crystallized grains, for example, an individual grain will contribute to the diffraction only if its orientation corresponds to the Bragg condition, whereas all grains, whatever their orientation, scatter at small angles.

It must be noted that the scattering efficiency increases linearly with the atomic number for X-rays when the wavelength is far from the edges of the constituting atoms of the sample. Now, with synchrotron radiation facilities, it is possible to modify this linearity and consequently to vary the contrast on the same sample, this is called "*anomalous scattering*".

It is possible to study a large variety of materials by small-angle scattering: phase-separated alloys, glasses, polymers, colloids, proteins.....Distinction must be made between monodisperse or polydisperse systems, between diluted systems and concentrated ones. For dilute systems, size and shape of the scatterers can be determined quite easily. When the system is more concentrated, the SAS experiment is influenced by the structure of the "**particles**" and their spatial arrangement: form factor and structure factor. If the system is non-particulate, as porous silicon for example, another approach is necessary with the concept of "**chords**" and "**characteristic length**" of the phases. At last, but not least, SAS is an appropriate technique to characterize "**fractal**" objets.

2. THEORETICAL ASPECTS OF SMALL-ANGLE SCATTERING

This section will first give the general principles of the theory of SAS; the emphasis will be more on the appropriate use of the essential formulae rather than on an exact derivation, which can be found in the textbooks referred in the bibliography.

In this course we are concerned with X-rays only, but, as indicated in the introduction, all equations may be applied with slight modifications to neutron, light scattering or electron diffraction as well.

2.1. The scattering problem

The basic principle of the scattering theory is shown in Fig. 1, where a monochromatic plane wave of X-rays, represented by $\psi = \exp(ikz)$ illuminates a small object, scattered waves of variable intensity are emitted in all directions of space. All these waves have the same intensity given by the Thomson formula:

$$I_e(2) = I_o \, r_e^2 \;\; \frac{1}{D^2} \;\; \frac{1 + \cos^2 2\theta}{2} \tag{1}$$

Figure 1 : scattering geometry

Where I_0 is the primary intensity, D the distance from the object to the detector, r_e^2 the square of the classical electron radius ($r_e^2 = e^4/m^2c^4 = 7.95 \times 10^{-26}$ cm^2). As indicated in Fig. 1, the scattering angle 2θ, is the angle between the primary beam and the scattered beam which can be described by their wave vectors k_0 and k, respectively, of modulus $2\pi/\lambda$ and oriented perpendicular to the wave front. The last term in equation (1) is the polarization factor and is practically equal to 1 for small-angle scattering problems.

We can define a vector q such that $k = k_0 + q$, where q is the scattering vector which defines the geometry of the experiment. For small angles 2θ, its modulus is $4\pi\theta/\lambda$. Then we have:

$$q = |q| = \frac{4\pi}{\lambda} \sin \theta \qquad (2)$$

In SAS, sin may be replaced by θ.

As indicated before, small-angle scattering measures spatial correlations in the scattering density, averaged over the time scale of the measurement. Because the energy transfer is small compared with the incident energy of the photon ($\approx$ 10 keV), inelastic effects can be neglected.

Scattering can still be coherent or incoherent. In SAS we shall consider only coherent scattering, where scattered photons from different atoms interfere, and consequently provide all information on the spatial distribution of atoms. The main source of incoherent scattering is fluorescence. Very often, now with the synchrotron radiation, it can be easily avoided by choosing a photon energy below (or far from) the edge of the species constituting the sample. Another source of incoherent scattering is the Compton scattering, but it is negligible in the angular range of the SAS.

2.2. Particle scattering

In this part we shall discuss scattering from monodisperse systems, i.e., all particles in the scattering volume have the same size, shape and internal structure.

2.2.1. Dilute system : form factor $P(q)$

A system is considered as a diluted system when particles are isolated, that means with no interaction between them as those in the box in figure 2.

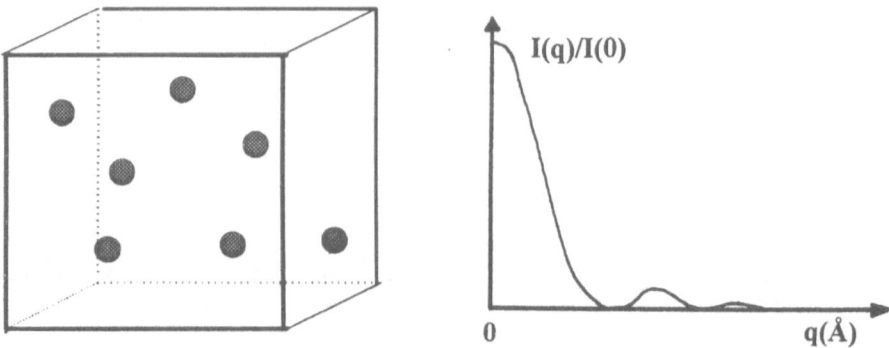

Figure 2 : Dilute system of particles (widely separated) and their scattering intensity given by formula (3)

The angular dependence of the scattering intensity can be calculated by analytical expressions. For example, for a homogeneous sphere of radius R, the scattering intensity is given by :

$$I(q) = I_0 \left| 3 \frac{\sin (qR) - qR \cos (qR)}{(qR)^3} \right|^2 \tag{3}$$

where I_0 is the intensity of the incoming beam. This function has a series of well-defined zeros corresponding to $qR = 4.52$, 7.72, etc.

The shape analysis of cylinder, ellipsoids, flat particles, etc. has also been calculated and can be found in textbooks (see bibliography).

2.2.2. Dense packing : structure factor S(q)

When the scattering particles are numerous, they are close to each other and consequently they interact. The observed intensity reflects both their geometry and the interactions between them.

For N identical, randomly distributed and centrosymmetrical particles, the scattered intensity is the product of the form factor of the particle P(q) and the structure factor of the assembly S(q)

$$I(q) = N \ P(q) \ S(q) \tag{4}$$

The general trend of such a dense packing of particles is shown in Fig.3 with its scattering curve showing a peak related to an average distance between the particles. In general, this peak appears in the structure factor S(q) and is related to the presence of correlations between the positions of the scattering centers. However the separation between the two contributions, P(q) and S(q), is generally very difficult to solve because they interfere on each other.

There are different approaches to find a solution to this problem:
- Dilution method : When the particles are identical and dilution is possible, extrapolation to zero concentration yields P(q).
- Labeling method : If one can label only a point in each particle, the contrast of the rest being

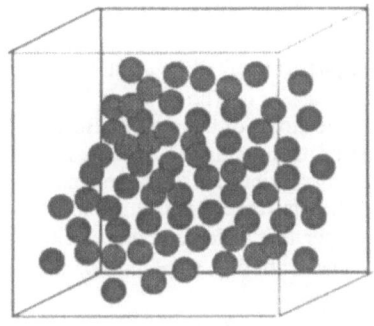 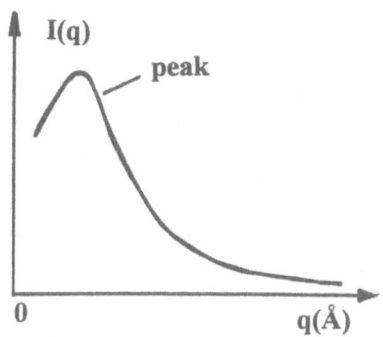

Figure 3 : Dense packing of identical particles and their corresponding scattering intensity

- Simulation : It is possible to calculate the scattering profile in a model when assuming size, shape and the interaction potential of the particles.

2.2.3. Two-phase system

The two-phase model is one of the most commonly used approximation. It assumes that the scattering is due to a particulate system which can be described by a collection of particles (of different sizes) with uniform electron density ρ_1, embedded in a matrix with uniform electron density ρ_2, as shown on the scheme of figure 4.

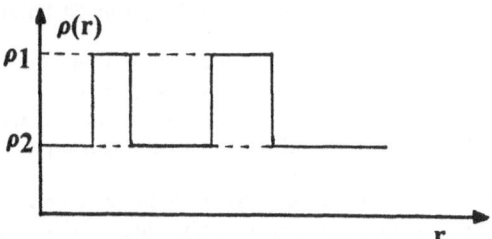

Figure 4 : Scheme of the two-phase system

Here, it is difficult to solve the problem, because the particles are not identical, and equation (4) is no longer valid. It must be replaced by:

$$I\,(q) = \ P(q) \ S(q) \tag{5}$$

P(q) is now an average structure factor. Each particle has its own scattering profile as shown in figure 2 for a sphere. The scattering profile is size and shape dependent. Consequently, the polydispersity of the particles smears out the characteristic damped oscillations of the scattering curve.

More complicated is when the particles have an internal structure. Here it is quite impossible to solve the problem by a conventional small-angle scattering experiment.

2.3. Correlation function

The superposition of waves of different amplitudes and phases results in a scattering intensity depending on the scattering angle 2θ and the momentum transfer q. The total amplitude at q is the sum of the waves scattered by all the electrons of all the atoms of the sample. Because SAS deals with characteristic lengths that are large compared to interatomic distances, it is not possible to separate the contributions of individual atoms as point-like scatterers and the sum over discrete atoms can be replaced by an integral. For this reason, we introduce now the electron density ρ (r). This is the number of electrons per unit volume at the position r. A volume element dV at r contains ρ (r) dV electrons. The scattering amplitude of the whole volume V is given by :

$$A (q) = \iiint \rho (r) \exp (-iqr) \, dr \tag{6}$$

We see that the amplitude A is the 3-d Fourier transform of the electron density distribution. The scattering intensity is the product of A (q) by its complex conjugate A*(q) :

$$I (q) = |A(q)|^2 = A(q) \, A^*(q) = V \iiint G (r) \exp (-iqr) \, dr \tag{7}$$

with
$$G (r) = \frac{1}{V} \iiint \rho (r_0) \, \rho (r_0 + r) = \langle \rho (r_0) \, \rho (r_0 + r) \rangle \tag{8}$$

G (r) is the autocorrelation function. All information is included in G (r) which expresses the correlation between the electron densities measured at two points separated by a vector r, averaged over the total irradiated volume V.

All information is indeed in G (r), but experiment gives I(q). Although the inverse Fourier transform is mathematically possible to extract ρ (r), too much information is lost for the following reasons:
- The scattered intensity is the result of averaging over all sizes and distances.
- The squaring of the amplitudes causes phase information to be lost.
- The cut-off value of q can give rise to artifacts in the inverse Fourier transform.
- The spatial and orientationnal correlations of the particles (for a particulate system) have to be considered.

These are the reasons why, instead of performing the inverse Fourier transform of the scattering intensity, it is preferable to look at the limit values of G(r) and I(q): for $r \to 0$ one can determine the invariant Q_0, for $q \to 0$ the size will be obtained through the radius of gyration R_g, and for $q \to \infty$ the specific surface S_p could be reached.

2.3.1. The invariant Q_0

If we assume spherical symmetry, equation (7) becomes:

$$I (q) = 4 \, V \int_0^\infty G(r) \, \frac{\sin qr}{qr} \, r^2 \, dr \tag{9}$$

The correlation function is obtained by a Fourier transform of the intensity:

$$G(r) = \frac{1}{2\pi \, V} \int_0^\infty I (q) \, \frac{\sin qr}{qr} \, q^2 \, dq \tag{10}$$

If $r = 0$, we have:

$$G(0) = \frac{1}{2\pi\ V} \int_0^\infty I\,(q)\ q^2\ dq \tag{11}$$

where the quantity $\quad \int_0^\infty I\,(q)\ q^2\ dq \quad$ is called the invariant and noted Q_o

Q_o represents the total integrated intensity in the reciprocal space (in fact in its first cell or the first Brillouin zone).

Q_o is independent of the topology of the sample. That means that in a two-phase system, Q_o gives an evaluation of the total volume of the scattering entities in the sample, whatever the size of the scattering objects.

It is important to note that there is the same integrated intensity around each Bragg's peak, including the (000) one, which gives the small-angle scattering. As the intensity of the direct beam is much more intense than any Bragg's peak, the intensity around the angular origin is the highest and the easiest to measure.

In a two-phase system, if c is the volume fraction in % of the scattering objects, one has:

$$Q_o = \int_0^\infty I\,(q)\ q^2\ dq\ = 2\pi^2\ V\ c(1\text{-}c)\ \Delta\rho^2 \tag{12}$$

2.3.2. Radius of gyration

Let us recall equation (9) for an isotropic object, when $q \to 0$, the expansion of $\dfrac{\sin qr}{qr}$ can be done, leading to the following formula:

$$\lim I(q) = 4\pi\ V \int_0^\infty r^2\ G(r)\ \exp{-1/3}\ (\,q^2\ R_g{}^2\,) \tag{13}$$

where R_g is the radius of gyration, as shown by Guinier [1] in 1938. It means that a particle of any shape, randomly oriented, is described by a radius of gyration.

R_g is obtained by the plot of Log(q) versus q^2 which displays a linear region for small q values. The characteristic size of a particulate system can be easily obtained within this approximation. For instance, for spherical particles, one can relate the radius of particle R with the radius of gyration R_g by $R = \sqrt{5/3}\ R_g$. However, the interparticle interference effects should modify the scattering curve, leading to wrong value of R_g. Nevertheless R_g is very often a good information upon the size determination of the scattering particles.

2.3.3. Porod law [2,3]

Porod analysis or Porod approximation is a common technique used in the characterization of two-phase systems. It is based on the so-called Porod law which shows that when interfaces between phases are sharp, the scattering intensity decreases asymptotically with q^{-4} when q goes to infinity. One has:

$$\text{Lim } q^4\ I(q) = \quad \text{Constant} \ = \ 2\pi\ S_p\ (\Delta\rho^2)\ V \tag{14}$$

S_p is the specific surface or the surface area per unit volume S/V.

Porod law is valid at all distances for which the boundaries between the two phases are seen as sharp. As for the invariant Q_o, the validity of Porod law is independent of the geometry and topology of the sample.

However, let us note that great care must be taken with Porod law. Deviations from Porod law, which were observed in the past and attributed to some particular interface, are now attributed to the "fractal" behavior of the scattering objects.

2.4. Non-particulate system

2.4.1. Case of a porous sample

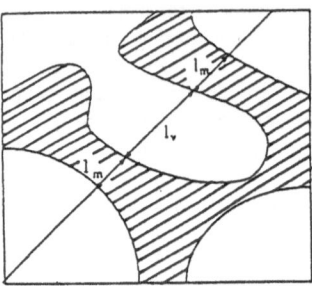

Figure 5 : Schematic drawing of a non-particulate two-phase system, with chords l_m and l_v in matter and voids respectively.

Small-angle scattering is nor restricted to the presence of well defined particles. It can be applied to any system where electron density heterogeneities of any kind are present. The only condition is that these heterogeneities are with respect to the interatomic distances.

A porous sample, as for example porous silicon with an open porosity, is a non-particulate two-phase system of matter and voids. If c is the porosity of the sample (the volume fraction of voids), then (1-c) is the fraction of matter. Such a system is schematically represented in Fig. 5.

One can see that, in a porous sample, the invariant is given by:

$$Q_o = \int_0^\infty I\,(q)\ q^2\ dq\ = 2\pi^2\ V\ c(1\text{-}c)\ \rho_{Si} \tag{15}$$

where ρ_{Si} is the electron density of silicon. It represents here the contrast between silicon and voids.

2.4.2. Chord distribution

According to Porod [3], a chord is formally an "intercept length" as schematically shown in figure 5 for a non-particulate two-phase system as porous silicon. A line crossing through the system will cut out alternating chords l_m in matter and l_v in voids. An average intersect length <l> can be defined by:

$$1/\text{<}l\text{>} = 1/\text{<}l_m\text{>} + 1/\text{<}l_v\text{>} \tag{16}$$

<l> verifies the following relationship:

$$\text{<}l\text{>} = c\ \text{<}l_m\text{>} = (1\text{-}c)\ \text{<}l_v\text{>} = 4c\ (1\text{-}c)\ V/S \tag{17}$$

V/S is the inverse of the specific surface S_p that we find also in relation (14).

So, when the classical Porod power law is obeyed, the intercept length <l> is given by the following equation when q → ∞ :

$$\lim_{q \to \infty} [q^4 \, I(q)] = 8\pi/<l>$$ (18)

2.4.3. Characteristic dimensions

In a non-particulate two-phase system, with equations (17) and (18), we have the relationship between the outer part of the intensity function, the specific surface and the concept of chord distribution. Furthermore, this chord distribution is very sensitive to the surface structure [4]. The advantage of such a treatment for porous silicon which is a non-particulate two-phase system is that the concept of size and shape (with a different meaning than in particles) may still be applied. So, we can get averaged values given by a small-angle scattering experiment: average intersect length <l>, specific surface S_p and information about interface between matter and pores; however different cases have to be considered here about the nature of the interface:

- *Sharp interface*: As we have already seen, this is the case where Porod law is obeyed. The average intersect length <l> is given by equation (18).

- *Rough interface and fractal surface*: The limit of $q^4 \, I(q)$, when q → ∞ is not a constant. The plot of Log I(q) versus Log q gives an exponent which is not an integer. Then a roughness of the interface, or a fractal dimension of the surface, can be deduced [5].

- *Diffuse interface*. For example, when porous silicon is oxidized, the resulting thin oxide layer constitutes an interface between the pores and the silicon skeleton. For X-rays, a diffuse interface corresponds to a continuous variation of the electronic density at the pore boundary. Such an interface has also been called "fuzzy" [6] and leads to a power-law-scattering exponent higher than 4.

In a two-phase system, when the transition between both phases is not sharp, the electronic density q(r) can be written by a convolution product between a theoretical sharp electronic density $q_{the}(r)$ and a smoothing function h(r) which takes into account the width of the interface (Fig. 6).

$$q(r) = \rho_{the}(r) \otimes h(r)$$

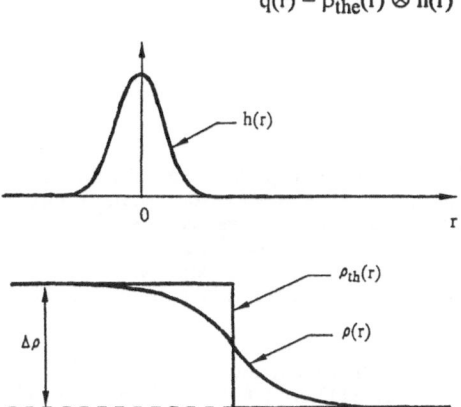

Figure 6: Smoothing function h(r) and electron-density ρ(r) with the theoretical sharp interface $\rho_{the}(r)$.

the intensity is then given by: $I(q) = H(q) I_{the}(q)$

$I_{the}(q)$ being the scattering intensity that would be obtained with a sharp interface.

If h(r) is assumed to be a gaussian [7], the Fourier transform of the convolution product h(r) ⊗ h(r) is then:

$$H(q) = \exp(-q^2\sigma^2) \tag{19}$$

σ^2 is the variance of h(r) perpendicular to the interface.

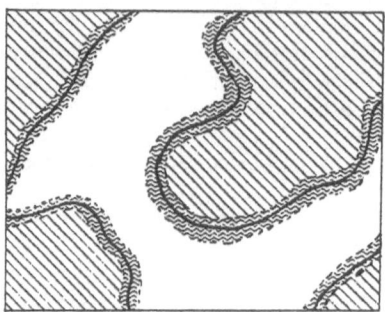

Figure 7: Schematic drawing of a non-particulate two-phase system with a diffuse interface.

So, the classical Porod law becomes:

$$\lim_{q \to \infty} \{\text{Log} [q^4 I(q)]\} = - q^2\sigma^2 + \text{Log } 8\pi Q_o/<l> \tag{20}$$

The plot of Log $[q^4 I(q)]$ versus q^2 gives a straight line with slope of $-\sigma^2$. In our case, the value of σ, which can be experimentally determined, constitutes a good approximation of the mean thickness of the oxide formed during the anodic oxidation process. A diffuse interface is schematically shown in figure 7. The average intersect length <l> can also be deduced with formula (20).

3. EXPERIMENTAL TECHNIQUE

3.1. In a laboratory

The prerequisites of a useful small-angle X-ray scattering apparatus are a small divergence of the incident beam and a large X-ray flux. Although the best way to satisfy these requirements is to take advantage of X-ray beams at synchrotron radiation facilities (as we shall see in the next paragraph), a laboratory instrument is also possible with a conventional X-ray source. The one we use in our laboratory [8] is described hereafter. It employs a rotating-anode X-ray source, an asymmetric monochromator and a position-sensitive detector. It is represented in Fig.8.

The rotating anode A delivers CuK_α X-rays with a linear focus size of 3 x 0.3 mm^2. The monochromator is an asymmetric, bent germanium (111) crystal. The polished surface is at an angle of 9.1° to the reflecting (111) planes. The crystal is held curved in a polished glass press which has a radius of 2600 mm and the focal distances are 200 and 1000 mm respectively.

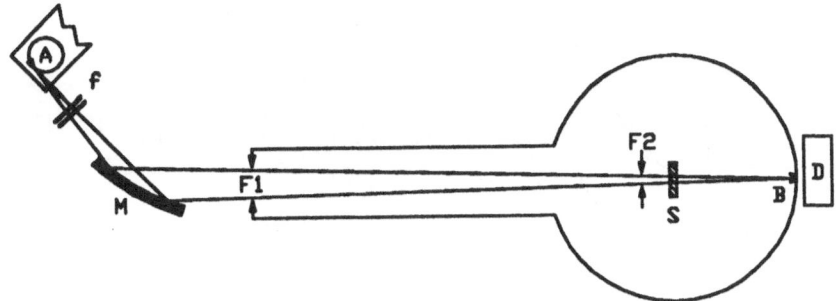

Figure 8: Top-view of the experimental X-ray set-up: A = rotating anode, f = filters, M = monochromator, F_1 = entrance slit, F_2 = collimation slit, S = sample and D = detector.

Such a large distance has the advantage of producing a beam which is geometrically very well defined by the slits F_1 and F_2 (Fig.8) and converges to a fine line focus with enough intensity. F_2, which can be moved independently, acts also as an antiparasitic slit, and its aperture which is always small, delineates the area of the surface layer illuminated by the X-ray beam.

The sample holder is in the middle of a cylindrical chamber under a vacuum of 1 Pa. This chamber is linked by a pipe to the entrance slit F_1, so the X-ray beam travels under vacuum from F_1 to the beam-stop B. The detector D is a one-dimensional position-sensitive proportional counter filled with a Xe-CH_4 mixture, with a spatial resolution of 150 µm. The fixed distance between the sample and the detector is 160 mm. However this distance is too small to obtain low q values to look at the large-sized scattering heterogeneities of porous silicon samples. It is the reason why we do perform experiments at the synchrotron radiation.

3.2. At a synchrotron radiation source

We describe here the spectrometer we use at LURE [9] on beamline D22. It is shown in Fig. 9.

A double-crystal constant-vertical-deviation monochromator, equipped with two Si (311) slabs selects a narrow energy band width ($\Delta E/E \approx 10^{-4}$) in the synchrotron beam emitted by the storage ring LURE-DCI running at 1.85 GeV. The energy resolution is only 10^{-3} with Ge (111) crystals. The rotation of the crystals is controlled by a stepping motor and the Bragg angle read on an encoder with a precision of 2.5".

Two sets of horizontal and vertical slits are used to defined a point-shaped beam (cross-section less than 1 mm^2). The incident X-ray intensity is monitored by a NaI scintillator placed before the sample. A second similar scintillator, placed after the sample-holder, constantly measures the sample absorption. A sample-holder with six positions allows to measure automatically five different samples, and also the parasitic scattering (without sample or with a non-porous silicon sample).

Monochromator, slits and sample-holder are all maintained in a vacuum of 1.3 Pa. SAXS profiles are recorded using a one-dimensional position-sensitive counter. The sample-counter distance can be varied with a modular assembly of evacuated pipes up to a value of 2 meters giving $q_{min} = 4 \times 10^{-3}$ Å^{-1} with $\lambda = 1.54$ Å.

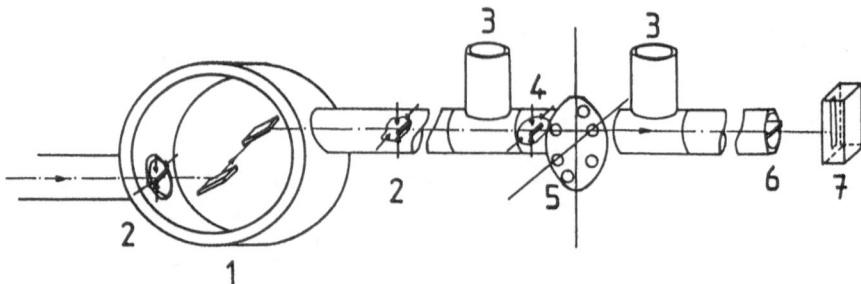

Figure 9 : SAXS experimental set-up at LURE showing the two-crystal monochromator (1) with entrance and exit tantalum slits (2); NaI scintillator detectors (3) with antiparasitic slits (4), sample holder (5), beam stop (6) and the position sensitive detector (7).

4. APPLICATION TO POROUS SILICON

4.1. Doping influence on the scattering

As small-angle scattering of X-ray can lead to a better understanding of the microstructure of porous silicon, we performed experiments on porous layers prepared on lightly doped and heavily doped silicon wafers according to the dopant type .

The dependence of the microstructure of P-type porous silicon on its doping level is the following [10]: In nondegenerate P-type silicon, the porous structure is an apparently random distribution of voids. The scattering pattern obtained in that case is isotropic, that is to say that the scattering spectra are the same whatever the tilt of the sample with respect to the direction of the incident X-ray beam. In degenerately doped P-type silicon (P^+), the small angle scattering peak is more or less accentuated, with an angular position dependent on the tilt of the sample. This feature suggests that the scattering pattern around the angular origin is not isotropic. One can say that the porous film structure consists of many long voids running perpendicular to the surface, with small "buds" on their sides.

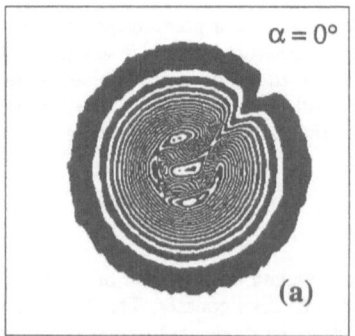
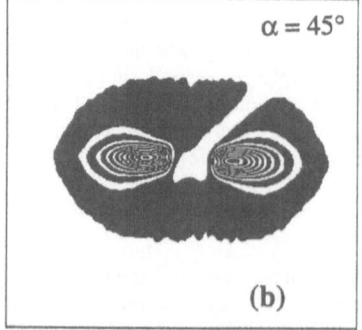

Figure 10: Evolution of the 2D scattering pattern with the tilt of the sample surface. (a) 0° and (b) 45°

SAXS experiments carried out at the synchrotron radiation source with a two-dimensional detector revealed the anisotropic character of the P^+ microstructure. The evolution of the 2D scattering pattern, according to the tilt of the sample, is illustrated in figure 10 for a P+ sample obtained with a current density of 80 mA/cm^2. Such an evolution remains the same whatever the forming current density.

4.2. Photoluminescent P silicon

Since photoluminescence and electroluminescence phenomena in the visible range and at room temperature have been recently obtained with high porosity (greater than 80%) in porous silicon layers [11,12,13], it was necessary to improve the understanding of the physics involved in these luminescence phenomena with more structural characterization of the high-porosity porous silicon layers (PSL).

One characterization technique which can be used is small-angle X-ray scattering (SAXS). In fact, we demonstrated in a previous paper [14] that this technique is well suited for the investigation of microstructures such as porous silicon. Here we present SAXS characterization of PSL when the porosity is increased from 55% till a porosity value around 80% for which the luminescence of PSL is observed at room temperature.

The porous silicon samples were prepared on P-type silicon wafers, (100) oriented, boron doped, with a resistivity of 1 Ω.cm. In order to minimize the X-ray absorption in the transmission mode, the wafers were chemically thinned down to an uniform thickness of 100 μ μm ($\pm$ 2%) prior to porous silicon formation. All details are given elsewhere [15].

The small-angle scattering measurements were performed at the French Synchrotron Radiation Source: Laboratoire pour l'Utilisation du Rayonnement Electromagnétique (LURE), Orsay, France, at the D22 station [9]. Two sample-detector distances have been chosen: D_1 = 540 mm and D_2 = 1540 mm, so the q values range from 3 to 3 x 10^{-2} nm^{-1}. As the cross-section of the X-ray beam is only 1 mm^2, it is possible to work in point-like collimation. Furthermore a small area of the samples can be easily analyzed.

Figure 11 shows three characteristic scattering curves, in a double $\log_{10}$ presentation, relative to three different porosities: (a) 55% ; (b) 68% and (c) 85% (the photoluminescent sample).

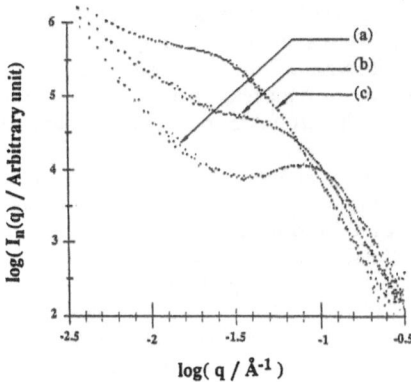

Figure 11 : Log-log representation of the X-ray small-angle scattering curves of porous silicon layers having three different porosities: (a) 55% ; (b) 68% and (c) 85%.

First of all we point out that when tilting the sample with respect to the X-ray beam, the scattering profiles remain approximately the same, confirming the isotropic behavior of the PSL even for the highly porous layer.

The three characteristic curves in figure 11 have a general trend. Two linear parts are separated by a curved region which is a shoulder for the (b) and (c) samples and a clearly visible maximum for the (a) sample (porosity 55%). In the curved region an arbitrary q_c value can be defined for each curve in order to make easy the interpretation of the spectra. So a distinction can be made for each side of the q_c value which separates the region of interfaces $(q > q_c)$ from the region of the bulk $(q < q_c)$.

For $q > q_c$: This is the region linked with small objects and with the interfaces between matter and pores. The log-log plot gives a linear part with a slope of about -4 for the three samples, indicating that the Porod power-law [2] holds in this region. Consequently the boundary between matter and void is sharp (well defined) for the three freshly prepared samples.

With the chord distribution analysis, it is possible to calculate chords in matter and chords in voids, $<l_m>$ and $<l_v>$ respectively, given by relations (16) and (17). Knowing the porosity of the samples, c, determined by weight measurements in separate wafers, we can calculate the variation of $<l_m>$ and $<l_v>$ with the porosity. They are indicated in Table I.

Table I
Geometrical parameters $<l_m>$, $<l_v>$ and the specific surface S_p
deduced from SAXS measurements in 3 different PSL.

c	$<l_m>$ (Å)	$<l_v>$ (Å)	S_p (m²/cm³)
55%	25	22	825
68%	29	63	435
85%	40	227	150

One can observe a large increase of the size of voids for the porosity 85% associated with a relatively small increase of the size of matter. This large increase of the size of voids $<l_v>$ with increasing porosity can be associated with the fact that there are communications between nearest-neighbor pores, because there is dissolution of silicon and the pores can coalesce. At the same time, dissolution of matter leads to a progressive disappearance or a decrease of some "walls" of silicon, but there is also an increase of some segments in matter (and consequently a small increase of $<l_m>$) because we have to remember that a SAXS experiments gives always an average value. An important result is the strong decrease of the specific surface S_p when the porosity increases.

For $q < q_c$: For small q regime there is an upturn in the scattering curves shown in figure 10. This upturn is obviously due to the existence of much larger objects than those which contribute to the observation of the Porod power-law. In fact the small q regime is related to the bulk which is in the present case the silicon skeleton.

The log-log plot in figure 10 indicates a linear part, at least for the (a) and (b) samples. Taking into account the high value of the porosity of the samples, such a slope could be interpreted as being a mass fractal dimension D_m which would characterize the structure of the Si skeleton [16]. The value of the slope decreases from -3 for curve (a) to about -2 for curve (b) and still less (around -1) for curve (c). According to Schaeffer [17], such an evolution of the slope means that the mass fractal dimension of silicon skeleton diminishes from a structure having 3 dimensions (when the volume of matter as compared with void is still important) to another one which tends to be linear when the volume of matter is very low (noodle-like structure). A possible explanation of our results is that the porosity increase is associated with an electrochemical dissolution of silicon "walls" separating nearest-neighbour pores. Such a dissolution would lead to a pore-pore overlap and dramatic thinning of the silicon crystallites. This is in agreement with the evolution of $<l_m>$ and $<l_v>$ with the porosity as shown above.

The q_c region : This region of the shoulder is more complicated to understand because it may have different possible explanations.

For the (a) sample, a well defined peak is observed. It is clear that the angular position of this maximum in intensity could be interpreted as a diffuse interference effect, in the same manner as for liquids or Guinier-Preston zones [18] or small precipitates [19] in alloys. Furthermore TEM observations [20] of P-type porous silicon samples having a porosity of about 50% show an isotropic network of small size pores.

In conclusion, a small-angle scattering analysis, based on the concept of chord distribution which is appropriate for non-particulate two-phase system as high porosity PSL, gives relevant geometrical parameters indicated in Table I. Although a continuous evolution of the scattering curves is observed, there is a large increase of the pore size when the porosity reaches a value of about 85%. At the same time the silicon skeleton becomes thinner and thinner, while keeping an isotropic morphology.

Our results suggest that the appearance of the photoluminescent phenomenon is in fact due to three-dimensional structure because of the isotropic behavior of the scattering curves. Considering that the phenomenon has been associated with quantum size effects in the silicon skeleton it would be more appropriate to describe the microstruture of the high porosity luminescent porous silicon by a 3D network of quantum segments in a "noodle-like" or "wool-ball" structure leading to a local one or two-dimensional quantum size effect in a disordered structure.

4.3. Blue-shift after anodic oxidation

We have already seen that when porous silicon is oxidized, the resulting thin oxide layer constitutes an interface between the pores and the silicon skeleton. For X-rays, such an interface leads to a power-law-scattering exponent higher than 4, and the thickness σ of the interface can be deduced from relation (20).

It is known [21, 22] that the photoluminescent (PL) intensity is substantially enhanced when the porous silicon layer is anodically oxidized, while the PL peak is shifted towards higher energy (blue shift) compared to the non-oxidized layer. During the anodic oxidation process, there is formation of a thin oxide film at the inner surface of the PS layer that decreases the size of the crystalline Si domains and acts as an efficient passivation layer that increases the possibility of radiative recombinations [22]. In the quantum confinement model, the size reduction of the crystalline Si domains by anodic oxidation is responsible of the PL blue shift [22,12].

As SAXS is a well suited technique to investigate the microstructure of porous silicon layers, we used this technique to study the microstructure of anodically oxidized PS layers [23]. It is possible to established a correlation between the oxidation level of the material, the size reduction of the crystalline Si domains, the thickness of the anodic oxide and the PL blue shift. Information on the sharpness of the Si/void interface are also obtained. The main results are listed in Table II together with the integrated intensity Qo (corrected from the "fuzzy" surface), the chord dimensions and the porosities of the samples as determined from weight measurements.

Table II

Structural parameters deduced from the SAXS data of the anodically oxidized samples, with the oxidation level C_0, the calculated porosity in %, the width σ of the diffuse interface, the integrated intensity Q_0 and the average chord dimension $<l>$.

Oxidation level	porosity (%)	σ (nm)	Q_0 (el/nm^6 x 10^4)	$<l>$ (nm)
0	70	0	7	2.00
$C_0/4$	66	0.25	8	1.70
$C_0/2$	64	0.33	8	1.50
C_0	55	0.39[a]	8	1.40[b]

[a] The true value is certainly higher
[b] The true value is certainly lower

The samples studied here correspond to different oxidation levels (C_0, $C_0/2$, $C_0/4$). The results were compared to a non-oxidized sample. C_0 is the exchanged Coulombic charge which corresponds to the end of the oxidation.

For measurements of the porosity of the oxidized porous layers, we assumed that the silicon dioxide and bulk silicon densities (2.27 and 2.32, respectively) are the same. For the same starting PS layer of 70% porosity we found that the porosity decreases with increasing oxidation level. This result is consistent with the well known volume expansion resulting from the transformation of silicon to silicon dioxide.

It appears that when the oxidation level increased, the average chord dimension $<l>$ decreased. When the oxidation level is increased from 0 up to C_0, the average chord length $<l>$ decreased from 2.00 nm down to 1.40 nm and thus the chord in matter $<lm>$ decreased from 2.86 nm down to 2.00 nm. This 30% decrease corresponds to the reduction of the silicon skeleton size and is in agreement with a model where the PL blue-shift is due to the thinning of crystalline Si domains.

5. DIFFRACTION OF X-RAYS

5.1. Small-angle scattering and wide-angle scattering

In a pure element, or in an alloy in solid solution, the atoms are distributed on a single lattice. When there are heterogeneities due to a non perfectly random distribution of the atoms, or holes, they give rise to a small-angle scattering response as we have seen before.

However, the predominance of one type of atom in a cluster (or its absence in a hole) can lead to a deformation of the lattice, the atoms being displaced more or less from the lattice points of the average lattice. These defects in periodicity, arising either from the positions of the atoms (or their absence in holes) or from both the nature of the atoms and their positions (clusters in a solid solution) modify the crystalline reflections and cause diffuse scattering outside the positions of the Bragg reflections. So, it appears a diffuse scattering around each Bragg peak, including the (000) one which is called small-angle scattering.

The general treatment of scattering (see bibliography) indicates that the small-angle region is sensitive only to the substitution disorder, while the large-angle scattering is sensitive to both substitutional and topological disorder. This is represented in figure 12 which shows that the scattering appears around each Bragg peak for a polycrystalline material.

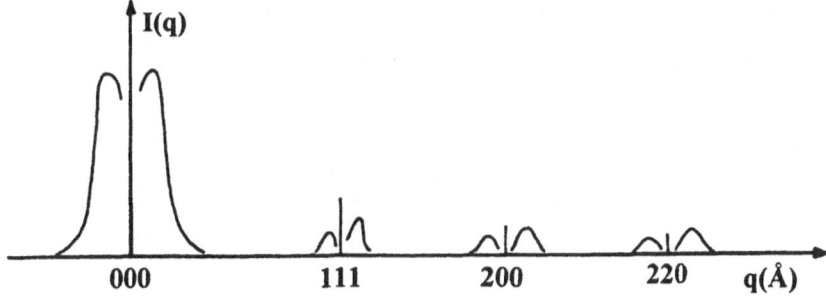

Figure 12 : Scattering around each Bragg peak, including the origin, for a fcc polycrystalline material

The scattering is symmetrical and the more intense around the origin (the small-angle scattering). Due to the elastic strains, it is non symmetrical around the Bragg peaks and the more when the Bragg peak order increases.

It can be concluded that the knowledge of a real structure requires the complete pattern of scattering and diffraction. The small-angle region gives the size and morphology of the scattering centers; the wide-angle region brings information on the interatomic spacing and the stresses in the material.

5.2. Lattice parameter of porous silicon

It was shown by X-ray topography [24] that porous silicon is a monolithic single crystal but that certain lattice planes of the PSL are misoriented with respect to the same type of planes in the substrate. These misorientations are in fact the result of a difference in lattice parameter which was determined on a double crystal diffractometer [25] for a series of porous samples. Two distinct diffraction peaks occurred for each X-ray reflection when studying a porous layer on the substrate: one from the substrate, the other from the PSL at a slightly smaller Bragg angle as shown in figure 13.

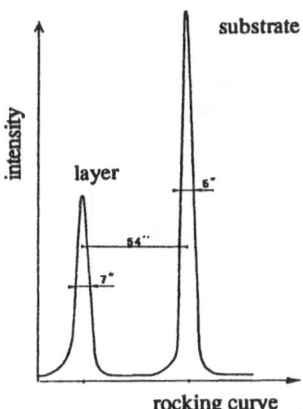

Figure 13 : Double crystal rocking curve of porous silicon layer on a silicon substrate. (400) reflection, porosity 34%. [from ref.25]

The measured width of the reflection curves indicates that the PSL appears to be a perfect crystal in spite of its porosity. Its lattice parameter is slightly larger than that of silicon; it increases with porosity and the mean pore radius. There is also a curvature of the sample which is due to the difference in lattice parameter between the layer and the substrate.

The elastic properties of the porous layer are different from those of the silicon substrate and also depend on details of the porous structure. The lattice expansion a/a of an electroluminescent P^+ porous silicon layer of 60% porosity was recently followed by double crystal diffraction during an anodic oxidation treatment [26].

5.3. Scattering around Bragg's peaks

The coherent diffraction Bragg peaks give the mean lattice of the porous silicon layer while the diffuse scattering around any Bragg peak is obviously related to the porous structure, as it is shown in figure 14.

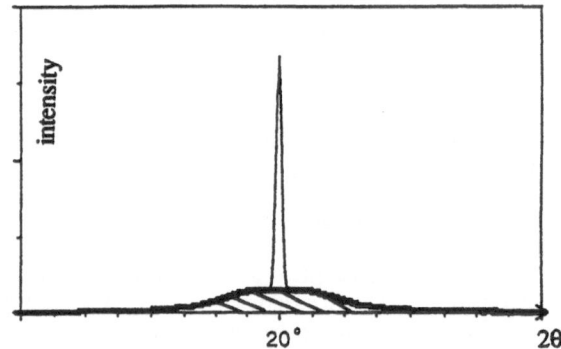

Figure 14 : Typical diffuse scattering around (111) Bragg reflection: P^+ sample with a PSL of 2 μm [from ref.27]

The (111) reflection profile obtained on P^+ porous silicon is superposed to the (111) reflection of bulk P^+ silicon which is always very sharp. The observed diffuse scattering has the same origin as the small-angle scattering which can be observed on a similar sample. It is centered on the Bragg peak, indicating that the strain effects are very weak.

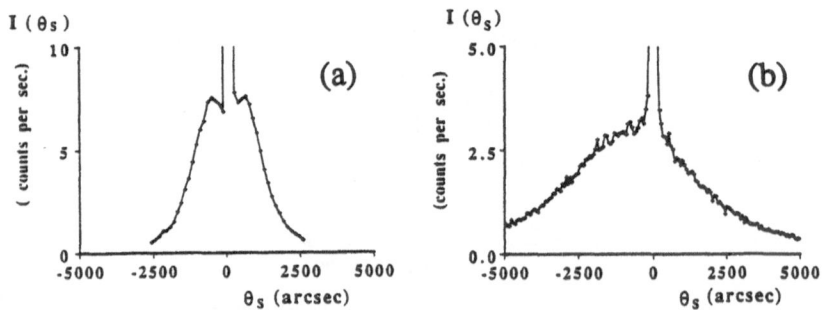

Figure 15 : Two-crystal X-ray rocking curves of a P^+ porous silicon sample of 47% porosity.
(a) before and (b) after anodic oxidation [from ref.27]

This diffuse scattering around the (111) Bragg peak is symmetrical, as is the small-angle scattering curve in a similar sample; that means that the distortions of the lattice planes are small.

The topological disorder is higher in the case of a PSL which is submitted to an anodic oxidation [28], as it can be seen in figure 15 where the diffuse scattering becomes asymmetrical after the oxidation treatment.

6. CONCLUSION

Porous silicon is often characterized by its porosity as measured by gravimetric methods, but porosity values alone are not enough to characterize the material. Other relevant parameters such as pore size, pore size distribution and shape of the pores must also be considered. These have been obtained by gas absorption measurements (with the BET technique), and by transmission electron microscopy -TEM-, in different types of porous layers.

Small-angle scattering of X-rays (SAXS) is a useful technique for learning about the structure of materials on a scale from a few nanometers through a thousand of nanometers or more. Since much of the structure in the porous silicon lies in this range, SAXS can also lead to a better understanding of the porous structures. It can estimate the pore size and provide information about the boundary surfaces of the pores with the concept of "chords". In fact the SAXS technique can provide more direct information on the structure related to geometrical parameters.

A better knowledge of a real structure requires the complete pattern of scattering and diffraction: scattering around the angular origin (SAXS) which gives the size, the morphology and the distribution of the scattering entities, and wide-angle scattering which brings information on the interatomic distances and the stresses in the material.

REFERENCES

1. Guinier A., *Nature*, **142** (1938) 569.
2. Porod G., *Kolloid Z.* **124** (1951) 83.
3. Porod G., Small-Angle X-Ray Scattering, edited by O. Glatter and O. Kratky, (Academic Press, London, 1982) p.17.
4. Mering J. and Tchoubar D., *J. Appl. Cryst.*, **1** (1968) 153.
5. Bale H.D. and Schmidt P.W., *Phys. Rev. Letters*, **53** (1984) 596.
6. Schmidt P.W., Avnir D., Levy D., Höhr A., Steiner M. and Röll A., *J. Chem. Phys.* **94** (1991) 1474.
7. Roberstein J.T., Marra B. and Stein R.S., *J. Appl. Cryst.* **13** (1980) 34.
8. Naudon A., Goudeau P. and Slimani T., *J. Phys. I, France* **2** (1992) 1083.
9. Dubuisson J.M., Dauvergne J.M., Depautex C., Vachette P. and Williams C.E., *Nucl. Instrum. Methods*, **A246**, (1986) 636.
10. Vezin V., Goudeau P., Naudon A., Halimaoui A. and Bomchil G., *J. Appl. Cryst.*, **24** (1991) 581.
11. Canham L.T., *Appl. Phys. Lett.* **57** (1990) 1046.
12. Bsiesy A., Vial J.C., Gaspard F., Herino R., Ligeon M., Muller F., Romestain R , Wasiela A., Halimaoui A. and Bomchil G. , *Surface Sci.*, **254** (1991) 195.
13. Halimaoui A., Oules C., Bomchil G., Bsiesy A., Gaspard F., Herino R., Ligeon M., and Muller F., *Appl. Phys. Lett.*, **59** (1991) 304.

14. Goudeau P., Naudon A., Herino R. and Bomchil G., *J. Appl. Phys.*, **66** (1989) 625.

15. Vezin V., Goudeau P., Naudon A., Halimaoui A. and Bomchil G., *Appl. Phys. Lett.*, **60** (1992) 2625.

16. Schmidt P.W. "Proc. of the IUPAC Symposium on the Characterization of Porous Solids", Bad Soden, FRG, April 1987, edited by K.K.Unger, J. Rouquerol, K.S.W. Sing and H. Kral (Elsevier, Amsterdam, 1988) p.35.

17. Schaeffer D.W. and Keefer K.D., Fractals in Physics, L. Pietronero and E. Tossati eds. (Elsevier Science Publishers, Amsterdam, 1986) pp. 39-45.

18. Gerold V. Small-Angle X-Ray Scattering, edited by H. Brumberger (Gordon and Breach, New-York, 1967) p.277.

19. Goudeau P., Naudon A. and Welter J-M., *Scripta Metall.* **2**, (1988) 1019.

20. Bomchil G., Halimaoui A. and Herino R, *Microelectron. Eng.* **8** (1988) 293.

21. Vial J.C., Bsiesy A., Gaspard F., Herino R., Ligeon M., Muller F., Romestain R,. Waslela A, Halimaoui A. and Bomchil G., *Surf. Sci.*, **254** (1991) 195.

22. Vial J.C., Bsiesy A., Gaspard F., Herino R., Ligeon M., Muller F., Romestain R and Macfarlane R.M., *Phys. Rev.*, **B 45** (1992) 14171.

23. Naudon A., Goudeau P., Halimaoui A., Lambert B. and Bomchil G., *J. Appl. Phys.*, **75** (1994) 2.

24. Barla K., Bomchil G., Herino R., Pfister J.C. and Baruchel J., *J. of Crystal Growth*, **68** (1984) 721.

25. Barla K., Herino R., Bomchil G., Pfister J.C. and Freund A., *J. of Crystal Growth*, **68** (1984) 727.

26. Bensaid A., Patrat G., Brunel M., de Bergevin F. and Herino R., *Solid State Com.*, **79** (1991) 923.

27. Bellet D., Dolino G., Ligeon M., Blanc P. and Krisch M., *J. Appl. Phys.* , **71** (1992) 145.

28. Bellet D., Billat S., Dolino G., Ligeon M., Mayer C. and Muller F., *.Solid State Communications*, **86** (1993) 51.

BIBLIOGRAPHY

Scattering Theory and Methods:

James R.W., The optical principles of the diffraction of X-rays (G. Bell & Sons Ltd. London, 1965).

Guinier A. and Fournet G., Small-Angle Scattering of X-Rays (John Wiley & Sons Inc., New-York, 1955).

Gerold V., Small-Angle X-ray Scattering, Edited by H. Brumberger (Gordon & Breach, New-York, 1967) pp 277-317.

Porod G., Small-Angle X-Ray Scattering, edited by O. Glatter and O. Kratky (Academic Press, London, 1982) p.17-51.

Kostorz G., Physical Metallurgy, 3rd edition, ed. by R.W. Cahn and P. Haasen (Elsevier, Amsterdam, 1983)

Neutrons, X-rays and Light Scattering, Edited by P. Lindner and T. Zemb (Amsterdam, North-Holland, 1991).

HERCULES: Neutron and Synchrotron Radiation for Condensed Matter Studies, J. Baruchel, J.L. Hodeau, M.S. Lehmann, C. Schlenker, J. Regnard, Eds. (Springer-Verlag, Heidelberg, 1993).

Scattering of porous systems:

Schmidt P.W. "characterisation of porous solids" IUPAC Symposium, Bad-Soden, FRG Apr. 26-29. (Elsevier Sci. Pubisher, Amsterdam, Netherland, 1988).

Teixeira J. On Form And Growth, Edited By H.E. Stanley & N. Ostravsky, 1985 (Martinus Nijhof, Boston) pp.145-162.

Schaffer D.W. and Keefer K.D. , Fractals in Physics, Ed. by L. Pietronero and E. Tossati, (Elsevier, Amsterdam, 1986).

Mering J. and Tchoubar D. , *J. Appl. Phys.* **1** (1968) 153.

Lewitz P. and Tchoubar D. , *J. Phys. France*, **2** (1992) 771-799.

X-ray photoemission spectroscopy

H. Münder

Institut für Schicht- und Ionentechnik, Forschungszentrum Jülich GmbH, 52425 Jülich, Germany

1. INTRODUCTION

The reports about the observation of intensive photo- [1] and electroluminescence [2] motivated scientists all over the world to look for the basic mechanisms of the luminescences. During the following 2 years a lot of different models were developed which might explain the luminescences. A high number of those models were chemically based. Therefore, different analysis techniques were used to study the chemical composition of porous Si layers. X-ray photoemission spectroscopy has been applied to investigate freshly etched porous films and to study post chemical treatments like thermal or anodic oxidation of the porous Si samples.

The principle of this method is based on the photoeffect first described by Einstein. During photoemission experiments the sample is illuminated with light. Electrons from electronic states are excited and emitted into the vacuum. The energy distribution of the emitted electrons is analyzed. Depending on the energy of the excitation light and therefore on the type of the used light source the technique is called *x-ray photoemission spectroscopy* (XPS) or *ultraviolet photoemission spectroscopy* (UPS). In literature XPS is also called *electron spectroscopy for chemical analysis* (ESCA).

In the first part of this paper a brief description of the electronic structure of solids is given and basic features which can be observed by XPS are explained. In the following sections the photoemission process will be discussed and the typical instrumentation will be described. In the last part of the paper experimental results obtained on clean Si and porous Si surfaces will be shown. In addition, the influence of post chemical treatments will be presented.

2. ELECTRONIC STRUCTURE OF SOLIDS AND SURFACES

In 1928 F.Bloch published a theoretical description of electrons in a solid based on quantum mechanics. He has shown that the solution of the Schrödinger equation of electrons in a periodical potential, as it is the case in a solid, behave nearly like plane waves. During the years several different ansatzes were developed to describe theoretically the electronic band structure of a solid including surface states [3].

To understand the basic ideas of the band structure models let us first assume free atoms. In this case the electrons are bounded in discrete atomic levels. The transition towards a solid can be made by decreasing the distance between these free atoms. The outmost wave functions will start to overlap first which will result in a nearly planewave behaviour if the distance between the atoms becomes the same as the interatomic distance in the solid. These wave functions describe the delocalized valence band electrons which do not feel any binding potential. The opposite extrema are the completly localized wave functions of the core levels. The electrons described by these wave functions feel a strong binding potential. Compared to the case of free atoms their energy is only slightly changed due to charge transfer processes between neighbouring atoms in the solid. Of course the binding energies E_b of electrons bounded in valence band states or core levels can be quite different (1...1000 eV).

For free Si atoms the outmost wave functions correspond to the 3s and 3p eigenstates. In the solid these wave functions are mixed and result in sp^3 hybrides which are responsible for the atomic binding. The corresponding wave functions exhibit a binding character (valence band) or an antibinding character (conduction band). Both bands are separated by the fundamental energy gap E_g. At zero temperature the valence band states are filled and the conduction band states are empty. At non zero temperatures the energtical distribution of the electrons is described by a Fermi–Dirac statistic. In this theory all states up to a certain energy–level which is called Fermi–energy E_F are filled. For undoped Si the E_F is at midgap position for T = 0 K but shifts towards the conduction band for higher temperatures.

A doping of the Si results in a shift of E_F closer to the band edges. The exact value of E_F at room temperature where all the XPS measurements were performed can easily be calculated [4].

The above discussion about the position of E_F in respect to the bandedges is only right for infinite single crystals. A real solid exhibits surfaces. Keeping in mind that we are interested in porous Si the surfaces become very important, because the specific surface of porous Si can be higher than $500 \, m^2/cm^3$.

At the surface the translation symmetry, which is a fundamental property of all models describing the electronic band structures, breaks down. This results in the formation of new electronic eigenstates at the surface which are called surface states[5]. Typically, the energy levels of those states are lying in the fundamental band gap. Therefore, at the surface the position of E_F is not only given by the doping level and the temperature but

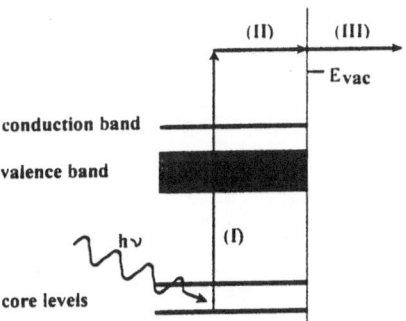

Figure 1: In a solid the XPS process can be described by 3 steps: the excitation of the electrons (I), the transport of the electrons to the surface (II), and the emission of the electrons into the vacuum (III).

also by the density of surface states and their corresponding energies. The surface states will be partly filled with charge carriers. This accumulation of carriers in surface states must be compensated by a space charge layer with opposite sign. This results in a band bending close to the surface which means that the energy difference between the valence band maximum and the E_F depends on the distance from the surface. The actual value of the potential corresponding to this band bending is given by the density of surface states and the doping level of the substrate. Typically, a density of surface states of $10^{12}\,\mathrm{cm}^{-2}$ is high enough to fix E_F to a midgap position which is called Fermi–level pinning.

The breakdown of the translation symmetry is not the only reason for surface states. The relaxation of atomic positions and the reconstruction of the surfaces due to the minimaziation of the surface energy also results in surface states [6, 7, 8]. In addition, adsorbates like oxygen can induce states which are inside the fundamental gap.

All surface states exhibit a translation symmetry parallel to the surface and are damped perpendicular to the surface. The extention length of those states are of the order of 10 to 15 Å. This shows how important these states are for the description of porous Si. The nanocrystals in these porous layers exhibit diameters ranging from more than 100 Å to less than 20 Å [9, 10].

3. PHYSICAL PRINCIPLES

The XPS process can be described in a 3 step model of independent events (Fig. 1) [11, 12, 13, 14]. First the photoelectron is created by the interaction of a core level electron with a photon. The second step is the transport of the electron to the surface and the last step is the emission of the electron into the vacuum.

The interaction of a photon with a core level electron can be discussed in a single particle picture where the perturbation is described by the Hamilton operator H'.

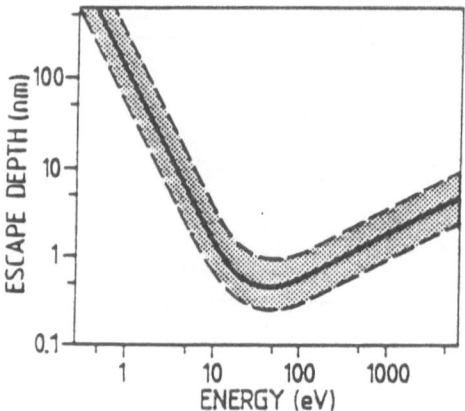

Figure 2: The escape depth of electrons depends on their energy.

$$H' = \frac{e}{2m}\{\mathbf{Ap} + \mathbf{pA}\} \sim \frac{e}{m}\mathbf{pA} \tag{1}$$

The transition propability between two Bloch eigenstates $|i>$ and $|f>$ is given by the golden rule.

$$W_{fi} = \frac{2\pi}{\hbar} m_{fi}\, \delta(E_f(\mathbf{k}) - E_i(\mathbf{k}) - h\nu) \tag{2}$$

$$\text{with } m_{fi} = \left| < f | \frac{e}{m}\mathbf{pA}\, |i> \right|^2 \tag{3}$$

m_{fi} is the transition matrix element which includes information e.g. about the oscillator strength and the energy of the transition. The δ-function represents the energy conservation in the process.

In the case of a solid the excited electrons must be transferred to the surface before they are emitted into the vacuum. On the way towards the surface the electrons can undergo scattering processes. Inelastic scattering processes, e.g. by the interaction with phonons and plasmons, will change the kinetic energy of the electrons. These scattered electrons are called secondary electrons. They will contribute to the spectra as a background signal and give no information about the sample properties. The subtraction of the background from the measured spectra will be discussed later.

The information depth of XPS is given by the mean free path of the electrons. This is the mean distance between two inelastic scattering processes. In Fig. 2 the mean free path or the escape depth of electrons λ is given as a function of the electron energy. The minimum is due to the excitation of valence band plasmons. For porous Si layers the information depth λ_{PSL} can approximately be assumed as

$$\lambda_{PSL} = \frac{\lambda_{Si}}{1 - P} \tag{4}$$

where P is the porosity and λ_{Si} the escape depth of electrons for Si. In the XPS experiments a typical value for λ_{PSL} is 10 nm.

Immediately after the emission of the electrons into the vacuum their kinetic energy E_{kin} is given by

$$E_{kin} = h\nu - E_b - e \cdot \phi \qquad (5)$$

where $e \cdot \phi$ is the workfunction of the sample. In the simplest model E_b is equal to the energy difference $E_b(atom)$ between E_F and $E_i(\mathbf{k})$. A more realistic definition of the measured binding energy is

$$E_b = E_b(atom) + \Delta E_{chem} + \Delta E_{Mad} + \Delta E_{relax} \qquad (6)$$

The measured binding energy has to be corrected due to the existence of statical and dynamical processes which change the electron energy. The dynamical processes are caused by the relaxation of the remaining N-1 electrons to minimize their total energy. This relaxtion energy ΔE_{relax} will change the binding energy of the emitted photoelectrons. The statical processes are represented by the Madelung term ΔE_{Mad}, which is important for ion crystals, and by the chemical shift ΔE_{chem}. In both cases the effective charge of an atom is changed due to the influence of neighbour atoms.

Mainly ΔE_{chem} is important for the investigation of porous Si. The oxidation of the porous layer results in a chemical shift of the core levels which can be used to follow the formation of suboxides. The chemical shift of Si 2p core levels as a function of oxidation state is given in Tab. I. The ionization probability or the cross section of a core level does not depend on the oxidation state. Therefore, it is possible to calculate the amount of oxidized Si by a comparision of the intensities of the chemically shifted core levels. Another possibility for the estimation of the oxygen content is to integrate the area below different core level peaks corresponding to Si and to oxygen. In this case the cross section of the different core levels must be taken into account.

If a photoelectron from a core level has been created the remaining Si atom is in an excited state. The deexcitation can occur by two processes. An electron from a higher level (e.g. valence band) can fill up the core level. The energy can be set free first by the emission of a x-ray photon or secondly, by the emission of another electron from an higher state. The latter process is called Auger–process. These Auger electrons can be seen in all XPS spectra. Their kinetic energy does not depend on the energy of the excitation x-ray light. Changing this energy will change the kinetic energy of all phototemission electrons but will not change the energy of the Auger electrons. Therefore, using x-ray tubes with different anode materials it is possible to distinguish between both contributions in the spectra.

4. INSTRUMENTATION

4.1 Light sources

Depending on the energy range different types of light sources are used for photoemission spectroscopy. For XPS the standard light sources are x-ray tubes with Mg (1253.6 eV) or Al (1486.6 eV) anodes. In some cases it is necessary to use different excitation energies for the experiments, e.g. to distinguish between Auger and core level electrons in the

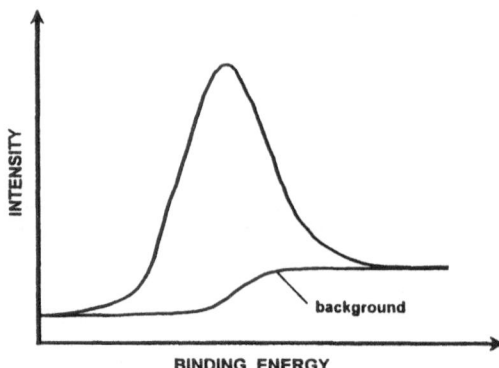

Figure 3: The background beneath a photoelectron peak shows a steplike behaviour.

recorded spectra. Therefore, twin anodes with two different metallizations are used. The anodes are water cooled to avoid a drastic heating, but still the temperature of the anodes increases which results in the desorption of adsorbed molecules. To protect the sample surface against this contamination the x- ray tubes are separated by an thin Al window from the analysis chamber.

More elaborated light sources are electron synchrotrons where the energy of the emitted light covers a wide range ($< 50 \ldots > 1000$ eV). The method based on these light sources is called *soft x-ray photoemission spectroscopy*. It is no longer possible to distinguish between UPS and XPS.

XPS has two inherent disadvantages. The first is the poor lateral resolution. Typically a large area of the sample is irradiated and the lateral resolution is given by the aperture of the analyzer. The second disadvantage is the full width at half maximum (FWHM) of the x-ray excitation lines . The typical value of 0.61 - 0.8 eV is much too large to be sensitive for small shifts of the core levels or of the valence band. Therefore, for such investigations monochromators in combination with x-ray tubes are used (FWHM ~ 200 meV). For small shifts in the energetic position of the valence band maximum it is better to use UV gas source lamps (FWHM ~ 3 meV).

4.2 Analyzers

There are several different methods to analyse the energy of the emitted electrons. Magnetic or electrostatic fields are used to disperse the electrons. Experimental setups for XPS include most often concentric hemispheric analyzers. Electrons emitted from a part of the sample ($\varnothing \sim 3 \ldots 5$ mm) are focused on the entrance slit of the analyzer. Depending on the potential of the inner and outer hemisphere electrons with a particular kinetic energy are focused on the exit slit. Behind the exit slit an electron multiplier (channeltron) is used for measurements because of the low number of the emitted electrons.

Table I: Chemical shifts of the Si 2p core level given for different suboxides.

oxidation state	chemical shift	suboxide type
Si^{+1}	1.0 eV	Si_2O
Si^{+2}	1.8 eV	SiO
Si^{+3}	2.6 eV	Si_2O_3
Si^{+4}	3.5 eV	SiO_2

5. LINESHAPE ANALYSIS

To obtain quantitative information from the measured XPS spectra it is necessary to do a careful lineshape analysis of the core level peaks. First of all the background due to inelastic scattering processes must be subtracted. The energy loss of the electrons due to these inelastic scattering processes results in a steplike increasing background beneath any photoelectron peak (Fig. 3) [15]. The peakform of the photoelectrons itself is best described by a Gaussian–Cauchy behaviour [16]. In addition, experimental broadening effects e.g. due to the spectral width of the used light cannot be ignored and are taken into account by a Gaussian contribution.

The experimental part of this paper will focus on results obtained by an analysis of the Si 2p core level. The Si 2p core level ($E_B \sim 99$ eV) is splitted into two lines, $2p^{1/2}$ and $2p^{3/2}$, because of the spin–orbit coupling. Their binding energies differ by 0.61 eV and their intensity ratio is 1:2. For the fits shown later it has been assumed that the FWHMs of the two peaks are the same.

By the lineshape analysis it is possible to follow the formation of suboxides. The different states of oxidation of the Si atoms result in chemical shifts ΔE_{chem} up to 3.5 eV (Tab. I) [18]. From the intensity ratio of the photoelectron peaks the thickness of the oxide layer can be calculated.

6. EXPERIMENTAL

The porous Si layers were formed in the dark on p-type doped substrates with a (100) orientation. The resistivity of the substrates were 0.01, 0.2, and 7 Ωcm. For the formation process a mixture (1:1) of ethanol and 50% HF was used. After the formation of the porous layers the samples were transferred into an ultra high vacuum (UHV) chamber operating at a basis pressure of $p < 10^{-8}$ Pa. The XPS measurements were performed by using an Al/Mg twin anode and a hemispherical analyzer (VG: CLAM2). The lateral resolution was about 5 mm.

7. RESULTS AND DISCUSSION

7.1 Si (100)

The qualitiy of the hydrogen passivation of the Si (100) surface has been investigated. The samples have been cleaned by a HF etch process. The obtained XPS spectra are

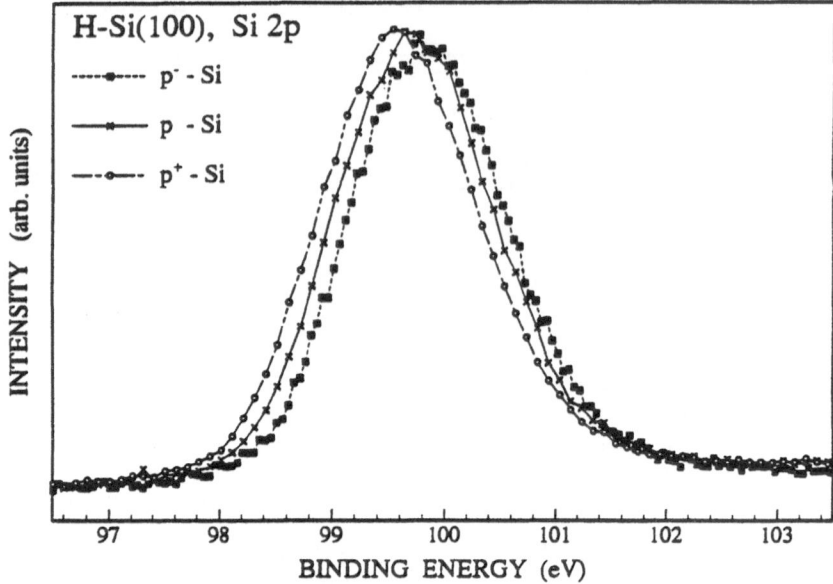

Figure 4: XPS spectra of clean single crystalline Si (100) surfaces. A shift in the Si 2p core levels due to the different doping levels of the samples is visible.

shown in Fig. 4. Clearly a shift of the core levels for the differently doped samples can be seen. The fits show a change in E_b from 99.34 eV for the highly doped substrates to 99.61 eV for the lowest doped material. The difference in E_b is of the same magnitude than the difference in the position of E_F relative to the valence band maximum. If there would be a high density of surface states the position of the valence band and therefore the position of the core levels would not be influenced by the doping level of the samples. Therefore, the shift indicates that there is a flatband situation and that no Fermi–level pinning occurs. There are no electrically active surface states which means a perfect passivation of the surface states by adsorbed hydrogen. In addition, this study is a control of the experimental procedure, because any oxidation of the sample would result in an additional chemical shift. Because no chemically shifted contributions are found the hydrogen passivation of the (100) surface should be perfect and stable at least on a 10 min scale.

7.2 Freshly prepared and aged porous Si layers

As already mentioned several models which could explain the luminescence behaviour were based on certain chemical compositions of the porous films. To investigate freshly prepared samples they were transferred into the UHV chamber in about 5 min after the formation of the porous layers. The XPS spectra are indicated as 0h in Fig. 5. Compared to crystalline Si the core levels are shifted towards higher binding energies by about 0.3 eV.

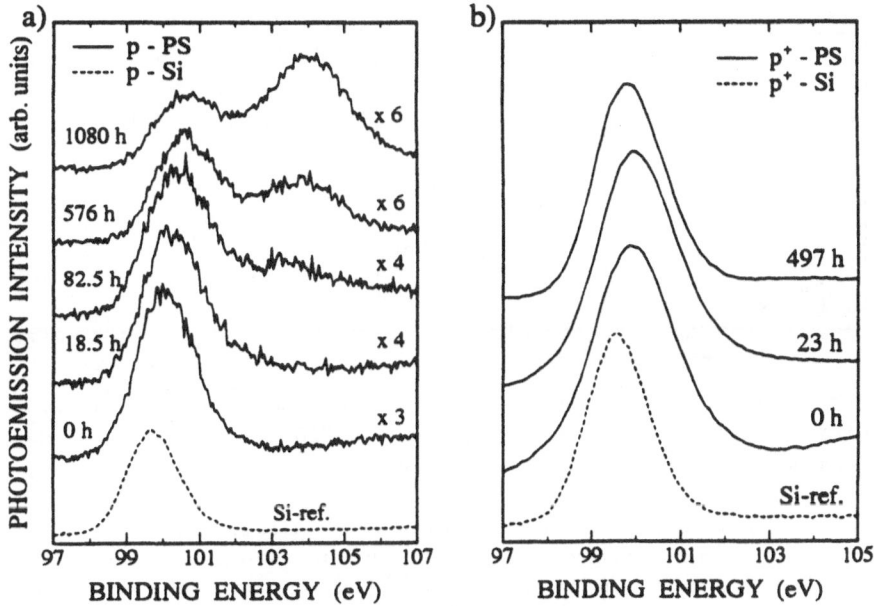

Figure 5: XPS spectra obtained on porous Si layers formed on differently doped substrates. The storing time in air and in the dark is indicated.

This shift does not depend on the doping level of the substrate. Possible explanations are an increased band gap due to quantum confinement effects [17] or strain effects. Changes of the atomic positions like it is the case in strained layers or surface layers will lead to different electronic environments of the atoms which will result in different E_b of the core levels [19]. A definite answer to the question, why the core levels are shifted, cannot be given.

A detailed lineshape analysis indicates that besides the Si^{+0} oxidation state further contribution occur in the spectra. For p^+-doped porous layers an additional peak which is shifted down in energy by 0.6 eV is found. This peak is due to the adsorbtion of hydrogen atoms. On the other hand for p-doped porous layers a contribution shifted up by 1eV is found. This corresponds to the Si^{+1} oxidation state. The existence of the Si^{+1} oxidation state can be explained by a fast oxidation of the samples or by the adsorbtion of other contaminants like C or F.

In literature the reports concerning contributions from F core levels are controverse. In some cases weak signals from F 1s are reported [20, 21, 22] whereas in this study and by some other groups [23] no F contributions to the spectra have been observed. There is also a controverse discussion concerning the amount of oxygen in freshly prepared porous Si layers. Results showing a high oxygen content might be due to a very fast native oxidation process [21]. In addition, the illumination of the samples during the electrochemical formation process results in an oxidation which depends on the light intensity [24]. Therefore, it is important to mention that the samples investigated here are formed in the dark.

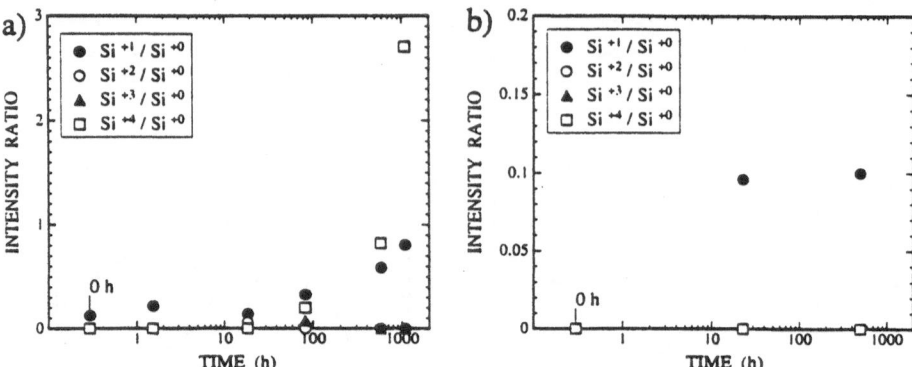

Figure 6: The formation of suboxides due to the native oxidation can be followed by a lineshape analysis of the Si 2p core level. The contribution from suboxides to the XPS spectra is shown as a function of storing time for p-type (a) and p+-type (b) doped samples.

XPS studies performed on p-doped porous Si layers are of great importance for the explanation of the basic luminescence mechanism. These porous samples show an intensive photoluminescence signal. In the siloxene model [25], e.g., this photoluminescence is explained by the presence of a certain molecule consisting of Si, oxygen, and hydrogen. Therefore, a high amount of oxygen should be present on freshly prepared samples which is definitely not found. In addition, the presence of hydrogen and oxygen in the siloxene molecule will lead to a chemical shift of the Si 2p core level of 0.5 eV [26]. As discussed before this has also not been observed. Both results in combination with results obtained from infrared measurements [27] definitely rule out the siloxene model.

To investigate the ageing behaviour of porous layers the samples have been stored in air and in the dark for a certain time before transfer into the UHV chamber. The ageing of the samples can clearly be followed on p- doped porous Si where pronounced changes of the lineshape of the peak occur. The peak position of the Si^{+0} shifts towards higher E_b by about 0.4eV. This shift can be explained by the arguments already discussed before – strain and quantum confinement. From Raman studies it is known that the strain of those samples increases with time. On the other hand it is obvious that an oxidation must reduce the size of the nanocrystals. In addition, the formation of electrical active surface states – due to the desorption of hydrogen – or the formation of interface states – due to a bad Si / SiO_2 interface – must be considered. Those surface or interface states will cause a shift of E_F towards a midgap position and therefore, the core levels would be shifted to higher binding energies. Again no final explanation can be given.

In addition, the lineshape analysis provides a detailed insight into the oxidation behaviour. The formation of suboxides can be followed (Fig. 6). During the first few hours after the formation Si^{+1} is the dominating contribution besides the Si^{+0}. This indicates the formation of Si_2O. For older samples contributions from suboxides corresponding to higher oxidation states occur. After 500h only the Si^{+4} and the Si^{+1} are present beside the Si^{+0}. This indicates the formation of a SiO_2 layer. For the sample stored in air for about 1000h the thickness can be determined to be about 3nm from the intensity ratio Si^{+4} / Si^{+0}.

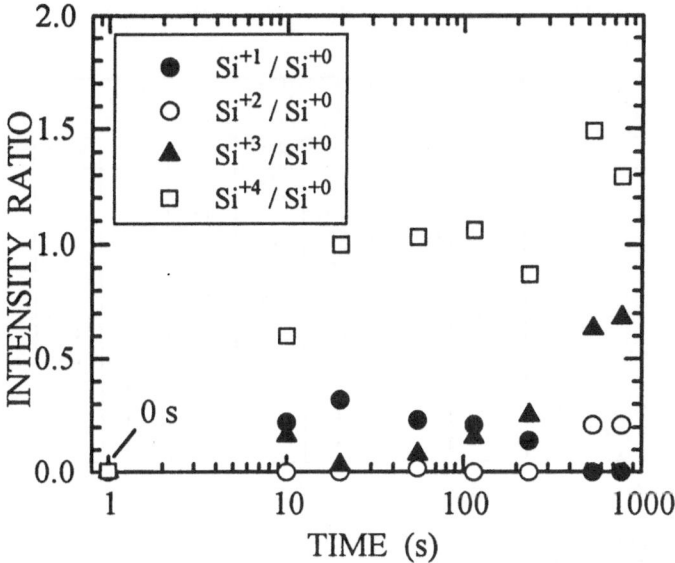

Figure 7: Fitting the XPS spectra taken on anodically oxidized porous Si layers indicate the formation of SiO_2 and a high amount of suboxides.

It should be mentioned that the oxidation does not stop even after 1000h. This indicates that the native oxidation is not a self–limiting process. Therefore, native oxide films can not be used to stabilize the porous structure.

For p^+-doped samples a different oxidation behaviour is found. Even after 500h the only contribution to the XPS spectra beside the Si^{+0} is Si^{+1}. This indicates that the passivation of the inner surfaces must be excellent. Comparing the results from the different doped samples the difference in the oxidation behaviour is obvious. For an explanation the difference of specific surfaces must be taken into account which is higher for the low doped sample by a factor of 3. In addition, it is known that the oxidation velocity depends on the doping level of the substrate. This should lead to a factor of 3 higher oxidation velocity for lower doped samples. Both together can not explain the difference which is more than a factor of 10. Therefore, an additional aspect must be considered – the orientation of the inner surfaces. For all times the Si^{+1} is the most contributing suboxide feature in the XPS spectra. A comparision of this result with studies performed on Si single crystal surfaces shows that this is the case for (111) oriented surfaces. In comparison to a (100) surface this will result again in a factor of about 3 different oxidation velocities. Therefore, the inner surface orientations must be most probably (111) for at least the p^+-doped samples.

7.3 Anodically oxidized porous Si layers

A main problem of a porous Si technology is the long term stability of those porous layers. Here, mechanical as well as electrical stability must be improved. Therefore, the controlled oxidation of porous Si layers has been investigated. One possible approach is

Table II: The deconvolution of the Si 2p core level shows the depth distribution of oxide after different anodical oxidation processes.

potential V	sputter time min	Si 2p composition %				
		Si^{+0}	Si^{+1}	Si^{+2}	Si^{+3}	Si^{+4}
1	0	13	29	5	10	43
	1	26	17	14	27	16
	30	33	14	14	22	17
	60	48	7	16	18	11
	300	46	14	12	18	10
3	0	11	10	1	14	64
	1	3	7	9	16	65
	1	3	2	3	13	79
	1	4	2	6	13	75
	1	4	3	4	13	76

the anodical oxidation of porous Si. The samples which have been investigated by XPS have been oxidized in 1N H_2SO_4 at a current density of $1mA/cm^2$. The galvanostatic oxidation process has been stopped after different oxidation times and the samples have been transferred into the UHV chamber.

The fitting results are shown in Fig. 7. Already after a few seconds a SiO_2 layer is formed at the surface. The thickness of the SiO_2 layer increases until 20s of anodic oxidation. It seems that the thickness is unchanged up to 300s and then increases again. In addition, an important result is the high intensity of suboxide contributions for times >100s. The contribution from Si^{+2} and Si^{+3} can be explained by the presence of oxygen vacancies in the grown oxide layer. This result suggests that the formed oxide layer and / or the interface between the oxide layer and remaining Si nanocrystals should exhibit bad electrical properties.

Due to the high surface sensitivity of XPS results concerning the quality of an oxide layer are always limited to the surface region of the wafer. To obtain depth resolved information a sputter depth profile can be recorded. Here, the 4KeV Ar^+ ions were used for the sputtering process. After certain sputtering times the process was interrupted and a XPS spectrum was recorded. The porous Si samples used for this depth profiling were oxidized up to 1 and 3V.

In Tab. II the results from the fit procedure are summarized. For the sample oxidized up to 1V the amount of non oxidized Si (Si^{+0}) increases with depth whereas the SiO_2 (Si^{+4}) contribution decreases. In the cases of the other sample the situation is totally different. Here, both contributions are homogeneous throughout the whole layer. Comparing the intensity ratios Si^{+4} / Si^{+0} the higher amount of oxide in the latter sample is obvious. In addition, the depth profile reveals also the continous distribution of Si^{+3}. This means that oxygen vacancies are present throughout the whole layer. Their existence might be correlated with the increasing voltage which is due to the decreasing electrical conductivity.

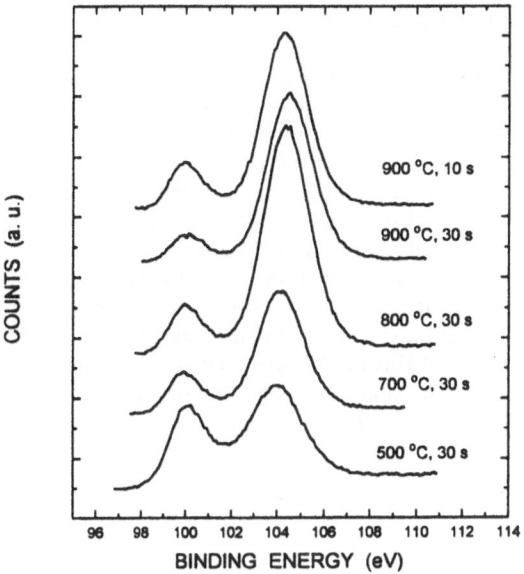

Figure 8: After the RTO process in the Si 2p core level regime two well defined peaks are present corresponding to Si^{+0} and Si^{+4} oxidation states.

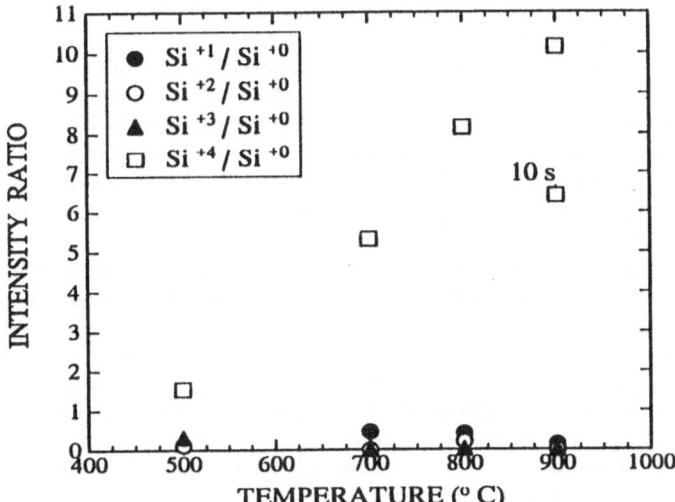

Figure 9: The lineshape analysis of the Si 2p core level (Fig. 8) shows that nearly no suboxides are formed. The main contribution is from the Si^{+4} oxidation state. Its intensity depends on the oxidation temperature and time.

7.4 Rapid thermal oxidized porous Si

Another technique for the formation of high quality oxide layers is the so called *rapid thermal oxidation* (RTO) process. During the RTO process the samples are heated within a few seconds to temperatures between 500 and 900°C. The temperature is kept constant for 10 to 30s before the samples is cooled down again. For these studies p$^+$-doped wafers have been used. The porosity has been 65% and the thickness 5μm.

The Si 2p core levels obtained on RTO samples are shown in Fig. 8. Clearly a two peak behaviour is found. The binding energy of the peaks correspond to the Si^{+0} and Si^{+4} oxidation states. The results from a detailed lineshape analysis are shown in Fig. 9. It is obvious that the Si^{+4} intensity increases with increasing oxidation temperature. Suboxide contributions from other oxidation states are missing indicating that a high quality oxide layer is formed. The thickness of this oxide layer does not only depend on the oxidation temperature but also on the oxidation time. The different Si^{+4} / Si^{+0} ratios for 10 and 30s oxidation time at 900°C indicate that the oxide layer depends on the oxidation time in a non linear way.

8. SUMMARY

X-ray photoemission spectroscopy (XPS) is a surface analysis technique with an information depth of about 10 nm. Using XPS occupied electronic states – valence band and core levels – can be investigated. The binding energies of the core levels (E_b) depend on the chemical and electronical environment of the atoms. E.g., a modification of the chemical composition of a sample will change E_b. In addition, the formation of surface or interface states can also be monitored by measuring E_b.

During the last years several models have been developed to explain the luminescence behaviour. One class of those models are chemically based. Therefore, the chemical composition of freshly prepared porous Si layers has been studied by XPS. No significant contribution from oxygen is found and therefore, models involving oxygen like the siloxene or a-SiO$_x$:H [28] model can be ruled out. On the other hand new results indicate a widening of the fundamental gap due to quantum size effects for freshly prepared porous Si samples [17, 29].

For possible device applications it is necessary to have a stable microscopic structure. Oxide layers can improve the stability of porous Si layers but may change other properties like the photoluminescence behaviour. Therefore, three different oxidation processes have been investigated. The native oxidation results in the formation of suboxides and the oxidation velocity depends on the doping level of the samples. For p-doped porous layers a SiO$_2$ layer is formed after 1000 h. Measurements performed on anodically oxidized porous Si layers show signals indicating the presence of a high amount of suboxides. By rapid thermal oxidation a SiO$_2$ layer is formed where the thickness of the layer seems to depend nonlinearly on the oxidation time.

Concerning the contact formation some work has already been done [30]. E.g., depth profile studies have been performed on porous Si layers after electroplating of a metal or the deposition of conducting polymers in the pores. Information about the formation process itself – how far does the metal penetrate into the pores – have be obtained. Those investigation will help to understand the electrical properties of porous Si nanocomposite structures.

REFERENCES

1. L.Canham, Appl. Phys. Lett. **57**, 1046 (1990).

2. A.Richter, P.Steiner, F.Kozlowski, and W.Lang IEEE Electron Device Letters **12**, 691 (1991).

3. N.W.Ashcroft and N.D.Mermin, *Solid State Physics*, Saunders College, Philadelphia, 1 edition, 1976.

4. Sze, *Physics of Semiconductor Devices*, John Wiley & Sons, New York, 2. edition, 1981.

5. W. Shockley, Phys. Rev. **56**, 317 (1939).

6. V.Heine, Phys. Rev. **138**, A1689 (1965).

7. J.Pollmann, R.Kalla, P.Krüger, A.Mazur, and G.Wolfgarten, Appl. Phys. A **41**, 21 (1986).

8. J.Pollmann, P.Krüger, A.Mazur, and G.Wolfgarten, Surface Sci. **152**, 977 (1985).

9. H.Münder, C.Andrzejak, M.G.Berger, U.Klemradt, H.Lüth, R.Herino, and M.Ligeon, Thin Solid Films **221**, 27 (1992).

10. H.Münder, M.G.Berger, S.Frohnhoff, M.Thönissen, and H.Lüth, J. Luminescence **57**, 5 (1993).

11. G.Ertl and J.Küppers, *Low Energy Electrons and Surface Chemistry*, VCH Verlagsgesellschaft, Weinheim, 2 edition, 1985.

12. M.Prutton, *Surface Physics*, Clarendon Press, Oxford, 2 edition, 1983.

13. M.Cardona and L.Ley, editors, *Photoemission in Solids I*, volume 26 of *Topics in Applied Physics*, Springer-Verlag, Berlin, Heidelberg, New York, 1978.

14. L.Ley and M.Cardona, editors, *Photoemission in Solids II*, volume 27 of *Topics in Applied Physics*, Springer-Verlag, Berlin, Heidelberg, New York, 1979.

15. D.A.Shirley, Phys. Rev. B **5**, 4709 (1972).

16. R.D.B.Fraser and E.Suzuki, Analytical Chemistry **41**, 37 (1979).

17. T.Buuren, T.Tiedje, J.R.Dahn, and B.M.Way, Appl. Phys. Lett. **63**, 2911 (1993).

18. G.Hollinger, J.F.Morar, F.J.Himpsel, G.Hughes, and J.L.Jordan, Surface Sci. **168**, 609 (1986)

19. F.J.Grunthaner, P.J.Grunthaner, R.P.Vasquez, B.F.Lewis, J.Maserjian, and A.Madhukar, Phys. Rev. Lett. **43**, 1683 (1979).

20. K.Murakoshi and K.Uosaki, Appl. Phys. Lett. **62**, 1676 (1993).

21. T.P.Kolmakova, V.G.Baru, B.A.Malakhov, A.B.Ormont, and S.A.Tereshin, JETP Lett. **57**, 410 (1993).

22. J.Terry, H.Liu, R.Cao, J.C.Woicik, P.Pianetta, X.Yang, J.Wu, M.Richter, N.Maluf, F.Pease, A.Dillon, M.Robinson, and S.George, Mat. Res. Soc. Symp. Proc. **259**, 421 (1992).

23. C.Lévy-Clément, A.Lagoubi, D.Ballutaud, F.Ozanam, J.N.Chazalviel, and M. Neumann-Spallart, Appl. Surf. Sci. **65/66**, 408 (1993).

24. C.Tsai, K.H.Li, J.C.Cambell, B.K.Hance, M.F.Arendt, J.M.White, S.L.Yau, and A.J.Bard, J. Electronic Materials **21**, 995 (1992).

25. M.S.Brandt, H.D.Fuchs, M.Stutzmann, J.Weber, and M.Cardona, Solid State Comm. **81**, 307 (1992).

26. J.A.Wurzbach, In *The Physics of MOS Insulators*, G.Lucovsky, S.T.Pantelides, F.L.Galeener eds., Pergamon Press, New York, 1980.

27. W.Theiß, P.Grosse, H.Münder, H.Lüth, R.Herino, and M.Ligeon, Appl. Surf. Sci. **63**, 240 (1993).

28. J.M.Perez, J.Villalobos, P.McNeill, J.Prasad, R.Cheek, J.Kelber, J.P.Estrera, P.D.Stevens, and R.Glosser, Appl. Phys. Lett. **61**, 563 (1992).

29. K.Inoue, K.Maehashi, and H.Nakashima, Jpn. J. Appl. Phys. **32**, L361 (1993).

30. M.Jeske, K.G.Jung, J.W.Schultze, M.Thönissen, and H.Münder, Surf. Interface Anal. (in press) .

Optoelectronic properties of porous silicon - The electroluminescent devices

W. Lang, P. Steiner, F. Kozlowski

Fraunhofer Institute for Solid State Technology Munich, Germany

1. INTRODUCTION

The discovery of photoluminescence[1] in porous silicon and the understanding of the growth of nanostructures[2] opened the field to a large amount of work on this material. For a practical application, electroluminescence (EL) is the crucial point. The development of EL devices in porous silicon technology is faced with some specific problems: The material has a large internal surface and therefore shows a tendency to undergo chemical change when exposed to air. Furthermore, nanoporous silicon shows a very low electrical conductivity, which causes problems for efficient EL. On the other hand, the EL with wet contacts[3,4] is very efficient. This shows that in principle porous silicon is a good material for EL.

A number of EL devices has been presented yet. Most of them show a simple structure of a porous layer and a contact layer on top. For contact, thin metals (Au), indium tin oxide (ITO), silicon carbide and conductive polymers are used. An advanced structure has a p/n junction within the porous region. This way the quantum efficiency is improved.

Within this paper, the devices are described. The technology and the characterisation of the devices is reported. The main problems and the basic tasks for future work are discussed.

2. BASIC DEVICES

For the technology of EL devices the capping layer is an essential point. It determines the efficiency of the device and its stability. Due to the large internal surface, porous silicon has the tendency to change its properties when stored in air. Furthermore, a current flow during EL will also change the material. Therefore, the capping layer must fulfil some very challenging demands:

- it has to be transparent;
- it has to be electrically conductive;
- it must not destroy the luminescence,
      better even increase it;
- it has to stabilize the surface against chemical
  and electrical degradation.

Up to now, four different approaches have been reported: Thin metal, ITO, SiC, conductive polymers.

Fig. 1 Photo of a light emitting pad with a needle
        probe.

2.1. Metal contacts

Porous silicon electroluminescence was shown in 1991 at the same time by A. Richter, P. Steiner, F. Kozlowski and W. Lang[5,6] in Munich and by N. Koshida and H. Koyama[7,8] in Tokio. Both groups used thin gold as a contact layer.

Fig. 1 shows the first electroluminescent active device. The description of the technology follows the process of the EL devices described in[5,6]. The electrochemical cell used is shown in Fig. 2. The wafer is immersed in a solution of 25% HF in ethanol. Two platinum electrodes on both sides of the wafer form the electrical contact. A halogen lamp is used to illuminate the wafer during the etching process. The material is 4 inch <100> n-type wafers, doped with phosphorus 1.4-2.3 Ωcm. When anodization is performed in the dark, n-type silicon will transform to a macroporous film with silicon skeleton sizes of more than 50 nm. This film is not luminescent. Nanostructures can be obtained from n-silicon by illuminating the wafer during the etching process. The wafer is illuminated from the anodic side. A current of 25 mA/cm^2 is applied for 20 minutes.

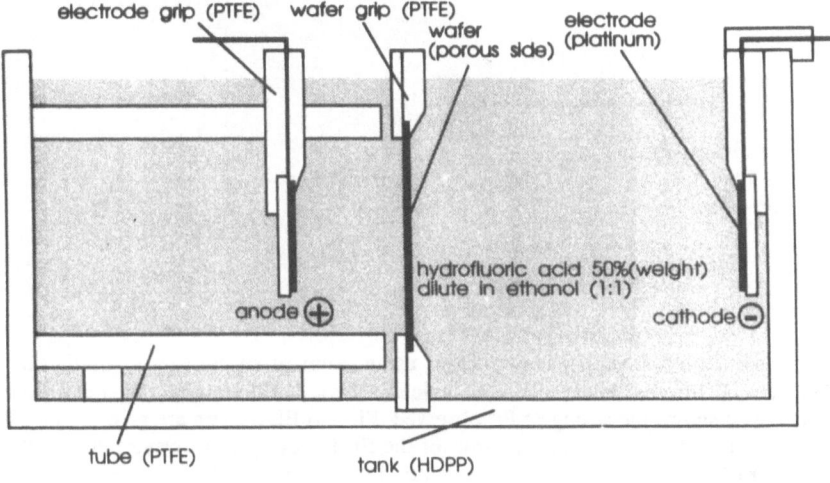

Fig. 2 The electrochemical cell used to etch porous
        silicon.

To make an electrical contact, a film of 12 nm gold is deposited on the wafer by sputtering or by thermal evaporation. The contact layer is structured using a shadow mask during deposition. This layer is used as an electrical solid state contact. It is thin enough to permit the light to pass. The transmission is above 80%. Nevertheless, the gold is thick enough to provide an electrical contact for the whole area of the pad.

It is very important to generate an ohmic contact from the back of the wafer to the electrochemical bath. This can be done either by a metallic film of gold or aluminum on the back of the wafer or by contact implantation.

Fig. 3 shows a surface of cleavage in an SEM view. Three structures can be observed: first the bulk silicon, then macroporous material, which is about 30 µm thick. This corresponds to n-type silicon etched in the dark. The pores are resolved by the SEM. This layer does not take part in luminescence. The top layer is nanoporous. It is about 6 µm thick. The pores are not resolved by the SEM and the film looks quite homogeneous. When the samples are stored in air for some days, the top layer of the nanoporous material is largely oxidized. Oxygen depth profiles are given in [6,9]. Based on spectroscopical data and on the electroluminescent observations, models of the internal structure have been developed[10,11].

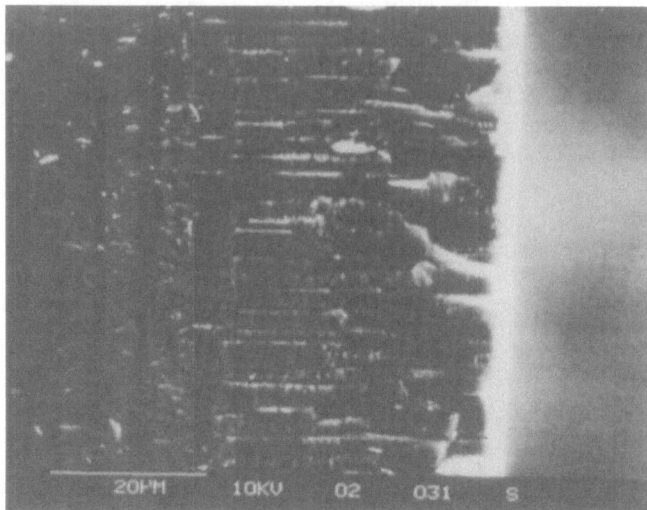

Fig. 3
Surface of cleavage
of porous silicon,
seen by an SEM.

Luminescence measurement is performed using standard spectroscopic equipment[12]. The comparison of different kinds of porous silicon shows that the n-type material anodized as just described shows the strongest PL observed. PL and EL spectra are given in Fig. 4. The spectra are quite similar, a small red shift of the EL is observed. However, this must not be the case; different colours of PL and EL are possible. I/V curves show a rectifying behaviour. At moment it is not clear, whether the rectifying characteristics is caused primarily by the Schottky contact from the gold to the nanoporous layer or by a homojunction from the nanoporous layers to the bulk silicon. Normally light emission is observed in the forward bias, but when a reverse bias is applied, the little current flowing will cause some EL light, too.

The intensity of the emitted light is linear to the current through the device. This is the same behaviour as that shown by LEDs in GaAs technology. Therefore the best value to characterize the efficiency of the EL is the quantum efficiency η_{EL}, defined as the ratio of the emitted photons to the electrons crossing the sample. The measurement of the quantum efficiency is described in[12]. We performed measurements with a number of different samples. The quantum efficiencies measured are in the region of 10^{-4} to $10^{-3}\%$. These are external quantum efficiencies. In every light emitting device there is a loss inside the structure. An estimation which is given within this volume[14] results in a loss of 97%. Therefore we suppose that the internal quantum efficiency is at least one order of magnitude larger than the external one. The onset voltage is less than 2 V.

Besides the n-type samples we also use p-type samples. The measured spectra, I/V curves and quantum efficiencies are similar to the ones described above.

The device demonstrated by N. Koshida and H. Koyama[3] is made from p-type silicon and has a contact layer of thin gold, too. Orange EL emission is measured under forward bias. The quantum efficiency is given as $10^{-4}\%$. Photoconductivity measurements of this device[8] show a peak at 500 nm. This device shows a characteristic quenching of the PL intensity, when an electric field is applied[13].

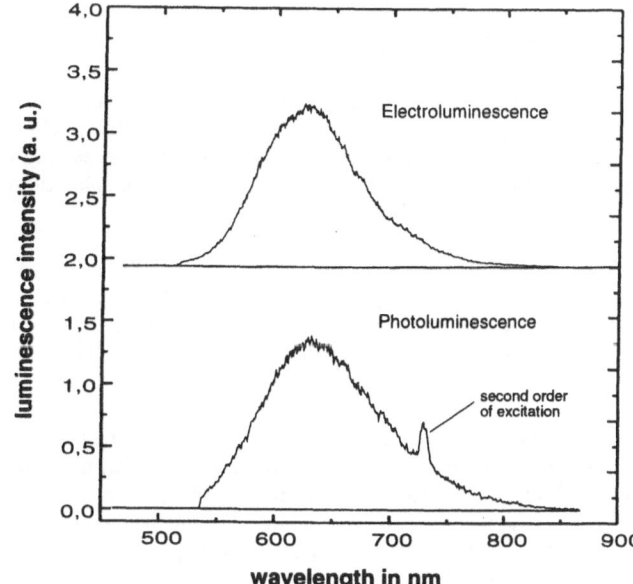

Fig. 4:
Photoluminescence (PL)
and
electroluminescence (EL)
spectra of porous silicon.

2.2. ITO and SiC contacts

ITO (indium tin oxide) on porous silicon has been presented by F. Namavar et al from SPIRE corporation[14,15] and by IFT[12]. ITO normally is deposited by sputtering. Possible alternativ ways are reactive evaporation, spray deposition and CVD. The EL, the I/V characteristics and the efficiencies of our devices with sputtered ITO are similar to the ones with gold samples. From that we conclude that the basic mechanism is the same.

The EL device made by SPIRE Corporation[14,15] is shown in Fig. 5. It is a heterojunction from porous silicon (p-type) to ITO (indium tin oxide). The ITO is 300 nm thick and made by sputtering. Under forward bias, a light emission around 580 nm is found with an onset voltage lower than 2 V. The I/V characteristics shows a diode with an ideality factor of 10.

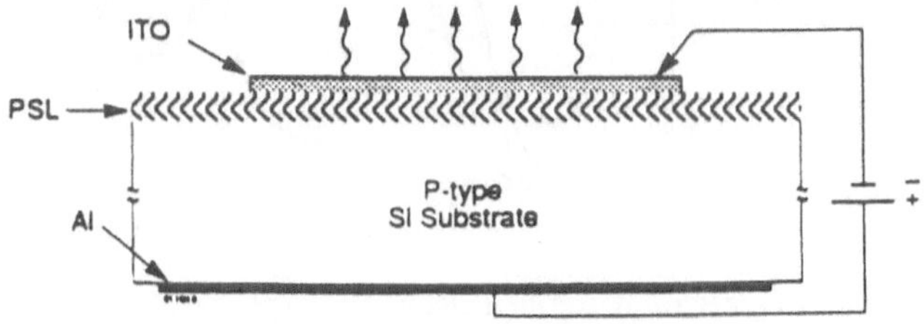

Fig. 5
The electroluminescent
device by SPIRE Corp.[14,15].

Figure from
N. M. Kalkhoran et al.[14].

A device using SiC as a contact layer is given by T. Futagi et al.[16,17]. The samples are etched from p-type silicon, a thin layer of SiC is deposited on top, then a contact layer of ITO is applied. The samples show a weak EL with different spectral distributions depending on the doping of the silicon used.

3. UV - TYPE DEVICES

All the samples described up to now show a red/orange EL light. In 1992 we found that using these basic structures, also green and blue EL could be obtained. To do this, we illuminate the wafer with ultraviollett light during the etching process. For this reason this type of samples is called UV-type, whereas the samples illuminated with visible light as described above are called VIS-type.

The idea of this experiment is to get even smaller structures by a change in the etching process. From the etching theory [18,19] it can be derived that smaller samples should be obtained by UV illumination. When short wavelength light is used, the structures should be smaller and the luminescence should shift to the blue. The details of the argument are given in [19].

To test this idea, a process is run using UV light. The material is n-doped Si (P, 1 - 2 Ohm cm). To get an ohmic contact on the back of the wafer, ion implantation is used (P, 100 keV, $5 * 10^{15}$ /cm^2). Etching is done in the double cell described in (1,2), 25% HF is used. 10 mA/cm^2 are applied for 10 - 60 minutes. During the etching the wafer is illuminated from the front with a UV glow discharge lamp. A nanoporous layer with a thickness of about 0.25 to 0.5 µm is formed. For contact, pads of 10 nm gold are deposited by evaporation.

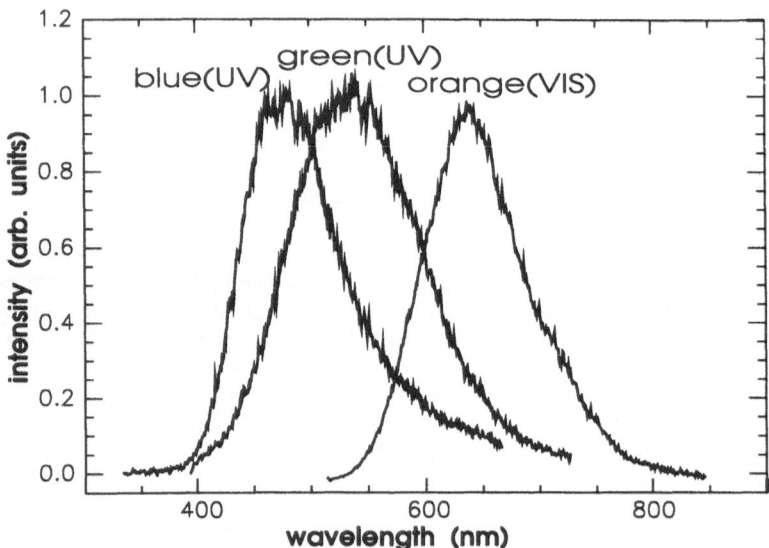

Fig. 6 EL-spectra of porous silicon, VIS-type and UV-type.

When electroluminescence is measured, the samples emit light at wavelengths between 480 nm (blue) and 560 nm (green). Fig. 6 depicts the EL spectra. The VIS-type samples were illuminated with visible light during etching as described above. For comparison, their spectrum is given in Fig. 6 (orange), too. The current voltage characteristic of the devices given in Fig. 7 shows a rectifying behaviour. Again, we find an onset of EL with some volts applied and a quantum efficiency in the region of 10^{-6} to 10^{-5}.

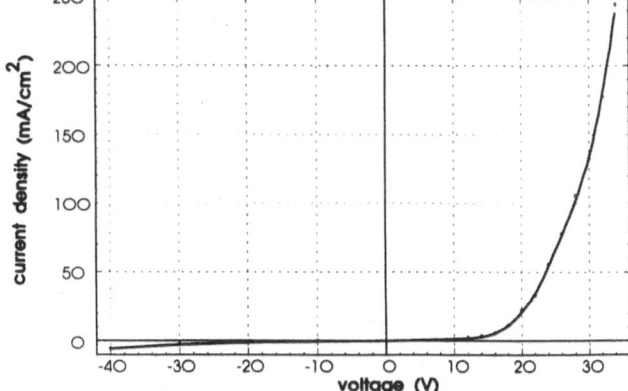

Fig. 7
Current voltage characteristics of a UV-type sample.

Spectral investigation of these samples show an interesting development of the EL spectrum due to the electrical current. A freshly contacted pad will first show red light for a moment, then the blue light will come up and stay. To measure this effect, time resolved spectroscopy is used. Fig. 8 shows the development of the spectrum with time. Time resolution is 1 second. At first, red light is emitted. This is seen at around 700 nm for about 10 seconds. Then the red emission vanishes and the green/blue emission rises and forms the large peak at 520 nm, which comes up after 40 seconds. The spectra show that there is not a shift from red to green, but a decrease of one mode and an increase of another. When the pad is contacted for a second time, the green/blue emission is seen from the beginning. This time dependent behaviour allows some conclusions on the EL mechanism [20] and we hope to get further insight into the basic principles of the luminescence by a detailed investigation of this phenomenon.

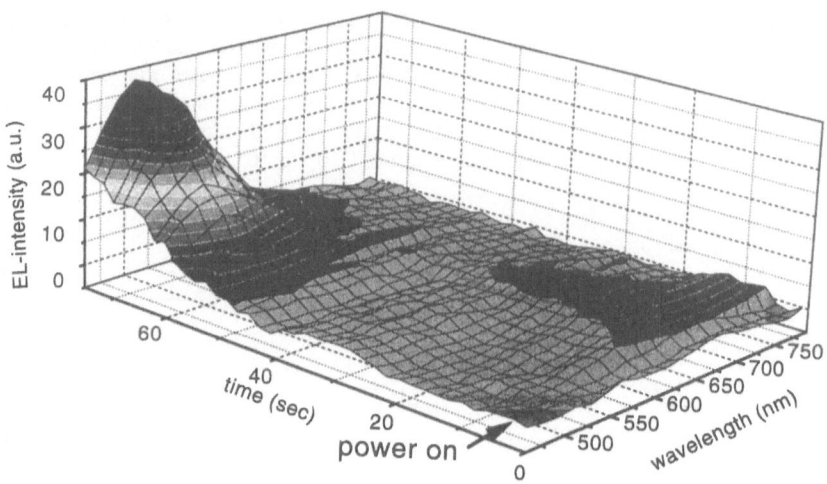

Fig. 8 The development of the EL-spectrum with time.

4. HOMOJUNCTION DEVICES

All the devices described till now show a junction from silicon to a metal or to a nonsilicon semiconductor. For efficient EL it seems most promising to exchange these interfaces for a homojunction in silicon. This way, the electrons and holes are supplied by the n- and p-silicon outside the nanoporous region. They recombine in the luminescent active layer. Moreover, the bad interface from the contact layer to the porous layer is one of the reasons for the low quantum efficiencies we have encountered so far. Therefore, a homojunction which makes a good interface may be better for efficiency.

A set of samples with p/n junctions has been processed. The flow chart reads as follows:

- The basic material is an n-type silicon wafer, 1-2 Ωcm.
- Doping is done by ion implantation. 5E15/cm^2 boron ions are applied with an energy of 150 keV. The back of the wafer is also implanted for an ohmic contact.
- Anealing with 1000^0 C for 60 min.
- Anodization is done with 25% (wgt) HF. A current density of 50 mA/cm^2 is applied for 10 min. The wafer is illuminated from the front during the etching with visible light.
- Deposition of 12 nm gold by thermal evaporation using a shadow mask.

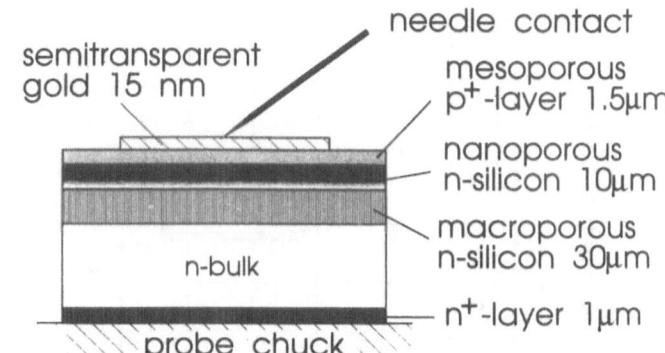

Fig. 9
A porous p/n junction.

A detailed description of this technology is given in [18,21,22]. A sketch of the structure is given in Fig. 9. After implantation and annealing, the top layer of 1.5 µm thickness is p+-doped. This layer is transformed to mesoporous silicon during etching. Mesoporous p+ material has structures with sizes of about 50 nm. These structures are too large to show luminescence due to quantum size effects. Thus, the top layer is not luminescent. Below the p+, there is a small region of p-silicon, which is nanoporous. Below that, there is the n-doped material, which is nanoporous as well due to the fact that the wafer is illuminated during etching. When an electrical current is applied, we find luminescence. The measured quantum efficiency of this device is 10^{-4}. The p/n junction allows an increase in quantum efficiency of one or two orders of magnitude as compared to the basic devices. The onset voltage is around 2 V.

5. INDIUM TREATED DEVICES

A lot of different materials is discussed which can be filled into the pores to increase quantum efficiency, to improve the electrical conduction and/or to stabilize the electroluminescence. Here, we describe experiments made with indium. An electrolyte of $InCl_3$ together with water and ethanol is used to deposit the material in the pores. To verify wether the metal did penetrate the pores, RBS depth profiling is used. It shows, that the indium is deposited mainly at the pore tips. In the lower half of the 500 nm thick porous film the In concentration is 5% (At). In the upper part, the concentration is low, but there is a surface peak near the contact layer.

The indium treatment causes a large increase in the quantum efficiency. Fig. 9 shows the electroluminescence intensity versus the current for samples with (upper curve) and without (lower curve) indium treatment. The quantum efficiency rises by two orders of magnitude. The experiment is made using UV-anodized material, thus the colour of the light is blue. The measured quantum efficiency is $0.5 \cdot 10^{-2}$ % for blue electroluminescence.

Indium is known to akt as an acceptor and causes p-doping. Thus, if the porous area can be doped by indium, a p/n-junction to the n-type bulk silicon can be formed. Under forward bias electrons from the bulk can recombine radiatively with holes in the porous layer. This way the recombination in the luminescent active region is enhanced.

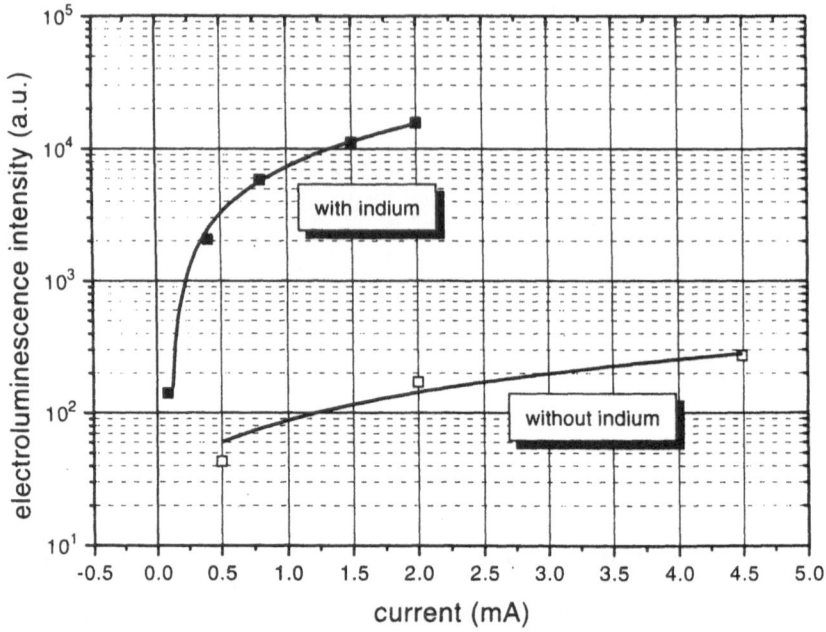

Fig. 10:   Electroluminescence intensity vs current for samples with (upper curve) and
without (lower curve) indium treatment.

6. MAIN TASKS FOR FUTURE WORK

Looking at the results of the first two years in EL research, there are three major tasks to work on in order to achieve an applicable device:

1.   Stability
2.   Quantum efficiency
3.   Time constants.

6.1. Stability

All electroluminescent structures in porous silicon investigated up to now exhibit degradation during electroluminescence. When no current is applied, the structures are stable. Porous silicon LEDs can be stored in air for two years and will not lose electro-luminescence; the luminescence may even become stronger during storage. This behaviour may be explained by the growth of a native oxide in the pores which passivates the structure. All the same, when a current is applied, the luminescence declines. When a sample is freshly contacted, the light intensity will drop to about 30% to 50% of the initial value within a rather short time. Then the sample stabilizes and a steady emission is observed. We made an experiment to test the long-term stability. We found continuous light emission for more than 100 hours in one case.

Porous silicon LEDs with ITO contacts show also 80 hours of continuous emission[23]. At the moment, the clue to stable EL seems to be a better capping layer and/or the filling of the pores.

To obtain stable devices, we have to work on passivation layers and contact layers. The degradation has its reason in the extremely large internal surface of porous silicon. Therefore, for its stabilization, it seems appropriate to fill the pores with a passivating substance. A large number of technologists think that conductive polymers are a very promising approach. This layer is deposited by a galvanic process. Therefore,the material can enter the pores. This way the layer may not only provide contact, but stabilize the internal surface. The deposition of conductive polymers on porous silicon is investigated by [24], first experiments on EL[13] show very promising results.

6.2. Quantum efficiency

An external quantum efficiency of $10^{-3}\%$ is found for simple porous silicon/metal structures; $10^{-2}\%$ are measured for the homojunction devices. When we calculate a loss of 97% within the device, the resulting internal efficiency is 0.3%. With the use of wet contacts, high efficiencies are reported. The understanding of the material is in the beginning. Research up to now has mostly focussed on the physical origin of PL. When this is better understood and when the basics of transport and EL injection are clearer, another improvement of the EL efficiency can be expected. All the same, the amount of light generated with the existing devices is sufficient for a number of applications.

6.3. EL time constants

The time constant of the luminescence is an essential quantity for the application as well as for the characterization of the material. If a device is used for data transmission, the time constant determines the maximum frequency which can be transmitted. On the other hand, the time constant is important for the experimental test of theoretical models on the luminescence mechanism. Time constants of photoluminescence have been published by several groups, but the results measured are quite different. Several distinct luminescent processes seem to be possible in porous silicon[25]. Each of them shows a characteristic spectral region and decay time. Experiments by different groups show that the green and blue photoluminescence light is correlated to fast processes and the red light is correlated to comparatively slow processes[26,27]. For blue emission, very short times in the nanosecond region are reported for PL[26,27]. If this is valid for electroluminescence, too, our blue and green samples could be one step towards fast devices.

Decay times of electroluminescence have not yet been measured with satisfying quality. The measurement is more complax than the photoluminescence decay time measurement. Our experiments made up to now show decay times in the region of microseconds. This value is composed of the luminescence decay time and the electrical decay time caused by the capacity of the device. Since the PL time constants are very short, we hope to get very fast devices, when the RC constant of the samples is reduced.

References

1. L. T. Canham: Appl. Phys. Letters, Vol. 57(10) pp. 1046-1048 (1990)

2. V. Lehmann und U. Gösele: Appl. Phys. Lett., 58, 8, 856 (1991).

3. J.C.Vial, A.Bsiesy, F.Gaspard, R.Herino, F.Muller, R.Romestain, R.M.Macfarlane: Phys. Rev. B, 45, 171 (1992)

4. A. Bsiesy, F. Muller, M. Ligeon, F. Gaspard, R. Herino, R. Romestain, J. C. Vial: Advanced Research Workshop, Meylan 1993.

5. A.Richter, P.Steiner, F.Kozlowski und W.Lang: IEEE Electron Dev. Lett., 12, No.12, page 691, (1991).

6. A.Richter, W.Lang, P.Steiner, F.Kozlowski, H.Sandmaier: Proc. MRS Vol 256, 1991

7. N.Koshida und H.Koyama: Appl. Phys. Lett., 60, No 3, 347, (1992).

8. N.Koshida und H.Koyama: Proc. MRS Vol 256, 1991

9. P. Steiner, J. Weidhaas, W. Lang: Mat. Res. Soc. Symp. Proc. Vol. 281. (1993) (MRS-Meeting Boston 1992)

10. F.Kozlowski, W.Lang: J.Applied Phys., 72 (11), 1. Dec 1992

11. F.Kozlowski, P.Steiner, W.Lang EMRS Meeting, Strasbourg, May 1993.

12. F.Kozlowski, M.Sauter, P.Steiner, A.Richter, H.Sandmaier, W.Lang: EMRS Strasbourg 1992, Thin solid films, 222 (1992)

13. N. Koshida: 8. Deutsch-Japanisches Forum, Weimar (1993)

14. N.M.Kalkhoran, F.Namavar, H.P.Maruska: Proc. MRS Vol 256, 1991

15. H. P. Maruska, F. Namavar, N.M.Kalkhoran: Appl. Phys. Lett. 61 (11), 14. September 1992, pp. 1338-1340.

16. T.Fugati, T.Matsumoto, M.Katsuno, Y.Ohta, H.Mimura, K.Kitamura: Jpn. Appl. Phys. Vol. 31, Part 2, Nr 5B, 15 May 1992

17. T.Fugati, T.Matsumoto, M.Katsuno, Y.Ohta, H.Mimura, K.Kitamura: Mat. Res. Soc. Symp. Proc. Vol. 283. (1993) (MRS-Meeting Boston 1992)

18. P.Steiner, F.Kozlowski, H.Sandmaier, W.Lang: MRS Meeting, Boston, Dec 1992

19. P.Steiner, F.Kozlowski, W.Lang Electron Device Letters, to be published June 1993

20. P.Steiner, W.Lang, F.Kozlowski, H.Sandmaier
    MRS Spring meeting, San Francisco, April 1993

21. P.Steiner, F.Kozlowski, W.Lang Appl. Phys. Letters, May 1993

22. W.Lang, F.Kozlowski, P.Steiner, H.Sandmaier,
    J. of Luminescence 57 (1993) 169-173.

23. N.M.Kalkhoran, F.Namavar, H.P.Maruska: Private Communication.

24. V. P. Parkhutic, J. M. Martinez-Duart, J. Diaz Calleja, E. Matveeva: J. Electrochem.
    Soc, to be printed.

25. F.Koch, V.Petrova-Koch, T.Muschik, A.Nikolov, V.Gavrilenko: MRS Meeting,
    Boston, Dec 1992

26. A.V.Andrianov, D.I.Kovalev, V.B.Shumann, I.D.Yaroshetskii, JETP Lett. 56, 236,
    (1992)

27. V.Petrova-Koch, T.Muschik, D.Kovalev, F.Koch,
    Mat. Res. Soc. Symp. Proc. Vol. 283. (1993)
    (MRS-Meeting Boston 1992)

Porous silicon luminescence under cathodic polarisation conditions

A. Bsiesy

*Laboratoire de Spectrométrie Physique, URA 08 du CNRS,
Université Joseph Fourier de Grenoble, BP. 87,
38402 Saint Martin d'Hères cedex, France*

---

1. INTRODUCTION

Since the first experimental results concerning the intense visible photoluminescence obtained on highly-porous silicon layers have been published[1], a great deal of work has been devoted to the understanding of this property. Different physical origins have been proposed. Among these hypotheses, the quantum confinement of free carriers taking place inside the silicon nanocrystallites[1,2], forming the body of the highly-porous silicon layers, seems at present to be the most supported one. Such nanometric structures can be obtained either by using electrochemical etching parameters producing high porosity [3] layers or by chemically etching layers of lower porosity[1]. Moreover, the broad luminescence line is commonly attributed to the size distribution of the silicon nanocrystallites which are at the origin of the light emission. An interesting consequence of the quantum confinement model associated with the distribution of crystallite size is the possibility to obtain a selective excitation of the inhomogeneously broadened optical transition. This gives rise to a tunable emission which, again, can be found for the PL [4,5] as well as for the EL [6] of porous silicon.

It will be shown that the association of the material inhomogeneity on the nanoscopic scale and the voltage-induced enhancement of charge injection into the silicon nanocrystallites provides a satisfactory description of the porous silicon luminescence tunability (the VTEL and the selective QPL) observed under an electrical excitation.

2. EXPERIMENTAL

Porous silicon is formed by anodisation of n-type (100) silicon samples ($810^{14}cm^{-3}$) in HF aqueous solution using a classical experimental setup. The electrolyte is composed of 2 parts DI water, 5 parts ethanol and 3 parts HF 50%. The anodisation is performed under illumination with a tungsten lamp in order to get the hole (h^+) supply required for silicon etching. The lamp power, the current density and the anodisation time are 40 mW/cm^2, 5mA/cm^2 and 5 mn respectively. The resulting layer is 1 μm thick with a porosity of about 80 %. In these conditions, no macroporous structure is obtained [7].

For QPL experiments, the freshly formed porous layer is put into a 1M H$_2$SO$_4$ aqueous solution whereas for the VTEL experiments various (NH4)$_2$S$_2$O$_8$ concentrations (0.05, 0.1 or 0.2 M) were added to this electrolyte. The cathodic polarisation is scanned between -0.4 and -1.8 V vs a saturated calomel electrode (SCE) at 25 mV/s. A forward (-0.4 V ---> -1.8 V) and a reverse scan (-1.8 V ---> -0.4 V) form a single run. The resulting current intensity is measured using an electrometer and the emitted EL and PL spectra are recorded and analysed by using an Optical Multichannel Analyser coupled to a cooled charge-coupled-device array detector. A whole spectrum can be recorded in less than 0.1 second.

3. THE VOLTAGE-TUNABLE ELECTROLUMINESCENCE

3.1 Why cathodic polarisation conditions ?

Bright visible light emission obtained by electrical excitation was first observed during anodic oxidation of highly-porous p-type silicon layers in contact with aqueous solutions [8]. This EL results from radiative recombination between the holes supplied by the bulk of the p-type semiconductor and the electrons injected from the electrolyte or from the crystallite surface. It has been proposed that the oxidation of the Si-H bonds which initially cover the porous silicon surface [9] might be responsible for the electronic injection [8]. However, such electroluminescence is accompanied by a rapid and irreversible modification of the porous material through its partial oxidation which finally results in the quenching of the electrical excitation [10]. In spite of its unstable character, the EL under anodic polarisation conditions has shown that porous silicon can emit an efficient visible light under an electrical excitation.

A similar and even more efficient electroluminescence has been observed on a cathodically polarised n-type porous silicon in contact with S$_2$O$_8^{-2}$ (persulphate) concentrated solutions [11,12]. The reduction of the S$_2$O$_8^{-2}$ ion leads to a very reactive intermediate oxidant SO$_4^-$[13] whose reduction results in hole injection into the crystallite where a radiative recombination takes place with the electrons supplied by the cathodically biased substrate. This light emitting junction is very attractive because in this case the oxidation of the material, and consequently its evolution, is almost totally avoided. This rather stable and very efficient light emitting system can also bring some improvements in our understanding of the luminescence properties of porous silicon.

3.2 Current and light intensity-potential characteristics

The current-potential (I-V) curve obtained in 0.2M (NH4)$_2$S$_2$O$_8$ solution is shown on figure 1a. The S$_2$O$_8^{-2}$ reduction current appears for a polarisation lower than -0.8 V which corresponds roughly to the flat-band potential of n-type Si in an aqueous solution [14]. For a potential around -1.2 V a current shoulder is observed which is found to be proportional to the S$_2$O$_8^{-2}$ ion concentration as shown by figure 2. For potentials more negative than -1.3V the current

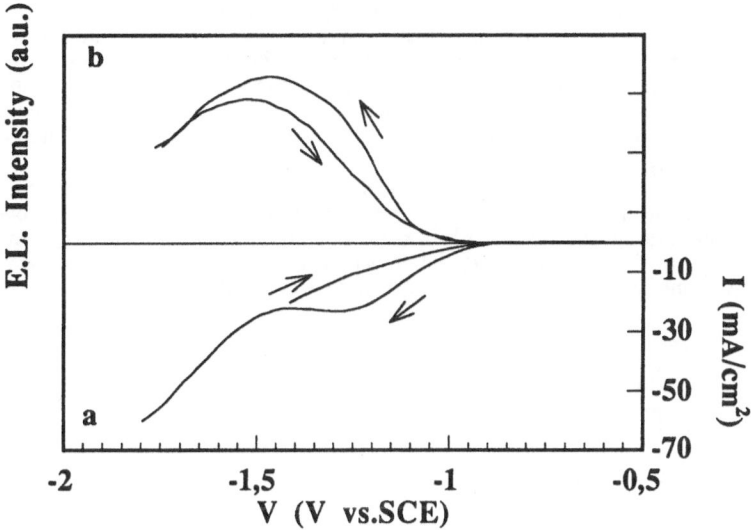

Figure 1: Cathodic current (a) and λ-integrated EL (b) obtained for 0.2M (NH4)2S2O8 as a function of the cathodic bias. The forward and the reverse scans are represented.

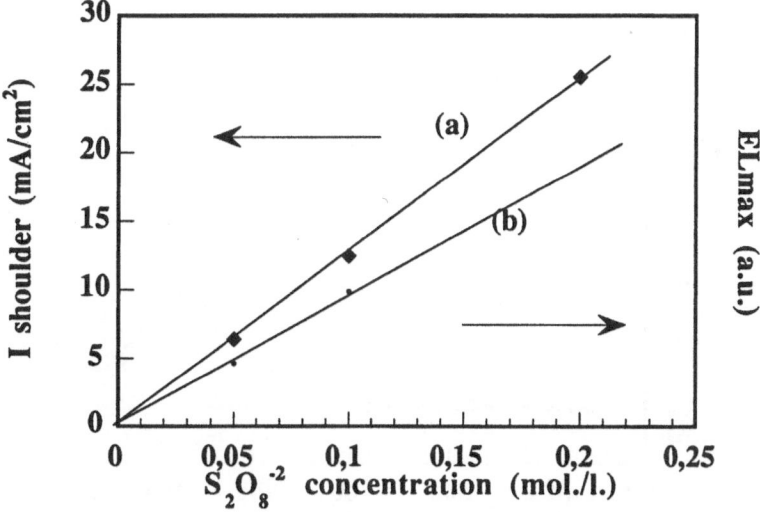

Figure 2: Current shoulder (a) et λ-integrated ELmaximum (b) as a function of $S_2O_8^{-2}$ ion concentration.

increases rapidly as a result of the H+ reduction which takes place by an electronic exchange with the conduction band. This reduction results in an intense hydrogen bubbling which can perturb the luminescence detection and therefore the cathodic bias cannot be further increased. We have thus chosen to stop at -1.8V. The corresponding λ-integrated EL intensity is recorded as a function of the cathodic polarisation (figure 1b). The EL signal appears at -0.9 V and the intensity increases rapidly, passes through a maximum (ELmax) and then decreases continuously . During the reverse scan, the current is found to be smaller than that observed for the forward scan while the EL starts to increase again and shows a new maximum which is obtained for nearly the same bias as the first ELmax. The same behaviour is obtained whatever the $S_2O_8^{-2}$ ion concentration, and ELmax is found to be proportional to the ion concentration in the studied range(Fig.2).

The evolution of the current clearly indicates the depletion of the oxidising agent at the electrode surface which should indeed take place in our conditions where a stationary electrode is used. This would well account for the first ELmax; the depletion of the oxidising agent results in a decrease of the hole exchange current leading to a decrease in the EL intensity. However, during the reverse scan, the EL intensity shows another Elmax whereas the reactive species depletion

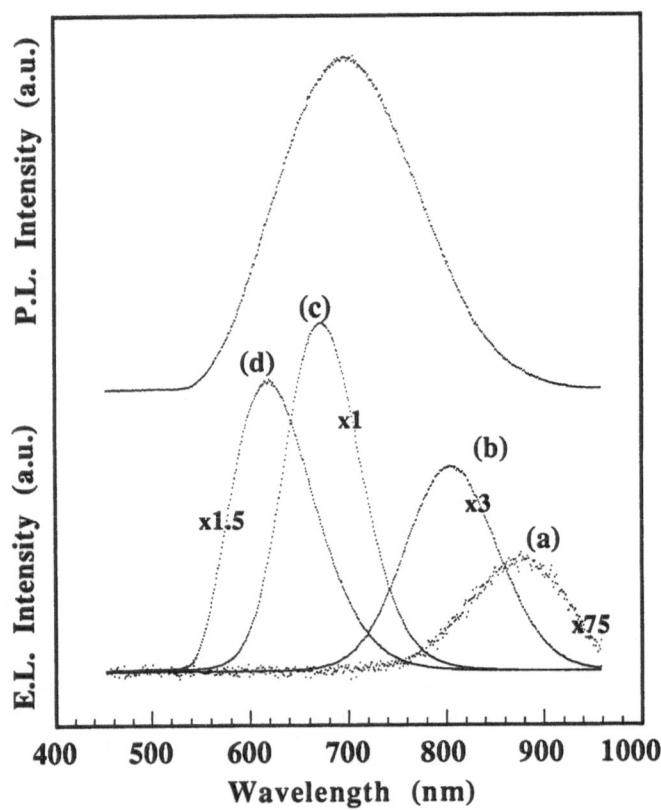

Figure 3: PL (open circuit conditions) and EL spectra obtained in 0.2M (NH4)2S2O8 for different cathodic bias : (a) -1V, (b) -1.2V, (c) -1.5V, (d) -1.6V.

would have induced a continuous decrease of the emitted light during this scan. This behaviour proves that the observed reversible EL modification is not related to a concentration effect but must be governed by the external voltage. On the contrary, the fact that the second ELmax is found to be of a smaller value is an oxidant depletion effect. A possible explanation of this bias-induced EL modification will be discussed in a following part.

3.3 Spectral shift of porous silicon EL, a <u>voltage</u>-induced effect

The EL spectra show a large reversible shift upon cathodic biasing. Figure 3 presents several spectra recorded at different voltages during a typical scan. A quite important peak shift, from 880 nm at -1 V to 610 nm at -1.6 V, is observed. This emitted light, whose colour is seen to change from dark red to bright green as the voltage is increased, is easily visible to the naked eye even in daylight. The peak energy is nearly proportional to the external voltage with a slope equal to 1 as illustrated on figure 4. On the other hand, the FWHM of the EL is found to be much narrower (0.25eV) than that of the PL signal (0.6eV). Furthermore, the observed PL can be considered to be roughly the envelope of all the emitted EL spectra as shown by figure 3. This strongly suggests that in this case we are in presence of a selective excitation of the porous silicon luminescence, which has been found, as already mentioned, to be inhomogeneous [3,4]. This selective excitation is induced by the cathodic bias. This spectral tunability could be compared to that obtained in the case of a semiconductor laser where the inhomogeneous luminescence of a multiquantum well InGaAs/InP structure has been used to scan the laser wavelength of about 200 nm in the infra-red range (1440-1640 nm) [15]. In this case, however, the tunability is obtained by modifying the external-cavity configuration.

In addition, one must note that the spectral position of the EL is determined by the cathodic bias but could be also affected by the value of the exchanged current through a band-filling mechanism. In order to check the dominant phenomenon, variations in the cathodic current, independently of the voltage, were obtained by changing the concentration of the electroactive

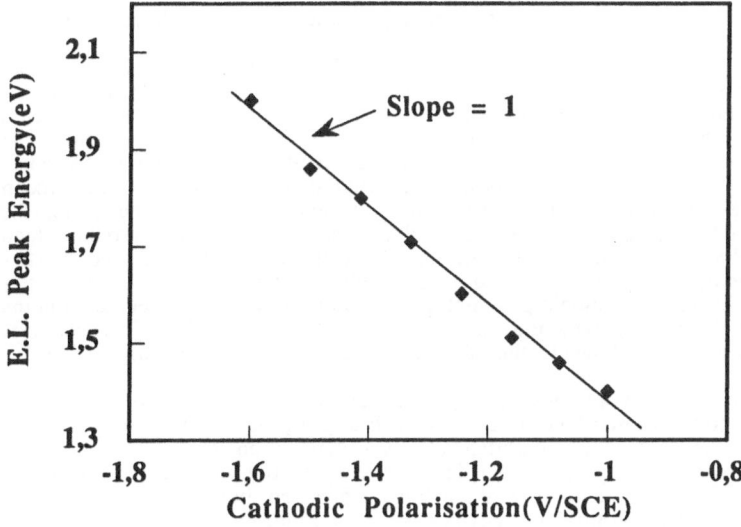

Figure 4: Emitted photon energy as a function of the applied voltage for a cathodically-biased n-type porous silicon layer.

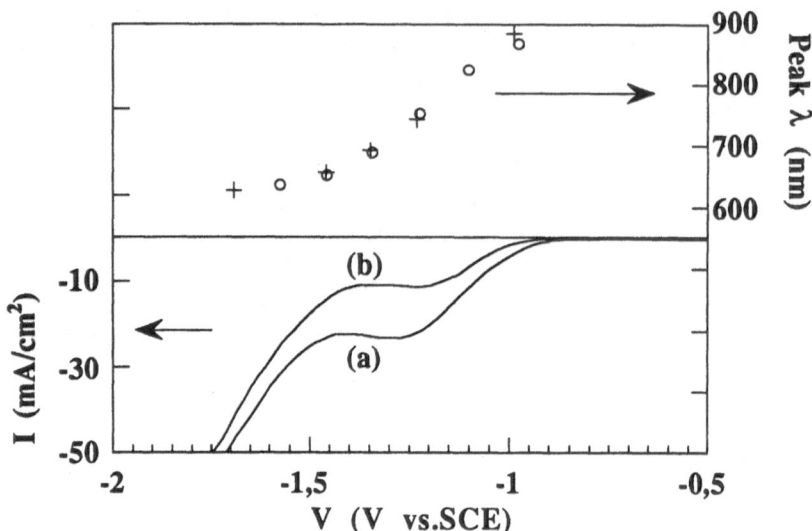

Figure 5: Cathodic current and EL peak wavelength obtained for 0.2M (a,o) and 0.1M (b,+) (NH4)2S2O8 as a function of the cathodic bias.

species. Figure 5 shows that when the $S_2O_8^{-2}$ concentration is doubled, the cathodic current is increased, but no differences are obtained in the potential dependence of the wavelength peak. In these conditions, using the "Voltage-Tunable" term is well justified.

3.4 EL voltage tunability : a consequence of a possible material inhomogeneity ?

Commenting this large porous silicon VTEL, L.T. Canham [16] wondered if the observed blue-shift does not rather result from a possible vertical inhomogeneity of the studied porous layers. Such porosity gradient with depth may exist on porous silicon structures [17] which would introduce a vertical inhomogeneity of the porous layer on the macroscopic scale. This is the reason why it is important to verify if this tunability can be related to a possible macroscopic inhomogeneity of the porous layer. In fact porous layers with relatively long fabrication time in the HF electrolyte (few minutes) might have such porosity gradient. This is due to the fact that the top of the layer spends more time in the HF electrolyte. Chemical dissolution produces then silicon nanocrystallites of smaller size on the top of the layer. This would give shorter-wavelength emission on the top. In these conditions, the blue-shift may seem to be due to this inhomogeneity; the increasing cathodic potential, putting progressively into accumulation the porous layer from the bottom up to the top, leads to the observed spectral shift. This scheme is indeed plausible but, as will be shown by the following results, this is not at the origin of the EL tunability.

In order to avoid any vertical porosity gradient, anodisation time as short as 30 seconds was used. It produces a 100 nm thick layer (measured with a step profiler). This layer must be vertically homogeneous because no important chemical dissolution takes place during 30 seconds. Figure 6 shows the EL spectra obtained on such samples. It is clear that, in spite of the absence of any porosity gradient, an important tunability is obtained similar to that observed on samples with longer anodisation time. Moreover, even if we admit the existence of a vertical inhomogeneity, the progressive charge accumulation in the layer does not explain the constant FWHM of the EL line. In fact, one would expect in this case a widening of the EL line along with

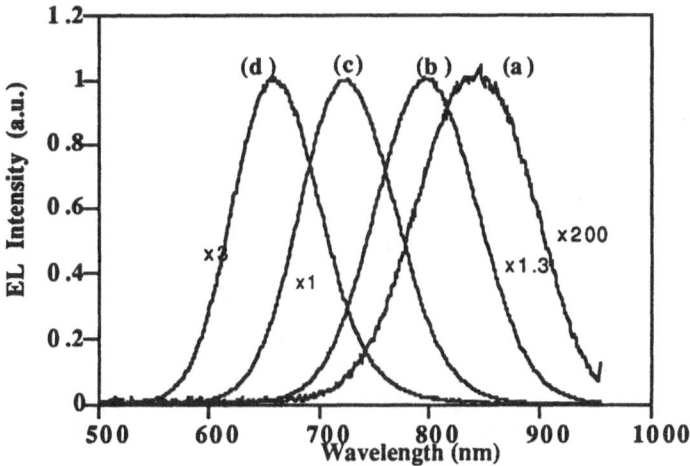

Figure 6: Electroluminescence spectra as a function of the cathodic polarisation at -1V(a), -1.2V(b), -1.35V(c) and -1.5V(d). The sample thickness is 100nm.

the blue-shift rather than a shift of the whole spectrum as is the case for example of EL spectrum (b) and (c) of figure 6. These EL lines are a good example because of their quasi-identical intensities.

3.5 Origin of the voltage-tunable electroluminescence

In order to better understand the origin of the spectral shift as well as the EL behaviour one must examine the charge exchange mechanism at the porous Si-electrolyte interface. As it has been already mentioned, it is widely accepted that the highly-porous silicon is formed of quantum-size crystallites of different dimensions in which a quantum confinement effect takes place. A large distribution of the confinement energy results from this size distribution which can be represented by a Gaussian law. In the electrolyte, let us assume in a first stage that there is a discrete hole donor level (which would be in our case that corresponding to the oxidising agent) and that all the external voltage is applied to the interface. An increase in the cathodic polarisation results in a shift of this energy level. In this case, an increase ΔV in the applied voltage will induce an energy shift $\Delta E_h = e\Delta V$ of the distribution maximum of the selected holes. Since for a given crystallite the electron and hole confinement energy (E_e, E_h) are roughly equal [18], the shift of the luminescence energy $h\nu = E_G + E_h + E_e$ is expected to be $\Delta h\nu = 2e\Delta V$. The experimental results (figure 4) give a factor equal to 1 which shows that in our case the applied polarisation is not totally applied at the porous Si-electrolyte interface but must be shared with the porous silicon bulk. In fact one should also take into account the finite distribution width of the electroactive species. This finite energy distribution is due to the solvatation of the electroactive species that is each ion in the solution is surrounded by the solvent dipoles which form the solvatation shell. This solvatation determines the energy of the ion in the solution. A fluctuation of this solvatation shell results in the fact that the electroactive species is characterised by an energy distribution. A typical value of the FWHM of such a distribution is 1 eV in an aqueous solution. Because no EL is observed when no electroactive species is present, it is likely that the EL observed in the cathodic range is governed by the hole exchange current between the electrolyte and the different crystallites. This current is determined by the product of the hole confinement energy distribution and the SO_4^- energy distribution shown in figure 7. It depends on the cathodic polarisation and

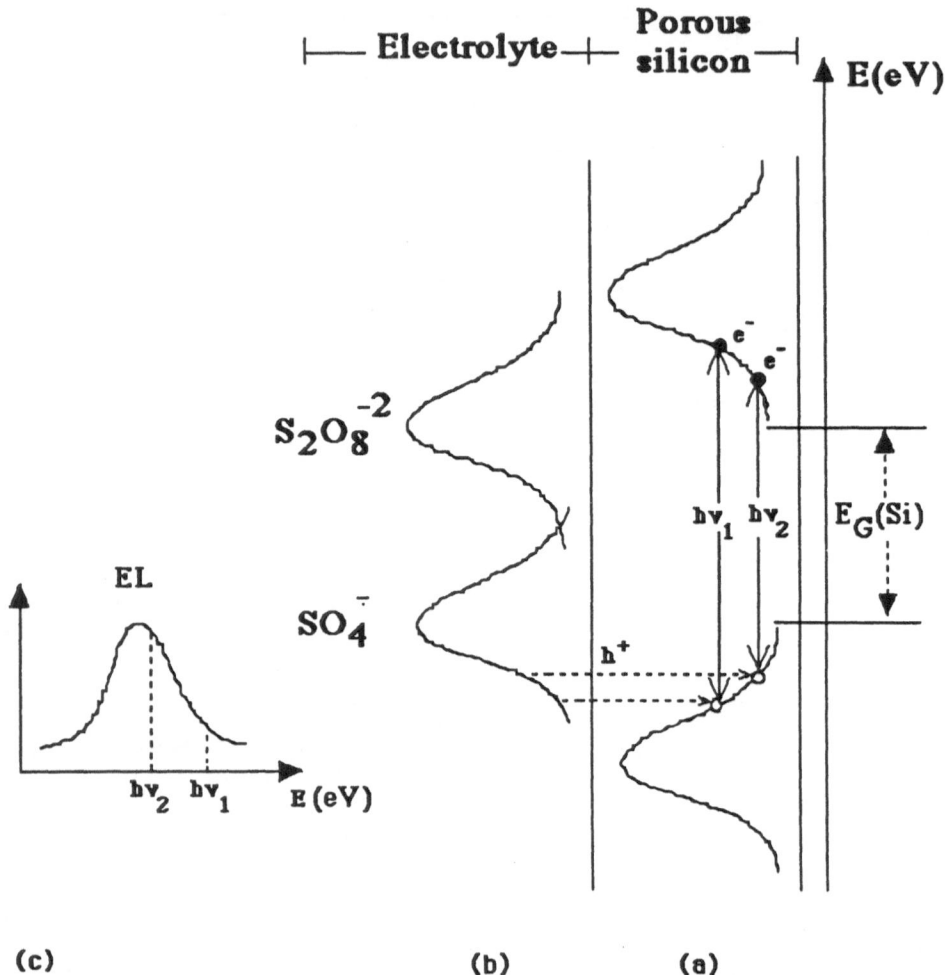

Figure 7: Energy band diagram showing: (a) Electron and hole energies and their distribution in the porous silicon and (b) Redox species ($S_2O_8^{-2}$ and SO_4^-) solvatation energy distribution (same energy scale for (a) and (b)). Scheme (c) represents the product of the energy distribution of holes by the solvatation energy distribution of SO_4^- which gives the EL spectrum. This product is shifted by the applied voltage.

can lead, for a certain bias, to a maximum in the h^+ injection current when the distribution maximum match. For this given bias, the emitted EL should be maximum which is in agreement with the experimental results reported above. And finally the proportionality factor equal to 1 between the EL energy peak and the applied bias results from the fact that this applied external voltage is shared between the injected holes and the injected electrons. These electrons have to experience the energy barrier which exists between the silicon bulk and the confined electron levels in the silicon nanocrystallites due to the quantum confinement effect.

This mechanism can also explain the observed spectral shift. The injection of holes and electrons with increasing energy attains progressively smaller silicon crystallites of higher luminescence energy. They begin to emit light while the red ones stop to emit because holes are no more injected there. A bias-induced blue shift is then expected, in agreement with the experimental results.

3.6 Relation between the porous silicon VTEL and its morphology :

One of the consequences of this description is that if the silicon nanocrystallite size distribution is modified, the VTEL behaviour would be also modified. The following discussion will focus on the verification of this expectation.

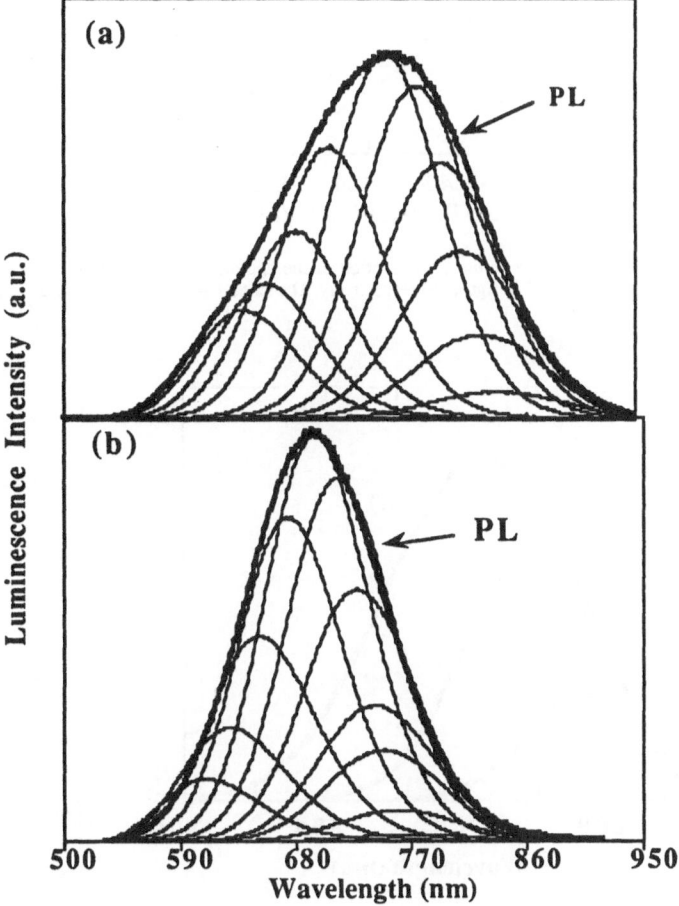

Figure 8: Comparison of PL and EL for two samples of different porosity. The bold lines refer to PL spectra. The non-normalised EL spectra are represented using a fine line. The change in luminescence energy of the two intensity peaks and in V_{ELmax} (see in the text) are 0.16eV and 0.15V respectively.

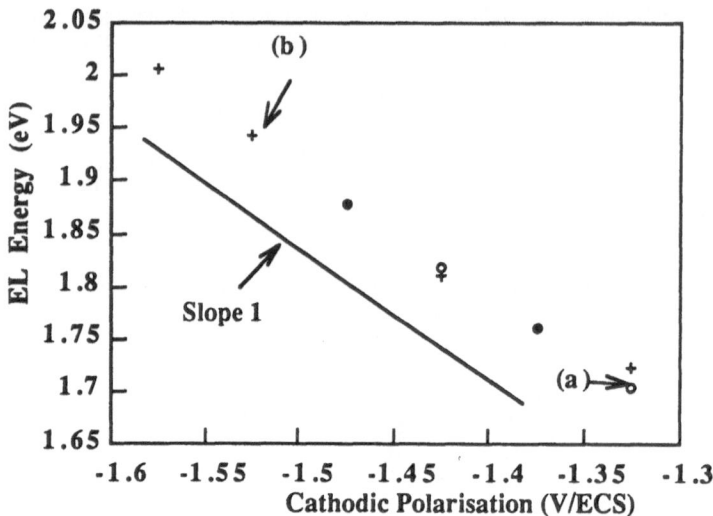

Figure 9: Electroluminescence energy maximum as a function of the applied voltage. The experimental points (a,o) and (b,+) refer to samples (a) and (b) of figure (8) respectively. The slope 1 is represented in order to guide the eye.

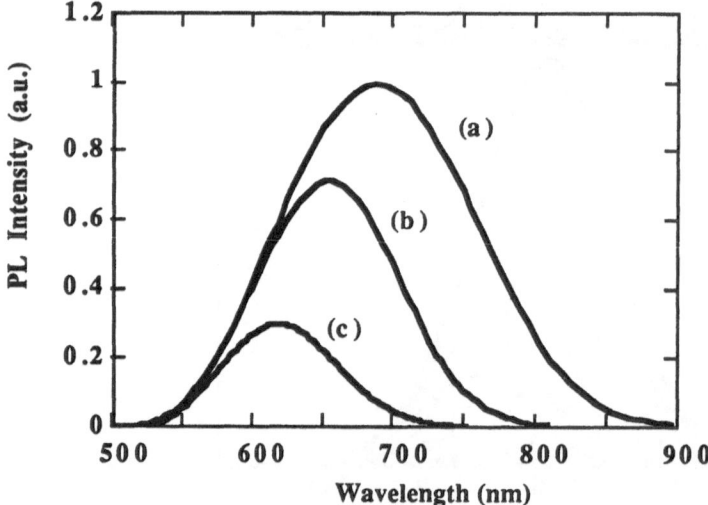

Figure 10:PL spectral evolution under cathodic polarisation. (a) rest potential to -1V, (b) -1.1V and (c) -1.2V.

It was shown that during a bias scan in the cathodic range the EL shows a maximum light emission, ELmax, obtained for a given bias: V_{ELmax}. According to the proposed description, this maximum light emission originates from the maximum fit between two energy distributions; the silicon crystallites confinement energy distribution and the SO_4^- solvatation energy distribution. In order to study the influence of the material morphology, two different types of samples were used. Figure 8a gives the photoluminescence (PL) of the first type of sample which is obtained in the above-mentioned conditions(anodisation time = 30 seconds). The second type of sample is obtained by further chemical etching performed in the fabrication electrolyte (during 15 mn. in dark). This leads to a blueshift of the PL emission (Fig.8b) as a result of the porosity increase. The samples thus obtained exhibit PL emission with maximum intensity peak shifted of about 60nm as illustrated by figure 8. Different EL spectra observed during a cathodic bias scan are also given by figure 8. It must be noted that these EL spectra are represented with the same scale. In both cases, the same tunability is obtained. This is characterised by the proportionality between the energy of the emitted light and the external applied voltage as illustrated by figure 9. The linear dependence shows a slope equal to 1 which corresponds to the fact that the applied voltage is shared by both electrons and holes which are injected selectively into the silicon crystallites. The PL line is found to be the exact envelope of all the emitted EL spectra regardless the kind of the sample. This implies that the EL line obtained at ELmax matches in both cases perfectly with the PL. In addition, the difference between the two V_{ELmax} (external bias at ELmax) is proportional to the luminescence energy difference. These two values are indicated in the legend of Figure 8. This result can be easily understood in the frame of our description. Since the increase in the porosity of the sample leads to a shift of the confinement energy distribution of both holes and electrons towards the higher values (of about 0.16eV in our case), it is fairly easy to deduce that a higher cathodic bias is necessary to have a maximum carrier injection probability.

4. THE ELECTRICALLY-INDUCED QUENCHING OF POROUS SILICON PHOTOLUMINESCENCE

4.1 Blueshift and narrowing of the PL line

Figure 10 shows the PL spectral evolution recorded as afunction of the cathodic bias. From the rest potential to a polarisation of V no modification is observed. Upon increasing the cathodic potential, a strong and progressive decrease of the PL intensity is observed leading to a complete

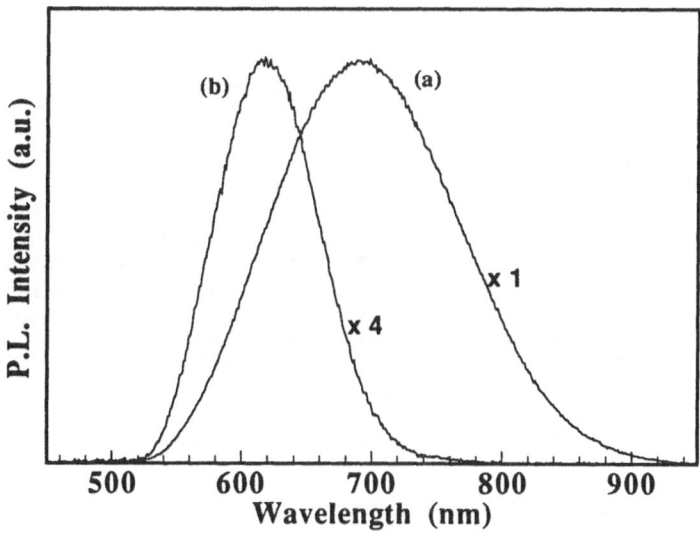

Figure 11:PL spectra obtained at -1V (a) and -1.3V (b) showing the PL line narrowing.

quenching of the emitted light. This quenching is accompanied by a blue shift of the peak wavelength. This blue shift is only due to the fact that the red part of the spectrum (low luminescence energy) is quenched first. Consequently, the increasing polarisation leads to the narrowing of the line width in such a way that it is possible to observe a clear green luminescence once the red-orange part of the spectrum has been quenched (Figure 11). This electrically-induced selective PL quenching appears clearly on figure 12 where the normalised PL intensity, recorded at different luminescence energies, is represented as a function of the cathodic polarisation. It appears obviously that the low luminescence energy (e.g.1.3 eV) is totally quenched before any important modification of the high luminescence energy. The insert of figure 12 shows that there is a linear dependence between the applied voltage cut-off variation and the luminescence energy. Furthermore, the PL quenching which is obtained for a bias variation of only about 500 mV is found to be completely reversible during the reverse scan (-1.8V------> -0.4V).

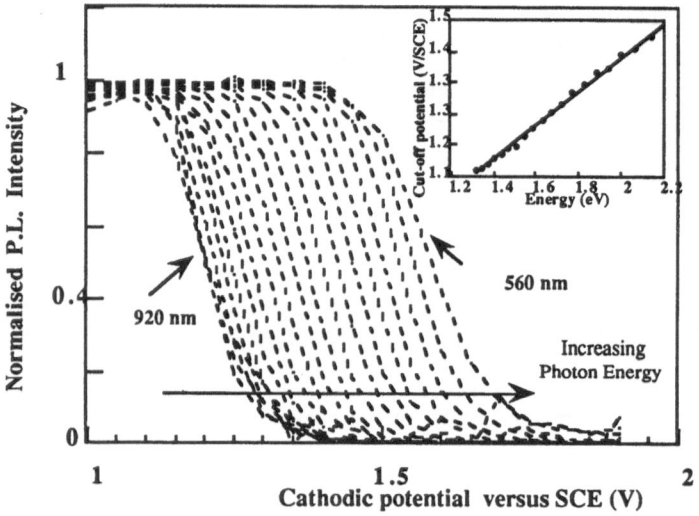

Figure 12: PL intensity for different luminescence energies as a function of the applied cathodic voltage and (insert) the cut-off potential versus the emitted photon energy.

Under cathodic conditions, a cathodic current is observed due to proton (H$^+$) reduction. This electrochemical reaction takes place by charge exchange with the semiconductor conduction band. Its intensity is proportional to the H$^+$ concentration. In order to investigate the influence of this current on the PL quenching, aqueous solutions containing different H_2SO_4 concentrations (1, 0.1, 0.01 M) are used. The PL evolution appear to be clearly independent of the cathodic current intensity. This same conclusion is also available if, instead of using H_2SO_4, a non electroactive species is used (e.g. K_2SO_4) as the supporting electrolyte.

4.2 Radiative and non-radiative recombination rates evolution under a cathodic polarisation

The emitted PL intensity (I$_{PL}$) can be given by the following expression[19]:

$$I_{PL} \; \alpha \; N^*. \; Wr/(Wr+Wnr) \tag{1}$$

where Wr and Wnr are the radiative and the non radiative photogenerated electron-hole recombination rates respectively. N* represents the number of the optically active silicon nanocrystallites. Since Wnr was found to be the dominant mechanism in the room temperature luminescence decay[19], the expression (1) can be written as follows :

$$I_{PL} \propto N^*.W_r/W_{nr} \qquad\qquad (2)$$

As I_{PL} is very strongly affected by the cathodic polarisation, the voltage-dependence of Wr, Wnr and N* has to be considered.

The radiative recombination rate (Wr) is proportional to the square of the overlap of the electron and the hole wavefunctions. An electric field is expected to spatially separate both carriers and thus reduce Wr [20]. However, numerical calculations assuming a voltage drop as large as 1 V across a 3 nm wide crystallite and an effective mass of $0.25m_0$ for both electron and hole would lead to a reduction of Wr by a factor smaller than 4. Moreover, this is accompanied by a red shift of the optical transition energy of about 250 meV opposite to the experimental evidence. It can therefore be concluded that Wr is in this case independent of the cathodic bias.

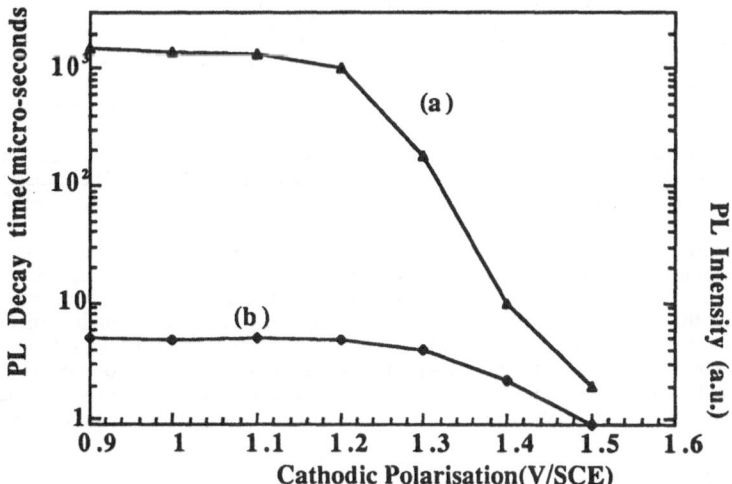

Figure 13: Evolution of the PL intensity (a) and decay time (b) at $\lambda_{luminescence}$ = 700 nm as a function of the cathodic polarisation.

The bias dependence of PL decay time (τ with $\tau = 1/W_{nr}$) has been studied. The results are given by figure 13. One can see that while the PL intensity shows a strong decrease by three or four orders of magnitude, there is only a slight decrease of the PL lifetime by nearly a factor of two upon the whole bias range. This proves that the I_{PL} decrease cannot be attributed to a modification of the non radiative recombination mechanism. In other words, one can consider Wnr to be constant.

This leads us to conclude that it is the modification of the number of the optically active silicon nanocrystallites, N*, which is at the origin of the observed QPL. In the following section different possibilities which can account for our experimental results through the evolution of N* will be considered.

4.3 Possible mechanisms at the origin of the decrease of the number of the optically active silicon crystallites

A first possible explanation of the decrease of the number of the optically active sources could be the result of the electron accumulation (driven by the cathodic bias) in the Si nanocrystallites which can make these entities transparent to the excitation wavelength. If we consider a Si nanocrystallite without an external bias, the confined electronic levels are unoccupied and the excitation absorption is possible giving rise to the observed orange-red PL. Once the electronic levels are occupied due to the electron flow coming from the substrate, it is no more possible to photogenerate any electron-hole pair, the excitation is no longer absorbed and this crystallite becomes optically inactive. This hypothesis could also explain the observed blueshift which accompanies the QPL. Let us consider a "red" (low confinement energy) and a "blue" (high confinement energy) quantum-size crystallites. The progressively increasing cathodic potential starts by putting into accumulation the "red" crystallite due to the weak difference between the energy of electrons in the bulk and their energy in this type of crystallite.

It should be noted that the coulomb interaction between the accumulated electrons leads to an increase of the confined level energy and thus increases the absorption energy threshold. Finally, according to this hypothesis, the PL decay mechanism is not supposed to be changed which is in agreement with the constant PL lifetime determined experimentally.

Another possibility is the enhancement of the escape probability of the non-thermalised carrier across the energy barriers surrounding the Si nanocrystallites. This barrier modification can be induced by the cathodic polarisation. This escape must concern the non-thermalised carriers, otherwise the luminescence decay rate should be affected by such a mechanism which is contrary to the experimental results.

The porous silicon surface depassivation could also explain the PL quenching. Smandek and Gerischer[21] have invoked this hypothesis to explain the EL quenching that they observed on the n-GaP/electrolyte system under cathodic polarisation conditions. They interpreted this EL quenching as a consequence of the reduction of the GaP surface which gives radiationless recombination centers. Such a mechanism could account for our experimental results as well as other process which lead to surface state modification resulting from the cathodic polarisation. However, these last mechanisms have to be able to kill the luminescence of some crystallites without affecting the other ones and have also to be completely reversible which is difficult to imagine in the case of a chemical depassivation of the silicon nanocrystallites. On the contrary various observations seem to support the absorption saturation hypothesis. Among them we have retained the photoluminescence saturation upon increasing laser excitation power which indicates that once a quantum dot is filled in with a hole-electron photogenerated pair it becomes transparent[22].

5. CONCLUSION

The porous silicon luminescence shows a drastic dependence upon the application of a small cathodic polarisation. It is found that the large EL blueshift cannot be attributed to a band filling effect or to a possible vertical macroscopic inhomogeneity of the porous layer. It originates in a local inhomogeneity on the nanoscopic scale. This inhomogeneity is due to the size distribution of the silicon nanocrystallites. The association of the confinement energy distribution, which results from the quantum confinement effect taking place in the silicon nanocrystallite, and the voltage-enhanced carrier injection from the solution and from the silicon bulk give a satisfactory description of the observed blueshift.

The quenching of the porous silicon photoluminescence under these cathodic polarisation conditions is found to be energy selective and strongly affected by a small polarisation variation. The selective injection of carriers into the silicon nanocrystallites allows in this case also to explain the large blueshift which accompanies the quenching of the PL. The strong PL quenching can be attributed to an excitation absorption saturation mechanism triggered by the carrier accumulation in the silicon nanocrystallites. Other possible mechanisms such as the electrically or the chemically-induced surface depassivation are less probable but cannot be ruled out.

Acknowledgements

The author would like to thank all the members of the porous silicon group of the "Laboratoire de Spectrométrie Physique" and especially : prof. F. Gaspard, prof. R.Herino, prof. M.Ligeon, prof. F.Muller, Dr. R. Romestain and Dr. J.C.Vial for their contribution in this work.

References

1. Canham L.T., *Appl. Phys. Lett.* **57**, (1990) 1046-1048.

2. Lehmann V., Gosele U., *Appl. Phys. Lett* **58**, (1990) 856-858.

3. Bsiesy A., Vial J.C., Gaspard F., Herino R., Ligeon M., Muller F., Romestain R., Wasiela A., Halimaoui A., Bomchil G., *Surf. Sci.* **254**, (1991) 195-200.

4. Calcott P.D. J., Nash K.J., Canham L.T., Kane M.J., Brumhead D., *J. Phys. Condens. Matter.* **5**, (1993) L91.

5. Bsiesy A., Muller F., Mihalcescu I., Ligeon M., Gaspard F., Herino R., Romestain R., Vial J.C., *J. Lum.* **57**, (1993) 29-32.

6. Bsiesy A., Muller F., Ligeon M., Gaspard F., Herino R., Romestain R., Vial J.C., *Phys. Rev. Lett.* **71**, (1993) 637-640.

7. Levy-Clement C., Lagoubi A., Ballutaud D., Ozanam F., Chazalviel J.N., Neumann-Spallard M., *Appl.Surf.Science* **65/66**, (1993) 408-414.

8. Halimaoui A., Bomchil G., Oules C., Bsiesy A., Gaspard F., Herino R., Ligeon M, Muller F., *Appl. Phys. Lett.* **59**, (1991) 304-306.

9. Venkateswara Rao A., Ozanam F., Chazalviel J.N., *J. Electrochem. Soc.* **138**, (1991) 153-158.

10. Bsiesy A., Gaspard F., Herino R., Ligeon M., Muller F., Oberlin J. C., *J. Electrochem. Soc.* **138**, (1991) 3450-3456.

11. Bressers P. M. M. C., Knapen J. W. J., Meulenkamp E. A., Kelly J. J., *Appl. Phys. Lett.* **61**, (1992) 108.

12. Canham L.T., Leong W. Y., Beale M. I. J., Cox T. I., Taylor L., *Appl. Phys. Lett.* **61**, (1992) 2563-2565.

13. Memming R., *J.Electrochem. Soc.* **116**, (1969) 785-789.

14. Ronga I., Bsiesy A., Gaspard F., Herino R., Ligeon M., Muller F., Halimaoui A., *J. Electrochem. Soc.* **138**, 1403-1408 (1991).

15. Lidgard A., Tnbun-ek T., Logan R. A., Temkin H., Wecht K. W., Olsson N. A., *Appl. Phys. Lett.* **56**, (1990) 816-817.

16. Canham L.T., *Nature* **365**, (1993) 695.

17. Inoue K., *Jpn.J.Appl.Phys.* **31**, (1992) L997.

18. Fishman G., Mihalcescu I., Romestain R., *Phys.Rev.B* **48**, (1993) 1464-1467.

19.   Vial J.C., Bsiesy A., Gaspard F., Herino R, Ligeon M., Muller F., Romestain R., Macfarlane R., *Phys. Rev. B* **45**, (1992) 14171-14176.

20.   Mendez E.E., Bastard G., Chang L.L., Esaki L., Morkoc H., Fischer R., *Phys. Rev. B* **26**,(1982) 7101-7104 .

21.   Smandek B., Gerischer H., *Electrochimica Acta.* **30**, (1985) 1101-1107.

22.   Mihalcescu I. et al. (to be published).

Interrelation between electrical properties and visible luminescence of porous silicon

N. Koshida

*Division of Electronic and Information Engineering, Faculty of Technology,
Tokyo University of Agriculture and Technology,
Koganei, Tokyo, 184, Japan*

---

1. INTRODUCTION

Studies on the electrical conductivity and photoconductivity of porous silicon (PS) [1-3] are very important for insight of the visible light emission phenomena, especially of the electroluminescence (EL) [4], since it provides direct information about the carrier transport and the interaction of light and electrons. Effects of electric field on the PL characteristics [5] should also be clarified in order to understand the luminescence mechanism.

In this article, the conduction and photoconduction of PS are discussed in relation to the PL and EL properties. It is shown that the band conduction with a thermal activation energy is dominant at room temperature, and that at low temperatures below 100 K the conduction mode becomes a tunneling type. The luminescent PS layers are sensitive for visible light, and the PC spectra are almost identical with the PL excitation spectra. Also there is an electrical quenching mode in the PL emission from PS. These results are consistent with the hypothesis that photoexcitation occurs in Si crystallites.

On the basis of these reuslts, the EL characteristics of PS diodes with an electropolymerized contact [6] are described. It is pointed out that the efficient and stable LED operation would be realized if the well controlled PS surface is appropriately combined with the large-area contact formation technique.

2. EXPERIMENTAL

2.1 Sample Preparation

Nondegenerate p-type or n-type (111) Si wafers (0.018-20 Ωcm) were cleaned, and then ohmic contacts were formed onto the back side. The PS layers were formed by anodizing these wafers in a solution of 50% HF or 50% HF:ethanol=1:1 at current densities of 10-100 mA/cm^2 for 5-60 min. The thickness of PS layers is 3 to 50 μm.

For n-type substrates, the anodization was carried out under illumination using a 500 W tungsten lamp from a distance of 20 cm. The p-type substrates were, on the other hand, anodized in the dark, in which immediately after anodization, the PS samples were illuminated in HF solutions by a 500 W tungsten lamp for about 10 min in order to enhance the PL intensity.

When it is required to control the PL emission wavelength more precisely [7], a 150 W halogen lamp equipped with an infrared blocking filter was used as an illumination light source, and the PS samples were illuminated through sharp-cut long-wavelength-pass filters placed between the light source and the samples.

2.2 Measurements

The experiments were carried out along the following direction.

2.2.1 Photoconduction, PL and EL

Anodized PS samples were immediately transferred into the vacuum chamber, and then semitransparent thin Au films were evaporated onto the PS layer surface. The photoconduction cell consists of thin Au film, PS, Si substrate and ohmic electrode. The PS layer was excited through the thin Au film using a 500 W Xe lamp as a light source, and vertical photoconduction was measured under the condition that a positive or negative bias voltage was applied to the Au contact with respect to the back electrode. The temperature dependences of the dark- and photocurrents and the spectral photoresponse were measured in the range of 10 to 300 K as a function of the bias voltage.

The PL and EL spectra of these diodes were also measured, and compared with the PC spectra. For PL measurements, the PS samples were excited with a 325 nm He-Cd laser or a 514 nm Ar laser through the thin Au film.

2.2.2 Electric field effects on PL

Effects of the external electric field on the PL characteristics were studied for PS diodes with the same electrode structure as the PC cells. The PL spectra and intensity were measured in a N_2 gas atmosphere as a function of positive or negative dc bias voltage. Dynamic response of the PL intensity was also measured under the pulsed bias condition. To avoid possible changes in the PL characteristics during measurements, the excitation intensity was adjusted to a minimum level acceptable for the instrument.

2.2.3 Formation of a large-area contact

After anodization, the PS samples formed on p-type wafers were rinsed in ethanol and then immersed in an electrochemical cell consisting of an electrolytic solvent and a counter electrode. Polypyrrole films were synthesized by galvanostatic polymerization in the same way as that described previously [6]. Some PS samples were lightly oxidized before immersion by a rapid thermal processing. After growth of polypyrrole films, thin Au films were evaporated in order to provide an electrical contact. The active area of the PS diodes was 6 × 6 mm.

The optoelectronic performance was studied at room temperature in terms of the voltage and current dependence of the EL intensity, the current-voltage (I-V) and capacitance-voltage (C-V) characteristics.

3. RESULTS AND DISCUSSION

3.1 Carrier Transport and Photoconduction

The PS diodes shows a definite PC effects independent of the polarity of applied bias voltage. In Fig. 1, a typical example of dark- and photocurrents for a positive bias are shown as a function of reciprocal temperature. It can be seen that at low temperatures below about 100 K the electrical conduction under either dark or illumination becomes independent of temperature, and that the behavior near room temperature is of a thermal-activation type. The tunneling and the band conduction are dominant at low and high temperatures, respectively. In the case of thin PS layers (e. g. 4 μm thick), the rectifying behavior of the I-V curve becomes apparent, and PC cells behave as photodiodes in which the photocurrent increases in proportion to the incident light intensity. There is also a possibility that this device also acts as a MIS tunnelling diode.

3.2 PC Spectra and Other Optoelectronic Properties

Figure 2 compares the PC spectrum of a PS diode formed on a heavily doped n-type Si substrate with the corresponding PL and EL spectra. The PS layer is sensitive for visible light,

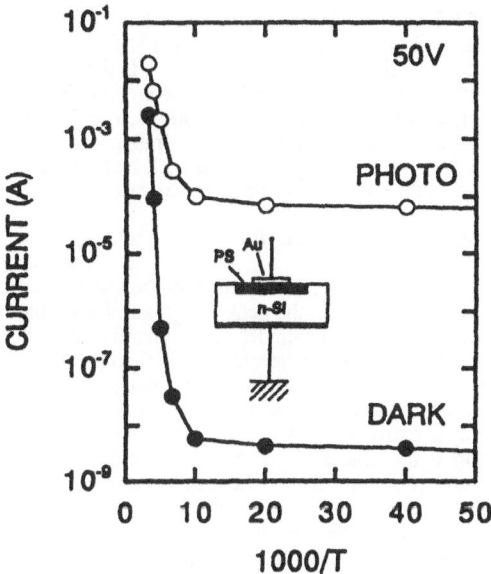

Fig. 1. Temperature dependence of the dark- and photocurrents for a thick PS layer cell at a negative bias voltage of 50 V. The PS layer (40 μm thick) was formed on a heavily doped n-type Si wafer. The light intensity was 100 mW/cm^2. The schematic structure of the sample is also shown.

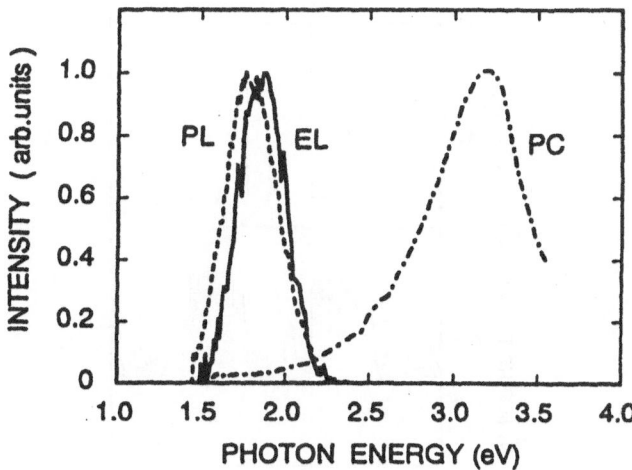

Fig. 2. Photoconduction spectrum of a PS cell at room temperature and corresponding PL and EL spectra. The sample is the same as in Fig. 1.

reflecting the property of PS as a semiinsulating semiconductor with a widened band gap [8,9]. It has been confirmed that the PC spectrum is very similar to the PL excitation spectrum. This is a support of the assumption that the photoexcitation occurs in the bulk of Si crystallites.

A significant difference between the PC and PL spectra implies the existence of a strong electron-phonon coupling before the radiative transition. The PC and PL spectra show a reasonable blue shift with decreasing temperature in a similar way to those of the conventional photoconducting semiconductors. The similarity of the PL band to the EL one suggests that the luminescence originates from the same transition in either case of optical or electrical excitation.

3.3 Electrical modulation of PL intensity

When a positive or negative bias voltage is applied to the Au contact of PS diodes, the PL intensity is significantly decreased, while the peak enegy and the width of the PL band remains almost unchanged [5]. As a typical example, the dynamic behavior of the electrical PL quenching effect is shown in Fig. 3. The external electric field produces a reversible modulation of the PL intensity with a response time similar to the PL decay time.

The quenching is independent of the excitation power and completely reversible. In addition, this effect is observed even at extremely low temperatures where the conduction is dominated by tunneling. The implication is that this PL quenching is not due to thermal effects, but due to electric field effect. The most probable effect is field-induced carrier sweeping-out from the bulk of Si crystallites. In fact, the degree of PL quenching is proportional to the net photocurrent induced by a He-Cd laser during PL measurements.

Similar electrical PL quenching effects have been reported for GaAs quantum wells by several authors [10,11]. Those were explained by the model of carrier leakage between adjacent quantum wells. This picture seems to be applicable to PS, because the luminescent PS layer is composed of a great number of Si nanocrystallites surrounded by a thin oxide layer. Anyway, the observation of the electrical effects on PL gives a further support of our central assumption that the electronic excitation takes place in Si crystallites.

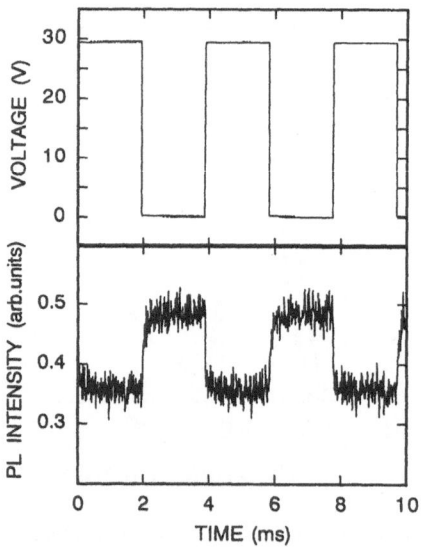

Fig. 3. Dynamic response of electrical modulation of PL. The sample was prepared under the same condition as in Figs. 1 and 2.

3.4 Improvement in the EL characteristics

It has been observed that all of the fabricated PS diodes with electropolymerized contact show a definite rectifying behavior as shown in Fig. 4. The forward bias condition in this diode corresponds to the case in which a negative voltage is applied to the polypyrrole contact with respect to the Al back electrode. Under the the forward-biased condition, therefore, electrons and holes are injected into luminescent elements from polypyrrole and substrate, respectively. As suggested from the result of Fig. 4, the n factor of the I-V curves for electropolymerized diodes was considerably improved in comparison to that for conventional Au contact PS diodes. It is evident that the carrier injection becomes more efficient. This is presumably caused by the impregnation of PS with conducting polypyrrole. This explanation has been supported by measurements of the impedance spectra, C-V characteristics and X-ray photoelectron spectra of the polypyrrole/PS/Si diodes. According to the results of the impedance spectra measurements [12], for instance, the frequency dispersion of impedance for the polypyrrole/PS interface behaves in the same way as that for the electrolyte/PS interface. It has also been confirmed by the C-V measurements that polypyrrole contact certainly penetrates into PS.

The improvement in the electrical properties produces desirable effects in the EL characteristics. In the forward bias region, visible light is uniformly emitted through the contact. The EL intensity is also plotted in Fig. 4 as a function of the forward bias voltage. Light emission begins at at about 3 V and increases more rapidly with voltage than for the conventional Au-contact diodes. The peak energy (2.09 eV) and the width (0.30 eV in FWHM) of the measured EL spectra were similar to those of PL.

Effects of the electrode penetration into PS were also observed with the current and input power dependences of the EL intensity. High efficiencies comparable to that of liquid junction cells [13,14] will be obtained by optimizing the process parameters of electropolymerization. The key factors are careful control of the PS surface and formation of a large-area contact without any structural damages.

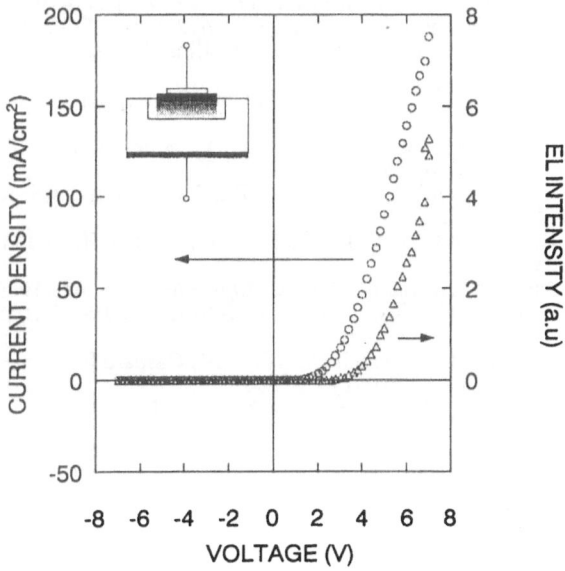

Fig. 4. Current-voltage curve of a PS diode with an electropolymerized contact. The corresponding behavior of the EL intensity is also shown.

4. CONCLUSIONS

Some concluding remarks can be deduced from measurements of the electrical and optoelectronic properties of PS as follows:
(a) The carrier transport in PS changes from the band conduction mode to the tunneling one at about 100 K.
(b) The PS layer is sensitive to visible light, and the relation between the PC, PL and PL excitation spectra suggests that the electronic excitation occurs in Si crystallites.
(c) The PL emission can reversibly be modulated by an external electric field, presumably because of field-enhanced escape of photogenerated carriers.
(d) The electrochemical approach like electropolymerization is promising for the formation of large-area contact to PS and for efficient injection EL devices.
(e) Complete surface passivation and uniform carrier injection are the key factors for development of stable EL operation.

Acknowledgements

The author would like to thank H. Koyama, T. Ozaki, M. Araki, Y. Yamamoto, T. Oguro and H. Mizuno for their collaboration. This work was partially supported by the Nissan Science Foundation, The Akai Foundation and a Grant-in-Aid from the Ministry of Education, Science and Culture of Japan.

References

[1] Koshida N., Kiuchi Y., and Yoshimura S., "Photoconduction effects of porous Si in the visible region", The 10th Symp. Photoelectronic Image Devices, London 2-6 September 1991, B.L. Morgan Ed. (Institute of Physics, Bristol, 1992) pp. 377-384.
[2] Koshida N. and Koyama H., *Mat. Res. Soc. Symp. Proc.* **283** (1993) 337-342.
[3] Koyama H. and Koshida N., *J. Luminescence* **57** (1993) 293-299.
[4] Koshida N. and Koyama H., *Appl. Phys. Lett.* **60** (1992) 347-349.
[5] Koyama H., Oguro T. and Koshida N., *Appl. Phys. Lett.* **62** (1993) 3177-3179.
[6] Koshida N., Koyama H., Yamamoto Y. and Collins G.J., *Appl. Phys. Lett.* **63** (1993) 2655-2657.
[7] Koyama H., Nakagawa T., Ozaki T. and Koshida N., *Appl. Phys. Lett.* **65** (1993) 1656-1658.
[8] Koshida N., Koyama H., Suda Y., Yamamoto Y., Araki M., Saito T., Sato K., Sata N. and Shin S., *Appl. Phys. Lett.* **63** (1993) 2774-2776.
[9] Suda Y., Ban T., Koizumi T., Koyama H., Tezuka Y., Shin S. and Koshida N., *Jpn. J. Appl. Phys.* **33** (1994) 581-585.
[10] Mendez E.E., Bastard G., Chang L.L., Esaki L., Morkoc H., and Fischer R., *Phys. Rev.* B **26** (1982) 7101-7104.
[11] Kash J.A., Mendez E.E. and Morkoc H., *Appl. Phys. Lett.* **46** (1985) 173-175.
[12] Koshida N., Mizuno H., Koyama H., and Collins G.J., *Jpn. J. Appl. Phys.* **33** (1994) (in print).
[13] Halimaoui A., Oules O., Bomchil G., Bsiesy A., Gaspard F., Herino R., Ligeon M. and Muller F., *Appl. Phys. Lett.* **59** (1991) 304-306.
[14] Beale M.I.J., Cox T.I., Canham L.T. and Brumhead D., *Mat. Res. Soc. Symp. Proc.* **283** (1993) 377-382.

Characteristics of porous n-type silicon obtained by photoelectrochemical etching

C. Levy-Clement

*Laboratoire de Physique des Solides de Bellevue, UPR 1332 du CNRS,
1 Place Aristide Briand, 92195 Meudon, France*

1. INTRODUCTION

The term porous silicon includes many different and complex materials whose physical properties depend critically on many parameters, such as the initial silicon properties or the preparation parameters. Most of the research performed on porous silicon has been dedicated to the p-type material and is well documented [1-4]. However it should be mentioned that a possible application to light emitting diode technology would be based on n-type substrates and consequently control of porous n-type silicon is a major issue. Porous silicon obtained by anodization in the dark of n^+-type Si, is a sponge like material with randomly-positioned pores and a sparse silicon network of columns in the nanometer dimension range. In the case of n^--type Si, due to the rectifying behaviour of the semiconductor/electrolyte junction, it is necessary to illuminate the electrode during the anodization process (i.e. photoelectrochemical etching, PEC-etching) [3, 5-9]. Similarly porous n^+-type Si can be formed under illumination. Comparatively little is known about these latter two cases [3, 5-16]. An essential result is that the porous n-type (n^- and n^+) silicon, obtained under illumination, generally consists of a superficial nanoporous silicon layer formed of silicon fibers [15, 16]. It covers a macroporous silicon layer containing pores in the micron size range and also coats the interior walls of the macropores [10, 11]. The room temperature visible luminescence originates only from the upper nanoporous layer [12].

In this article we review our recent results on porous n-type silicon obtained by PEC-etching. They concern the dependence of the morphology of nanoporous and macroporous silicon, on the doping density and on the crystallographic orientation of the silicon substrate. Details of the nanostructure, the nanochemical characterisation and the photoluminescence properties of the nanoporous Si layer are presented.

2. CONDITION OF FORMATION OF POROUS SILICON

Porous silicon is generally obtained by an electrochemical oxidation-dissolution (performed in the dark for p-type (and n^+ type) and under illumination for n-type (n^- and n+)) of the silicon in fluoride containing media, under a controlled current density inferior to a critical value (i_{crit}).

The cyclic voltammogram (i.e. current-voltage curve - CV) of a silicon electrode in HF, in the dark and under illumination reveals information on the domain of formation of porous silicon [6, 9]. This is thoroughly explained in this book in the chapter on the silicon/electrolyte interface. The fact that the light is an extra parameter that can be adjusted is illustrated by the experiments of Eddowes [9]. He demonstrated, by controlling the light intensity in order to maintain the anodic photocurrent below i_{crit}, that for n-type Si the two-hole process corresponding to the formation of the porous silicon is independent of the anodic potential.

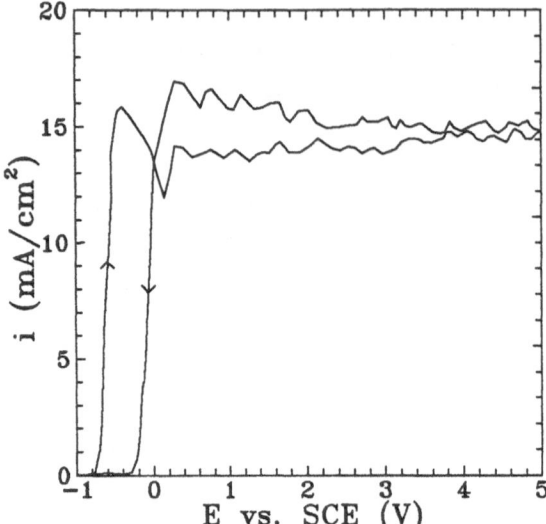

The photocurrent-voltage curve typically exhibits a light intensity limited plateau, as illustrated in Figure 1. For a weak illumination, the photocurrent is limited by hole generation and the light limited plateau observed in the photocurrent-voltage curve is associated with the formation of porous silicon. In this case the electropolishing regime is never reached due to the lack of carriers.

Figure 1 : Current-potential curves for n(100)LD Si under illumination in 5% HF versus SCE, scan rate = 100 mV/s. In the dark no current is observed.

Porous silicon was produced on n-type (n⁻ and n⁺, phosphorus doped) silicon single crystals with different doping densities and different orientations (see Table 1). n(100)LD and n(100)HD refer to lightly doped ($N_d = 10^{15}$ cm^{-3}) and highly doped ($N_d = 10^{18}$ cm^{-3}) n-type (100) oriented substrates respectively, and n(111)LD and n(111)HD to lightly and highly doped (111) oriented substrates respectively.

Table 1 : Characteristics of the silicon substrates studied.

Samples	Type	Orientation	Dopant (cm-3)	Resistivity (Ohm.cm)
n(100)LD	n	100	$1.2\ 10^{15}$	4.5
n(111)LD	n	111	2.10^{15}	1
n(100)HD	n	100	4.10^{18}	0.01
n(111)HD	n	111	$2.4\ 10^{18}$	0.02

The anodization was performed in a three electrode Perspex cell, under selected potentiostatic regimes, with the Si wafer as the anode, a Pt wire as the cathode and a calomel electrode (SCE) as the reference electrode as schematised in Figure 2. The geometry of the cell allows a 0.4 cm^2 circular shaped area to be horizontally exposed to the electrolyte for direct optical exposure. For PEC-etching under white light, the collimated beam of a tungsten-halogen lamp was used. The irradiance was set to a chosen value using neutral density filters. An ohmic contact on the back of the silicon was made by scratching the backside of the samples with a Ga-In eutectic.

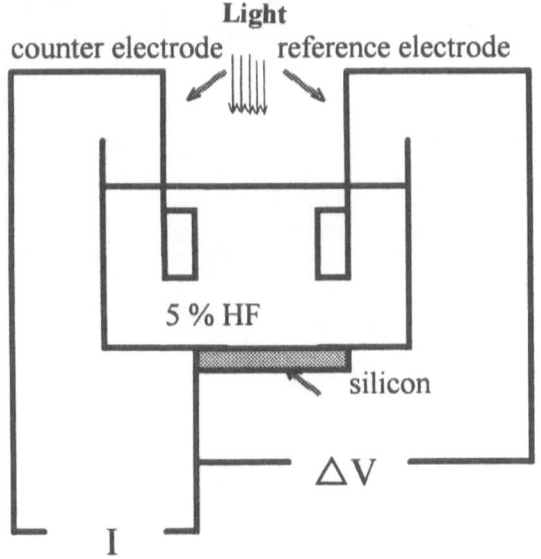

Figure 2 : Schema of the three electrode photoelectrochemical cell.

Porous n-type silicon was produced by PEC-etching in 5% aqueous HF, at a potential of 0.1 V, with the photocurrent held below i_{crit} by controlling the light intensity. The attack was continued until a certain amount of charge, Q, had passed through the cell. For example the (100)LD sample polarised at 0.1 V produced a photocurrent of 15 mA/cm^2. A charge Q = 4C/cm^2 corresponds to a PEC-etching duration of 4.4 min.

It was found, depending on the PEC-etching conditions, that during the formation of the double porous structure, an etch crater is also formed at the surface as schematised in Figure 3. The light emitting nanoporous Si layer not only covers the macroporous Si layer but forms inside the macropores as well, suggesting that this layer forms on all surfaces exposed to the electrolyte. The macroporous structure is revealed after removing the nanoporous layer by dipping in a KOH solution.

3. THE MACROPOROUS Si LAYER

3.1 Morphology

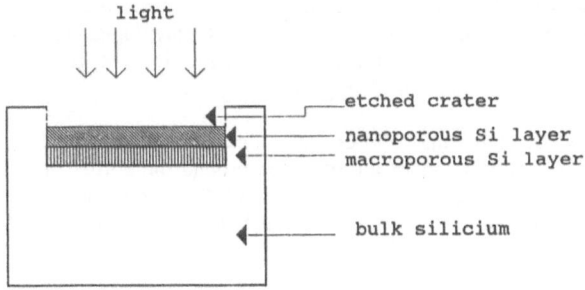

Figure 3 : Distribution of the different layers formed at the surface after PEC-etching

SEM plan view and cross sectional observations of the macroporous Si layer reveal the three dimensional structure of the macropores. The shape of the macropores [11] is found to be primarily dependent on the crystallographic orientation and on the doping density. The pore size depends on Q, the quantity of charge transferred.

<u>Crystallographic orientation</u> : The macropores formed on (100) substrates have a square sectional form which are round bottomed (Fig. 4a) and in the case of (111) substrate have an almost triangular section (Fig. 4b). In both cases the macropores delineate a regular network. This observation confirms results published elsewhere [7], i.e. that PEC-etching of Si in HF is anisotropic and occurs preferentially along the <100> direction as illustrated in Figure 5.

 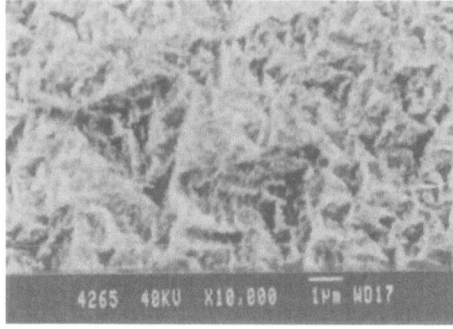

Figures 4a and 4b : Scanning electron micrographs of macroporous Si, plan views. Fig. 4a, on the left, n(100)HD (Q = 10 C/cm^2) and Fig. 4b, on the right, n(111)HD (Q = 10 C/cm^2).

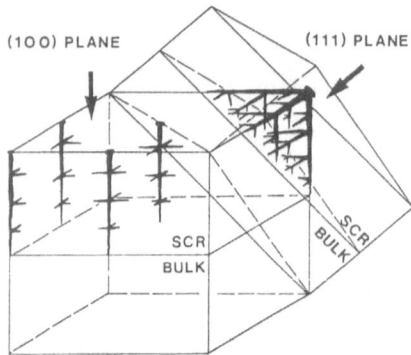

Figure 5 : Schema showing that the PEC-etching of Si occurs preferentially along the [100], [010] and [001] directions. Pores with square tubular shape and with triangular shape are expected to be formed for (100) and (111) oriented substrates, respectively, (from [7]).

<u>Doping density</u> : Two effects are noticed. For an equal amount of charge transferred, Q, the size of the macropores is slightly larger for lightly doped Si compared to highly doped (Fig. 4a and 6a and Fig. 4b and 6b). The square shape observed for the (100) samples is rather rounded for n(100)LD (Fig. 6a) and sharply defined for n(100)HD (Fig. 4a).

<u>Charge transferred</u> : The average width of the macropores varies linearly with Q, for (100) substrates. At high values of Q, small secondary pores are observed around the border of the large primary macropores (Fig. 7a). They develop along the <010> and <001> directions. For n(111)HD (Q > 4C/cm^2) and n(111)LD (Q = 50 C/cm^2), triangular macropores on at least

two distinct length scales are observed (Fig. 4 b and 7b).

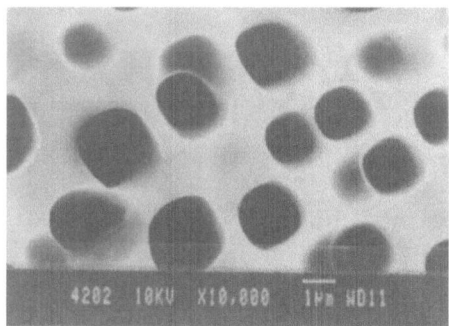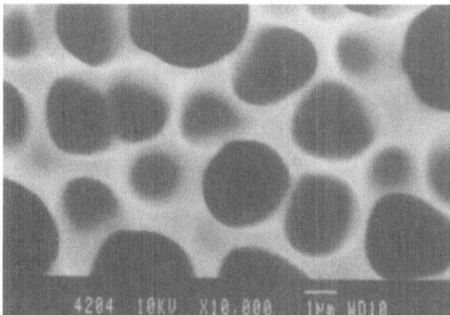

Figures 6a and 6b : Scanning electron micrographs, plan views of macroporous Si . Fig. 6a, on the left. n(100)LD (Q = 10 C/cm^2) and Fig. 6b, on the right, n(111)LD (Q = 10 C/cm^2).

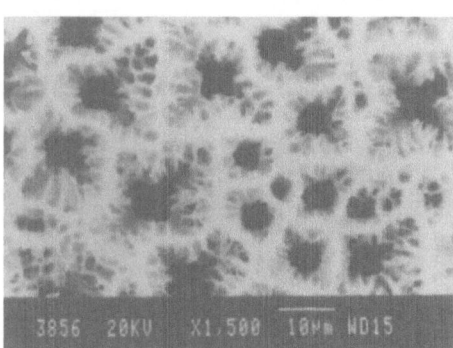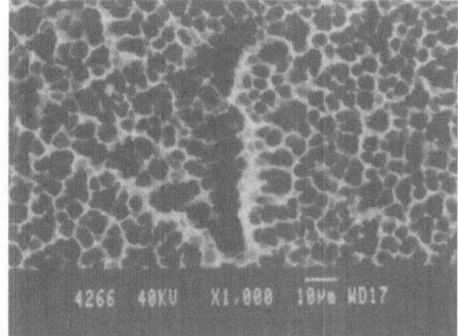

Figures 7a and 7b: Scanning electron micrographs, plan views of macroporous Si. Fig. 7a, on the left, n(100)LD (Q = 100 C/cm^2) and Fig. 7b, on the right, n(111)LD (Q = 50 C/cm^2).

Figure 8 shows the schematic evolution of the thickness of the different layers produced on the surface of a n(100) Si substrate during PEC-etching. The etch crater deepens regularly with the increase of Q transferred. The thickness of the nanoporous layer reaches a maximum value for Q values of about 50 C/cm^2, which for n(100)HD is equal to about 7 μm. At low values of Q (4C/cm^2), no macroporous layer is formed on lightly doped Si, while at high values of Q (50 C/cm^2) the macropores are longer (10 to 20 μm) than for highly doped Si (6 μm).

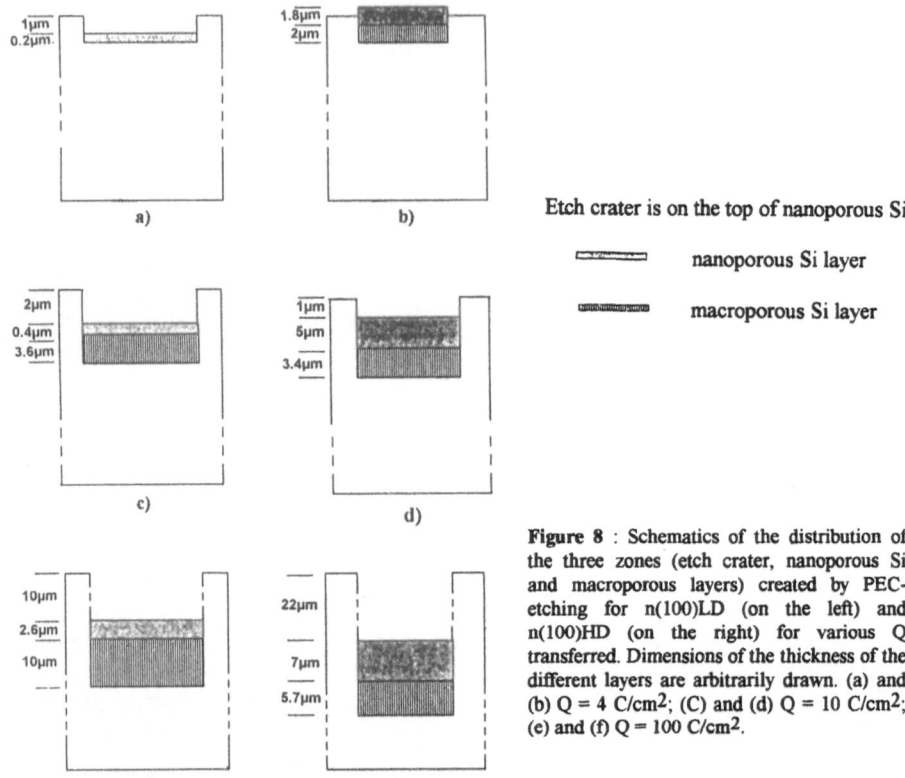

Etch crater is on the top of nanoporous Si

▭ nanoporous Si layer

▭ macroporous Si layer

Figure 8 : Schematics of the distribution of the three zones (etch crater, nanoporous Si and macroporous layers) created by PEC-etching for n(100)LD (on the left) and n(100)HD (on the right) for various Q transferred. Dimensions of the thickness of the different layers are arbitrarily drawn. (a) and (b) Q = 4 C/cm^2; (C) and (d) Q = 10 C/cm^2; (e) and (f) Q = 100 C/cm^2.

3.2 Optical properties of macroporous Si

No room temperature luminescence in the visible range is detected from the macroporous silicon [12]. For certain values of Q (Q $\leq$ 10 C/cm^2 for highly doped Si and Q > 10 C/cm^2 for lightly doped Si) the surface of the macroporous Si is black mat after the removal of the nanoporous layer using a KOH dip. The fact that such surfaces strongly absorb light is of potential interest to photovoltaic applications [13, 14]. Figure 9 shows the total reflectivity of the macroporous surface (n(100)HD, Q = 4 C/cm^2). Compared to the typically high 35% reflectivity for polished mirror like Si, a dramatic decrease to 2-9% in the visible part of the spectrum is observed for the macroporous Si. The increase above 1000 nm is in the vicinity of the bandgap energy.

The low value of the total reflectivity of macroporous silicon is due to the surface roughening leading to efficient light trapping on the surface. This effect is established over the entire visible spectrum, leading to the black appearance of treated samples. Use of PEC-etching as a new process of surface preparation to texturise the surface of the emitter of crystalline and multicrystalline silicon solar cells is thus expected to lead to an increase of the output of photovoltaic cells due to reflectivity decrease.

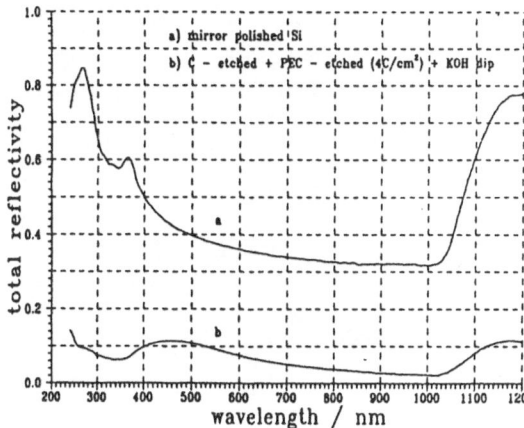

Figure 9 : total reflectivity spectra of n(100)HD : a) polished mirror like Si, b) macroporous Si (Q = 4 C/cm²).

4. THE NANOPOROUS Si LAYER

4.1 Morphology

The color of the surface of nanoporous Si depend on the doping density and Q. Low doped PEC-etched Si appears dull and exhibits interference colours. Surfaces of n(100)HD and n(111)HD PEC-etched Si are mat, with violet-green colours for low Q values and yellow-brown for high Q values.

Imaging of the nanoporous silicon morphology by transmission electron microscopy (TEM) has been reported elsewhere [15]. TEM images of specimens, both in plan and cross-sectional views were obtained using a JEOL-2000 FX operating at 200 KV. TEM samples were prepared by detaching the nanoporous Si layer by an electropolishing process and eventual chemical thinning, avoiding physical damage inherent in other commonly employed techniques. Cross-sectional specimens were prepared by ultramicrotomy.

For the same value of Q, the pore size is smaller in the lightly doped nanoporous Si compared to the highly doped one. This is the contrary to the size of the pores in the macroporous layer, discussed above.

The corresponding selected area diffraction patterns reveal the characteristic pattern of the single crystal bulk material, with the original orientation of the Si parent wafers conserved.

TEM clearly shows that the elementary building block of all nanoporous layers is a silicon fiber. The spatial distribution of these fibers results in structures with different length scales and depends on the initial substrate characteristics. Due to this common building block, fibrous Si may be a more precise term to include the various structural features of the nanoporous layer formed on n-type Si.

The plan-view bright field image of n(100)LD in Figure 10a shows a porous structure, with pore diameter of the order of 50 nm. Higher magnifications [15] show that the pore walls have themselves a fibrous structure and individual fibers can be visualised within the pores. Thus two length scales are observed.

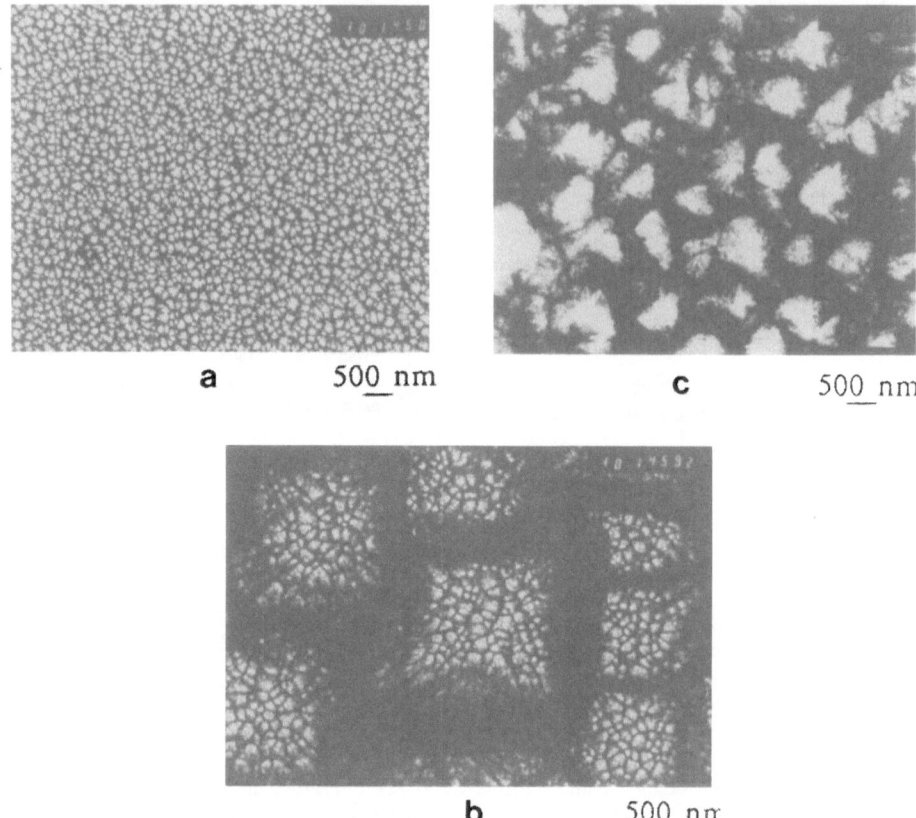

Figures 10a, 10b and 10c : TEM micrographs (plan-view) of nanoporous Si (Q = 10 C/cm²). Fig. 10a, on the left, n(100)LD; Fig. 10b, in the centre, n(100)HD; Fig. 10c, on the right, n(111)HD.

In the case of n(100)HD three length scales are observed (Fig. 10b). The largest length scale shows a square array forming pores of approximately 1 µm width, corresponding to the pores observed in the macroporous layer (Fig. 4a). Within this structure a secondary porous structure is observed of approximate 100 nm length scale. Higher magnifications show that the walls of both arrays, i.e. 1 µm and 100 nm arrays, consists of Si fibers.

The n(111)HD nanoporous layer exhibits two length scales (Fig. 10c). The walls of the triangular array, pore dimension 500 nm corresponding to the macroporous triangular structure of figure 4b, are fibrous [16].

For the highly doped specimens the largest structures are a fingerprint of the crystallographic orientation of the substrates. The square array corresponds to the (100) substrate and the triangular array to the (111) substrate. The rectangular and triangular structure of n(100)HD and n(111)HD respectively are also confirmed in the cross sectional views of Figures 11a and 11b. This is an indication that the pores are formed along the (100) planes as previously discussed by Chuang et al. for porous p and n⁺ Si obtained in the dark [17].

The fiber diameter is variable as illustrated in Figure 12a for highly doped Si, with wire diameters in the range 4.5 to 12 nm. In this case the diameter fluctuates along individual nanowires and has the required quantum size at certain positions [18-20]. For lightly doped Si the nanowires exhibit a more rigid structure, with smaller diameters in the range of 3 to 9 nm (Fig. 12b), but individual wires having a constant diameter along their length [15].

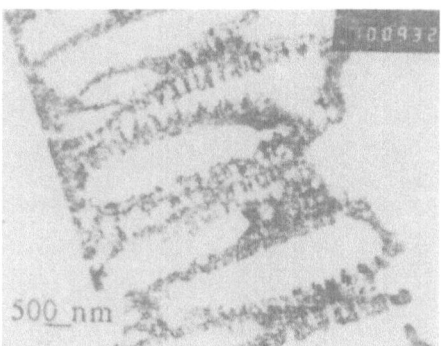 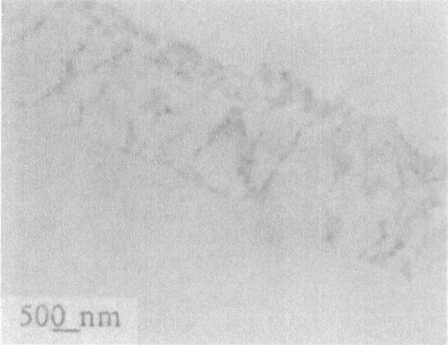

Figures 11a and 11b : Cross-sectional TEM micrograph revealing the typical internal structure through the thickness of the nanoporous Si (Q = 10 C/cm^2). Fig. 11a, on the left, n(100)HD; Fig. 11b, on the right, n(111)HD.

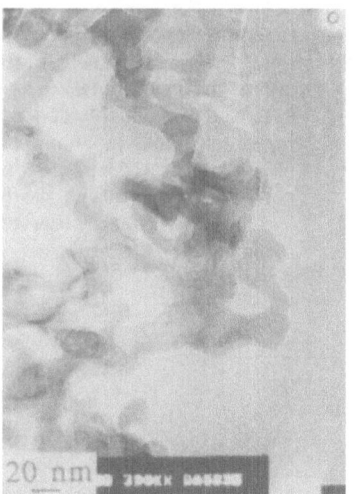 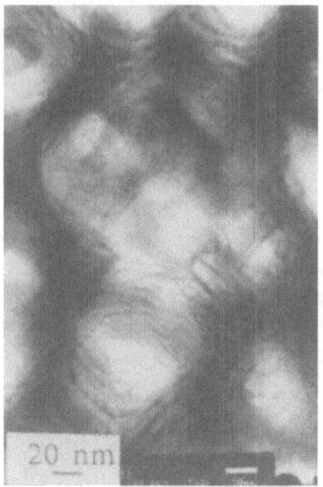

Figures 12a and 12b : TEM micrographs (plan-view) detail of the nanometer sized fibers. Fig. 12a, on the left, n(111)HD; Fig. 12b, on the right, n(100)LD.

4.2 Nanostructure of the Si nanowire

High resolution transmission electron microscopy (HRTEM) has been attempted to directly image individual fibers or nanowires, held in the pore, at atomic resolution. HRTEM was performed by C. Colliex group (Orsay-Paris Sud University). It was recorded with a Topcon 002-B EM fitted with the ultra high resolution pole piece and a top-entry tilting goniometer (±10° x, y). It was shown that all the nanowires exhibit a crystalline core surrounded by an amorphous layer [16, 21]. Figure 13 illustrates for a n(100)LD sample a nanowire oriented with the [100] axis parallel to the beam, showing a crystalline core with a square array of (220) lattice planes, inter reticular distance of 0.19 nm, surrounded by an amorphous skin (10 to 20 Å thick) .

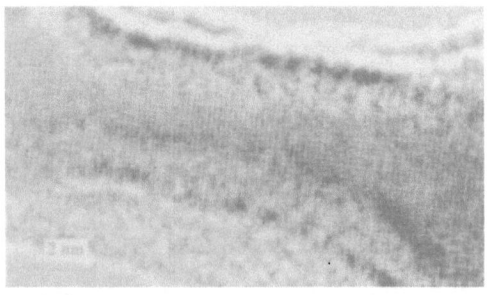

Figure 13 : HRTEM micrograph (plan view). Detail of a nanowire.

4.3 Photoluminescence

The as prepared nanoporous layer emits light at room temperature upon excitation with an argon ion laser. The photoluminescence spectrum of porous Si, formed on a lightly doped substrate shows a maximum at about 600 nm, and is shifted to the blue compared with the spectrum of the highly doped which has a maximum at about 700 nm [12], as illustrated in Figure 14. We correlate this shift with the thinner and more rigid fibers observed with the lightly doped specimen.

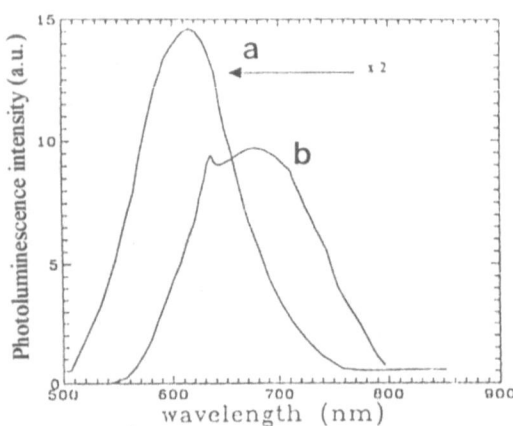

Figure 14 : Room temperature luminescence originating from the as prepared nanoporous Si layers :a) n(100)LD, Q = 100 C/cm^2, b) n(100)HD, Q = 10 C/cm^2.

4.4. Macroscopic chemical characterisation

In order to get a global insight into the chemical nature of the porous silicon, X-ray photoelectron spectroscopy, secondary ion mass spectroscopy and infrared spectroscopy have been employed.

The main result of previously carried out XPS studies [3] is that the nanoporous Si samples are free of SiO_2, as in the energy range 103 to 104.5 eV no peak characteristic of SiO_2 is detected. The broadening and the blue shift (c.a. 0.5 to 1 eV) of the Si 2p peak in the XPS (X-ray Photoelectron Spectroscopy) spectrum of the nanoporous silicon compared to bulk Silicon (Si 2p peak at 99.5 eV) (Fig. 15) leads to the assumption that substoichiometric silicon oxides are formed [3]. An alternative explanation is a quantum confinement shift of the conduction band as discussed by Van buuren et al. [22, 23].

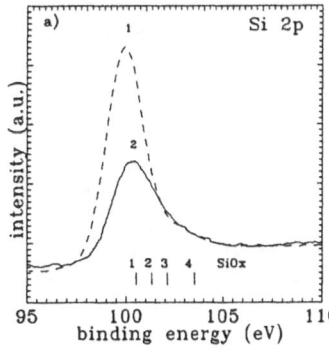

Figure 15 : X-ray photoelectron spectroscopy (binding energies) of the Si 2p peak for n(100)LD. 1) Si substrate, 2) nanoporous Si. The energy of the X-rays is 1486.6 eV. The expected Si 2p line positions for oxidation states 1-4 of Si in SiO_x are also indicated.

SIMS (Secondary Ion Mass Spectroscopy) measurements reported elsewhere [10] clearly indicate that the increased amount of oxygen detected in nanoporous silicon does not correspond to the formation of SiO_2. A noticeable increase in the concentration of H seems to be limited to the domain of the nanoporous silicon. A larger concentration (at least two times) of phosphorus is measured in the nanoporous and macroporous Si layers compared to the parent substrate. It is not clear if this correspond to a real increase in the impurity level or if the phosphorus is localised in certain areas.

Infrared vibrational absorption spectroscopy performed on lightly doped Si ($N_d = 10^{15}$ cm^{-3}) gives complementary insights [10]. A specially shaped electrode was employed. It was cut from a n(111)LD ingot obtained by zone fusion and shaped as an attenuated-total-internal-reflexion prism as reported elsewhere [24]. It is seen in Figure 16 that after PEC-etching a shift in the position of the Si-H stretching vibration occurs, together with a broadening of the peak which appears to involve three components. It has been shown that at single crystal surfaces the Si-H stretching vibrations depend strongly and systematically on the local bonding environment. Specifically, the frequency of the Si-H stretching vibration linearly shifts from $\approx$ 2090 cm^{-1} to $\approx$ 2270 cm^{-1} when the three silicon atoms backbonded to a SiH species are successively replaced by oxygen atoms [25, 26]. Fluorine substitutes could also raise the Si-H frequency to the same level, but, since no fluorine was detected by XPS, it is very unlikely that these three vibrational peaks are due to fluorine substitution. On the other hand, the amount of oxygen detected by XPS, as well as by SIMS, is consistent with the assignment of these vibrations to oxygen-backbonded SiH groups. The low frequency component at 2145-2155 cm^{-1} is primarily assigned to $(Si_2O)\equiv Si-H$, the 2200 cm^{-1} line to $(SiO_2)\equiv Si-H$ and the one at 2250 cm^{-1} to $(O_3)\equiv Si-H$ vibrations. Therefore the asymmetry in the XPS Si 2p peaks may be explained by the presence of SiH units backbonded to one or more oxygen atoms along the walls of the silicon pores.

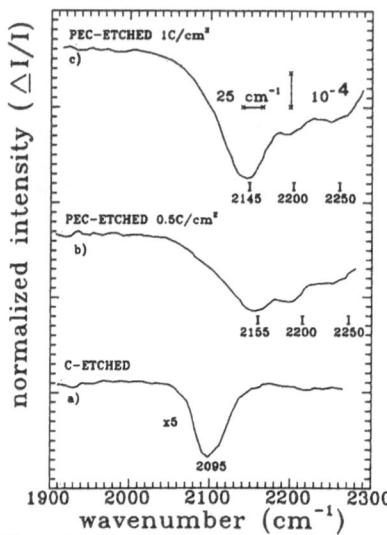

However. a small amount of substoichiometric silicon dioxide is also present at the interface as evidenced by the vibrational signal in the 1000-1200 cm^{-1} range. Previous Fourier Transform IR [27] studies of porous p-type silicon (5 Ohm.cm, (100) oriented) formed by anodization in 50% HF + ethanol (2:1 by volume) in the dark, showed that no such oxidised species were present but rather SiH, SiH$_2$ and SiH$_3$ surface groups. Apart from the difference in electrochemical conditions, the difference with the present experiments may be due to the fact that the experiments of Venkateswara Rao et al. [27] were performed in situ during the porous Si growth. Exposing the samples to air may lead to some oxidation of the surface.

Figure 16 : Multiple internal reflection spectra of the surface of n(100)LD Si : a) substrate; b) nano6porous Si (Q = 0.5 C/cm^2), c) nanoporous Si (Q = 1 C/cm^2). The FWMH resolution was » 25 cm^{-1} (horizontal bar). The vertical bar gives the absolute scale ($\Delta I/I$ scaled to one reflection).

4.5 Local chemistry of the nanowires

Electron energy loss scattering (EELS) fine structure of the Si nanofibers was recorded in order to identify the nature of the amorphous skin.

It is now well established that one can investigate atomic bonding, and both filled and empty electronic states in inhomogeneous materials with a sub-nanometer spatial resolution from EELS data. The useful information arising from the electronic structure can be identified in the low loss energy region, from about 2 to 30 eV, as well as in the fine structures occurring on core excitation edges, i.e essentially the Si L$_{23}$ edge corresponding to the excitation of the silicon 2p electrons to the empty states.

In the low energy-loss region the observed intensity distribution partially results from direct interband excitation. However the local change in both the valence band and conduction bands cannot easily be detected because this electron excitation domain is dominated by collective processes, i.e. bulk and interface plasmons.

Sequences of EELS spectra have been recorded by the Colliex group under the line spectrum mode that they have developed [28], when scanning the incident probe of a VG HB501 STEM across a nanowire. Typical experimental conditions are 0.5 nm probe size, step increments between spectra controlled in the range 0.3 to 0.6 nm and a typical recording time per spectrum of the order of 50 ms in the low loss range and a few 100 ms for the fine structures on the Si L$_{23}$ edge. The spatial resolution itself is mostly governed, for such thin nanowires, by the probe shape, i.e. also about 0.5 nm for the illumination parameters employed.

A 3D intensity-plot, as a function of the energy loss and the incident beam position of the Si l$_{23}$ edges during a line scan across a 30 nm Si wire is shown in Figure 17a. Figure 17 b corresponds to a line scan across two jointed Si wires recorded in the low-loss energy range. Careful comparison of the spectra at different positions across a single nanowire, especially at the external surface of the wire and in an intermediate domain close to the boundary between the crystalline core and the amorphous skin was made. A clear edge at 100 eV with a first maximum at about 101.5 eV is observed. No chemical shift is detected between the crystalline core and the amorphous skin. In the low energy-loss region the base of the zero-loss peak extends to higher energies and therefore hinders the visibility of weak features in the 2 eV energy loss domain (Fig. 17b). The energy of the different features are listed in table 2. No chemical shift is detected between the different features observed in the spectra of a crystalline core and that of the amorphous skin. To identify the chemical nature of the amorphous skin comparison has been made with EELS fine structure of reference samples, such as amorphous hydrogenated Si, bulk Si and amorphous SiO_2 (from aSi/SiO_2 interface) (Fig. 18).

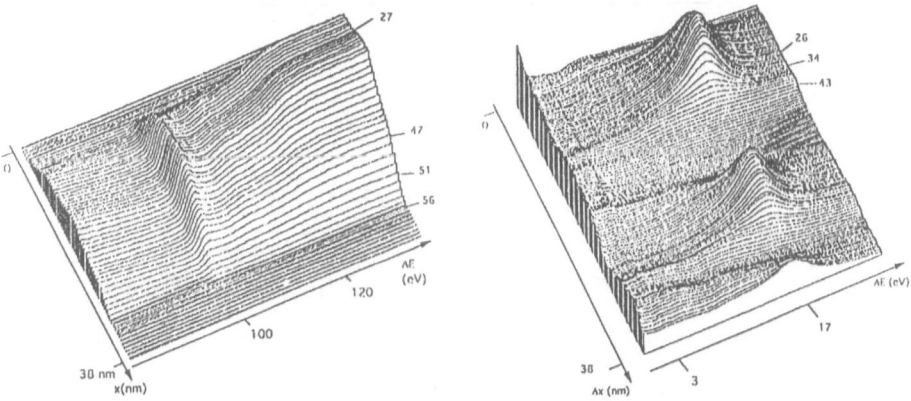

Figures 17a and 17b : EELS spectra across an isolated nanowire. Fig. 17a, on the left, in the 80-180 eV energy range showing detailed shape of the Si l_{23} edge (100 eV); Fig. 17b, on the right in the 0-30 eV energy range showing the dominant bulk Si volume plasmon peak loss at 17 eV.

The weak features at 10.5 eV and 106-108 eV visible across the whole nanowire could be interpreted in terms of native oxide (either due to the growth process of the Si substrate or to oxygen contamination when handling the sample in the air).

The volume plasmon is characteristic of the bulk crystalline silicon spectra. When moving the probe from the vacuum into the nanowire, two main features are observed : (i) the energy position of the volume plasmon and, (ii) the two first structures at the onset of the interband transitions, are quite identical across the whole nanowire. The position of the bulk plasmon exhibits weak shifts in opposite directions for hydrogenated and nanoporous Si with respect to crystalline Si. However due to the small amplitude of the shifts, the resulting information may be questioned. More informative is the double peak absorption behaviour near 3.2 and 4.7 eV exhibited by the porous Si which is not observed for the other materials [16]. This double peak constitutes a clear signature for the nanoporous Si, as it reveals a strong interband excitation process above the visible energy range.

Table 2 : Different features observed in the low-loss energy region spectra of the Si nanofiber, bulk Si, amorphous SiO_2 and amorphous SiH_x.

Samples	Transitions (eV)			weak edge (eV)	volume plasmon (eV)
	Onset	bump	bump		peak
Porous Si	2.8	3.7	4.8	10.5	17
Bulk Si					16.8
SiO_2				10.6	23.5
SiH_x			7		16.6

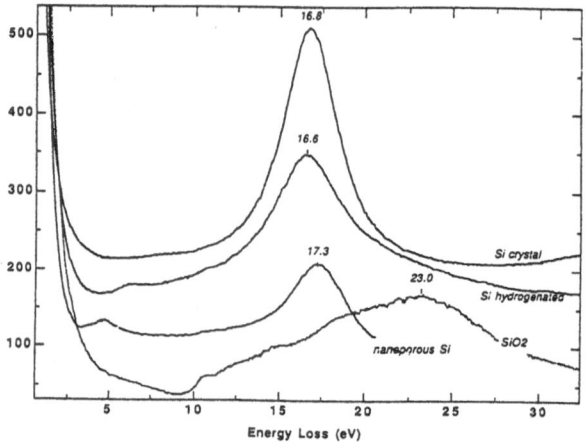

Figure 18 : Selected EELS spectrum in the low loss energy range of an isolated Si nanowire and EELS spectra of test samples : bulk Si, SiO_2 (from a Si/SiO_2 interface) and hydrogenated Si.

5. CONCLUSION

Illumination, is an important control parameter during the anodization of n-type Si which can be regulated to manipulate the morphology and hence the luminescent properties. In this paper we report details of the morphology and surface chemistry of photoelectrochemically etched (PEC-etched) n and n^+-type Si in HF and show that it leads to the formation of a surface with unique features, different from those obtained by the electrochemical process.

The HRTEM and spatially resolved EELS tools have proved to be very effective to investigate on a very local scale the structural, electronic and chemical properties of the nanoporous silicon. With the support of other high energy resolution spectroscopies, which average the information over wide areas, they help to corroborate the role of quantum sized crystalline Si in the interpretation of the luminescence of nanoporous Si. It should be noted, contrary to studies on p-type porous Si [29], that no nanosized crystallites embedded in the matrix have been observed in the n-type porous layers formed by PEC-etching. In the studies presented here only nanowires are observed attached to a fibrous Si network.

Acknowledgements: *The help of the following people in preparing the manuscript is warmly acknowledged : Ana Albu-Yaron, S. Bastide, Danièle Bouchet, Nathalie Brun, J.-N. Chazalviel, C. Colliex, E. Galun, A. Lagoubi, W.M. Shen, R. Tenne, M. Tomkiewicz, C. Vard, P. Williams. This work was supported by a CEE contract (JOUR2 - CT92-0179), the PIRSEM-CNRS through a Concerted Research Action on multicrystalline silicon and the CNRS through LUSIL a "groupement de recherche" on porous silicon.*

REFERENCES

1. R.L. Smith and S.D. Collins, J. Appl. Phys., 71, (1992) R1-R22 and reference therein.
2. P.C. Searson, J.M. Macaulay and S.M. Prokes, J. Electrochem. Soc., 139 (1992) 3373 and references therein.
3. C. Lévy-Clément, A. Lagoubi, R. Tenne and M. Neumann-Spallart, Electrochimica Acta, 37 (1992) 877 and references therein.
4. K.H. Jung, S. Sih and D.K. Kwong, J. Electrochem. Soc. 140 (1993) 3046.
5. R. Memming and G. Schwandt, Surface Sci., 4 (1966) 109.
6. V. Lehmann and H. Föll, J. Electrochem. Soc., 135 (1988) 2831.
7. V. Lehmann and H. Föll, J. Electrochem. Soc., 137 (1990) 653.
8. V. Lehmann, J. Electrochem. Soc., 140 (1993) 2836.
9. M.J. Eddowes, J. Electroanal. Chem., 280 (1990) 97, and J. Electrochem. Soc., 137 (1990) 3514
10. C. Lévy-Clément, A. Lagoubi, D. Ballutaud, F. Ozanam, J.-N.Chazalviel and M. Neumann-Spallart, Appl. Surf. Sc., 65/66 (1993) 408.
11. C. Lévy-Clément, A. Lagoubi and M. Tomkiewicz, J. Electrochem. Soc. 141 (1994) 958.
12. E. Galun, A. Lagoubi, R. Tenne and C. Lévy-Clément, J. of Luminescence, 57 (1993) 125.
13. A. Lagoubi, M. Neumann-Spallart, S. Bastide and C. Lévy-Clément, Proc. 11th European Photovoltaic Solar Energy Conference, Montreux, Harwood Academic Publishers, (1993) p.250-253.
14. C. Lévy-Clément, A. Lagoubi, Neumann-Spallart, M. Rodot and R. Tenne, J. Electrochem. Soc., 138 (1991) L69.
15. A. Albu-Yaron, S. Bastide, J.L. Maurice, C. Lévy-Clément, J.of Luminescence, 57 (1993) 67
16. A. Albu-Yaron, S. Bastide, D. Bouchet, N. Brun, C. Colliex and C. Lévy-Clément, J. Phys. I France, 4 (1994) to be published.
17. S.F. Chuang, S.D. Collins and R.L. Smith, Appl. Phys. Lett. 5 (1989) 675.
18. L.T. Canham, Nature, 353 (1991) 335.
19. L.T. Canham, Appl. Phys. Lett., 57 (1990) 1046.
20. V. Lehmann and U. Gosele, Appl. Phys. Lett., 58 (1991) 856.
21. N. Brun, C. Colliex, C. Lévy-Clément, P. Williams and A. Albu-Yaron, to be published in theProceedings of the 13th Intern. Congress on Electron Microscopy, Paris 17-22 July 1994.
22. T. Van Buuren, Y. Gao,T. Tiedje, J.R. Dahn and B. M. May, Appl. Phys. Lett., 60 (1992) 3013.
23. T. Van Buuren, T. Tiedje, J.R. Dahn and B. M. May, Appl.Phys. Lett., 63 (1993) 2911.
24. A. Venkateswara Rao,J.-N. Chazalviel and F. Ozanam, J. Appl.Phys., 60 (1986) 696.
25. J.A. Schaefer, D. Frankel, F. Stucki, W. Göpel and G.L.Lapeyre, Surf. Sci. 139 (1984) L209.
26. G. Lucovsky, Solid State Commun., 29 (1979) 571.

27.   A. Venkateswara Rao, F. Ozanam and J.-N. Chazalviel, J.Electrochem. Soc. **138**
      (1991) 153.
28.   C. Colliex, M. Tencé, E. Lefèvre, C. Mory, H. Gu, D. Bouchet and C. Jeanguillaume,
      Mickrochemical Acta Proc. 3rd Conf of the European Microbeam Analysis Society,
      Rimini (1993).
29.   I. Berbezier and A. Halimaoui, J. Appl. Phys., **74** (1993) 5421.

Porous Si: From single porous layers to porosity superlattices

M.G. Berger, St. Frohnhoff, W. Theiss*, U. Rossow** and H. Münder

Institut für Schicht- und Ionentechnik (ISI), Forschungszentrum Jülich GmbH, P.O. Box 1913, 52425 Jülich, Germany
**I. Physikalisches Institut, RWTH Aachen, 52056 Aachen, Germany*
*** Institut für Festkörperphysik, TU Berlin, Hardenbergstrasse 36, 10623 Berlin, Germany*

1. INTRODUCTION

Porous silicon has a manifold microscopic structure. Depending on the doping type and level of the substrate used for the anodization, pores with diameters up to 1 μm or down to a few nanometers can be formed [1]. Within this article the discussion will be limited to porous silicon formed on p-type doped substrates. This material has a spongelike microscopic structure which can be observed in transmission electron microscope (TEM) pictures. The typical dimensions of the pores as well as of the remaining silicon crystallites are in the nanometer range. The size of the so-called nanocrystals is of great importance in the view of the quantum confinement model which can explain the luminescence properties of porous silicon [2]. While TEM pictures are very suitable to get a visual impression of the microstructure they are not a good choice to obtain statistical values about the diameters of the silicon nanocrystals. This can be done by means of inelastic light scattering (Raman spectroscopy). The theoretical background mainly based on [3, 4, 5] will be discussed in the following section. Before actually presenting the results achieved by this technique some comments on the experimental conditions will be given.

The microstructure of porous silicon has a great influence on the dielectric function of the material. Due to the nanometer size of the particles porous silicon must be treated as an effective medium when dealing with light from far infrared to the UV. Therefore, samples of different porosities as well as samples with different microstructures have different effective dielectric functions [6]. Combining these layers in a multilayer system porosity superlattices are formed. They offer the possibility to design the optical properties of those samples which opens a completely new field of applications for porous silicon.

2. SINGLE POROUS SILICON LAYERS

2.1 Raman scattering from nanocrystals

First order Raman spectra from single crystalline silicon show contibutions from phonons near the Γ-point of the Brillouin zone. The reason therefore is the momentum conservation and the small momentum of the incident and scattered light. For nanocrystals – as in the case of porous silicon – momentum conservation is no longer valid. Now the wavevectors of the phonons which contribute to the Raman spectra are only given by the geometry, i.e. the size and shape of the nanocrystals.

In a perfect single crystal the phonon can be described by a wavefunction

$$\Phi(\mathbf{q_0}, \mathbf{r}) = u(\mathbf{q_0}, \mathbf{r}) \cdot \exp(i\mathbf{q_0} \cdot \mathbf{r}) \tag{1}$$

where $\mathbf{q_0}$ is the phonon wavevector and $u(\mathbf{q_0}, \mathbf{r})$ a Bloch function which takes care of the periodicity of the crystal.

For nanocrystals the phonons are localized to the dimensions of the crystallite. This can be taken into account by introducing a weighting function $W(\mathbf{r}, L)$ which determines the phonon amplitude at a certain location $\mathbf{r}$ in a nanocrystal with diameter L of either spherical or columnar shape [3].

The phonon wavefunction can then be written as

$$\Psi(\mathbf{q_0}, \mathbf{r}) = W(\mathbf{r}, L) \cdot \Phi(\mathbf{q_0}, \mathbf{r}) = u(\mathbf{q_0}, \mathbf{r}) \cdot \Psi'(\mathbf{q_0}, \mathbf{r}) \tag{2}$$

The wave function $\Psi'(\mathbf{q_0}, \mathbf{r})$ can be expressed as a Fourier series

$$\Psi'(\mathbf{q_0}, \mathbf{r}) = \int C(\mathbf{q_0}, \mathbf{q}) \, \exp(i\mathbf{q} \cdot \mathbf{r}) \, d\mathbf{q} \tag{3}$$

where the Fourier coefficients $C(\mathbf{q_0}, \mathbf{q})$ are given by

$$C(\mathbf{q_0}, \mathbf{q}) = \frac{1}{(2\pi)^3} \int \Psi'(\mathbf{q_0}, \mathbf{r}) \, \exp(-i\mathbf{q} \cdot \mathbf{r}) \, d\mathbf{r} \tag{4}$$

In equation (3) the wavefunction of the phonon in the nanocrystal $\Psi'(\mathbf{q_0}, \mathbf{r})$ has been written as a superposition of the eigenfunctions for infinite single crystals. $C(\mathbf{q_0}, \mathbf{q})$ is

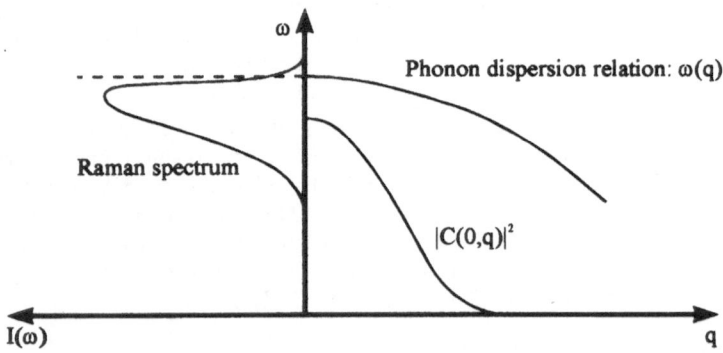

Figure 1: Raman scattering from nanocrystals causes a shift and an asymmetric broadening of the optical phonon peak. The reason therefore are scattering contributions from phonons all over the Brillouin zone weighted by $|C(\mathbf{q_0}, \mathbf{q})|^2$. The dashed line indicates the phonon frequency for bulk crystalline material.

the amplitude of the eigenfunction with wavevector $\mathbf{q}$. The Raman spectrum can be calculated as a superposition of wavefunctions with all possible wavevectors weighted by the square of their amplitude $|C(\mathbf{q_0}, \mathbf{q})|^2$ [3]:

$$I(\omega) \sim \int \frac{|C(\mathbf{q_0}, \mathbf{q})|^2}{[\omega - \omega(q)]^2 + (\Gamma_0/2)^2} \, d\mathbf{q} \tag{5}$$

The above equation is a weighted superposition of lorentzians where Γ_0 is the natural full width at half-maximum (FWHM) of the phonon and $\omega(q)$ is the dispersion relation. Figure 1 shows an illustration of equation (5). For the function $|C(\mathbf{q_0}, \mathbf{q})|^2$ a typical shape is shown. Due to the phonon dispersion relation, which is assumed to be parabolic for simplicity, the maximum of the phonon peak shifts to lower frequencies as compared to single crystals (dashed line). Furthermore the lineshape is asymmetrically broadened.

In order to calculate the function $|C(\mathbf{q_0}, \mathbf{q})|^2$ from equation (4) the weighting function for the phonon amplitude $W(\mathbf{r}, L)$ must be known. The function can only be chosen from physical arguments. In the majority of publications a gaussian function $\exp(-\alpha r^2/L)$ is used [7, 8]. Nevertheless this function is not a good choice because no physical reason exists to assume a gaussian lineshape. Furthermore, within this model the phonon amplitude is not zero outside the nanocrystal and the parameter α is arbitrary depending on the choice of the phonon amplitude at the boundary of the crystallite.

A better ansatz is to treat the phonon in the nanocrystal analogously to an electron in a potential well with hard boundaries (i.e. potential outside the nanocrystal is infinite) [4]. For spherically shaped crystals the phonon amplitude is given by $\sin(\alpha r)/\alpha r$ where $\alpha = 2\pi/L$. In contrast to the gaussian function, here the only free parameter is the size of the nanocrystal and the phonon amplitude at the boundary of the crystal is zero. In addition, for the calculation the phonon amplitude outside the crystal is set to zero. The function $|C(\mathbf{q_0}, \mathbf{q})|^2$ can be calculated to

$$|C(\mathbf{q_0}, \mathbf{q})|^2 = \frac{4L^4}{(2\pi)^4} \cdot \frac{\sin^2(qL/2)}{q^2(4\pi^2 - q^2L^2)^2} \tag{6}$$

2.2 Application to porous silicon

Due to the complicated microstructure of porous silicon it is not possible to calculate the phonon confinement for such a structure. Therefore, a simple model has to be chosen to describe the structure as adequately as possible. In literature spherical as well as columnar nanocrystals are used [9, 10]. In our opinion the topology is best described by spherical nanocrystals because a columnar structure assumes an infinite length in one direction which is not the case.

Nevertheless one cannot expect that porous silicon consists of nanocrystals of only one size. Therefore, we simulate the Raman spectra by assuming a size distribution of spherical nanocrystals [11]. Since a continous distribution cannot be calculated we use a set of sample points for which theoretical Raman spectra are calculated according to equation (5). The measured spectra are then fitted by a superposition of the normalized model spectra with the intensity as a free fit parameter. The size distribution fitted to the experimental spectra ranges from from 12.5 Å to 150 Å. For smaller crystals the model for the phonon confinement becomes questionable because it does not incorporate the discrete lattice structure of the material. To account for such small structures a contribution from amorphous silicon is also fitted. The upper limit for the size distribution is given by the fact that for crystals larger than 150 Å the theoretical Raman spectrum resembles that of bulk crystalline silicon.

It is well known that strain will result in a phonon shift. Because of the large strain values in porous silicon the frequency ω at the Γ point of the Brillouin zone must be changed. This improves the agreement between fit and spectrum. Using the fit procedure it is possible to separate these strain effects from the size effect [11].

2.3 Experimental

The porous silicon samples were prepared by a standard anodic etch process using a mixture of 50% HF with ethanol 1:1. Substrates were p-type doped (100)-oriented wafers with different (0.01 Ωcm, 0.2 Ωcm, 8 Ωcm) resistivities.

Prior to the etching the samples are cleaned in propanol in an ultrasonic bath. After this the samples are rinsed in deionized water. The etching process is carried out in the dark. After the etching the electrolyte is removed and the sample is rinsed in pure ethanol and dried with nitrogen gas. To avoid a native oxidation of the samples they are transferred into an ultra high vacuum (UHV) chamber within 10 minutes after preparation.

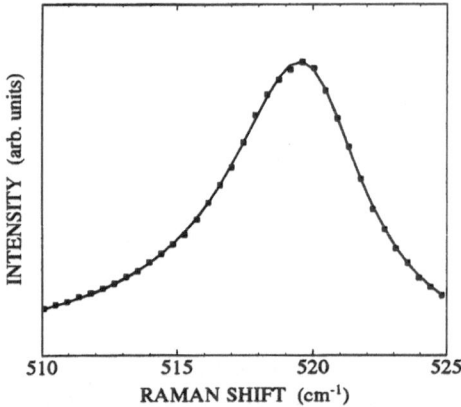

Figure 2: Raman spectrum from porous silicon 75% porosity formed on p-doped substrate 0.01 Ωcm. The agreement between experiment (points) and fit (line) is very good.

The Raman spectra are taken in the UHV chamber using the 457 nm line of an Ar^+ ion laser. The laser power was kept below 1 mW with a focus diameter of 100 μm. These experimental conditions are extremely important because higher laser power densities will result in an unwanted heating of the sample. On the other hand the UHV is necessary to avoid a photostimulated oxidation of the sample during the measurement. The scattered light is analyzed by a DILOR XY triple monochromator with multichannel detection.

The spectral resolution was $1.7\,cm^{-1}$. For all spectra the scattering configuration was $100(010,001)\bar{1}00$ for which the silicon optical phonon is symmetry allowed. The spectra were then fitted using the above described model.

The effective dielectric function was measured with spectroscopic ellipsometry using a rotating analyzer. The spectral range for the measurement was 1.5 to 5.5 eV. The reflectance spectrum of the samples was measured using a Perkin-Elmer Lambda 2 spectrometer in the range from 9000 to $50000\,cm^{-1}$ (1.1 to 6.2 eV). The TEM pictures were taken with a Jeol 4000FX and an electron energy of 400 keV.

2.4 Properties of single porous layers

In Figure 2 a Raman spectrum taken from a sample formed on $0.01\,\Omega cm$ doped substrate is shown. The porosity is 75% and the layer thickness is 11 μm. A perfect agreement between measurement and fit can be achieved. Figure 3(a) shows the size distribution functions obtained for a set of samples with different porosities (see also [12]). For the sample with the lowest porosity a very broad size distribution is found. Two distinct peaks occur around 15 Å and around 30 Å, but the dominating feature are crystals larger than 50 Å. It should be mentioned again that nanocrystals larger than 150 Å are represented by crystals around 150 Å and therefore might also be present.

With increasing porosity the number of crystals larger than 100 Å decreases and the center of the distribution shifts to smaller crystals. Furthermore the number of small nanocrystals around 15 Å and 30 Å increases.

(a)   (b)

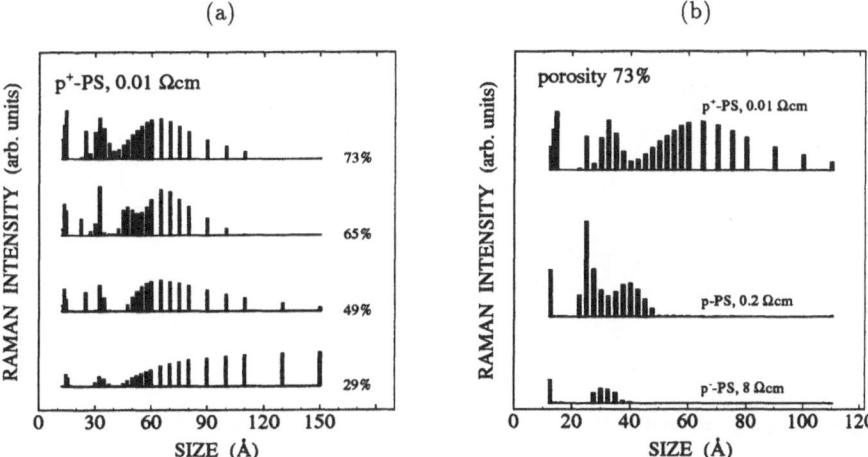

Figure 3: Nanocrystal size distributions for porous silicon formed on p+-doped substrate with different porosities (a) and formed on differently doped substrates with the same porosity (b).

Figure 3(b) shows the influence of different doping levels on the microstructure of PS. The current density has been chosen in a way that all samples have the same porosity (73%). As can be seen from the size distribution, the microstructure differs dramatically for the different doping levels of the substrate. For p+-doped substrates the nanocrystals show the above mentioned broad size distribution, whereas for lower doped substrates only small crystals are present. This is well known from TEM pictures but in contrast Raman is a non-destructive optical technique and is able to give a size distribution function. Although very small crystals are present for the samples formed on highly doped substrates they show only a very weak luminescence.

The difference in the microstructure between porous silicon layers formed on high and on low doped substrates results in a strongly different effective dielectric function of the layers. In Figure 4 the imaginary part of the pseudodielectric function $Im< \epsilon >$ determined by ellipsometry is shown [6].

In the frame of an effective medium theory the difference in $Im< \epsilon >$ means that the dielectric function of the particles, i.e. the nanocrystals, is different and/or the topology of the material differs. In reality both effects will occur. Due to the quantum confinement effect the band structure of the nanocrystals and hence the dielectric function is no longer identical to bulk crystalline silicon. Compared to bulk crystalline silicon the gaps of porous silicon are broadened and for p^- a completely new structure occurs.

For samples formed on the same type of substrate with different porosities the dielectric function is also influenced. The observed reduction in the height of $Im< \epsilon >$ with increasing porosity can be partially understood by the different volume fraction of Si but in addition topology effects must be assumed.

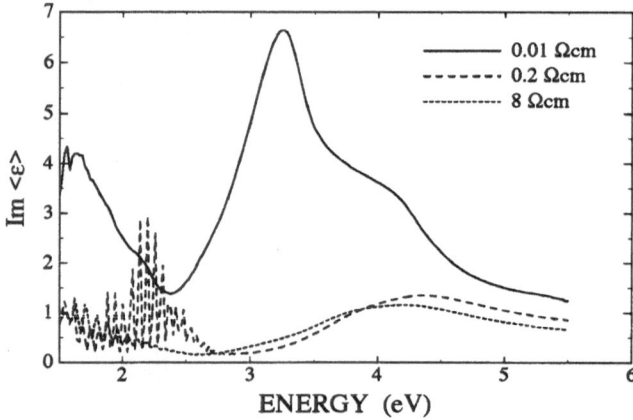

Figure 4: Imaginary part of the pseudodielectric function of porous silicon 75% porosity formed on differently doped substrates.

3. POROSITY SUPERLATTICES

The rather large differences of the dielectric function of porous silicon due to the microstructure has been mentioned above. If it is possible to form a stack of layers with different microstructures and/or porosities the optical properties of the samples can be designed. In this article the techniques for the formation of such porosity superlattices are presented. The strong modulation in the reflectance of porosity superlattices gives a first hint to think about applications like optical filters or waveguides. In fact results obtained on a filter made of porous silicon layers using a fabry perot type of structure will be presented.

There are two different approaches to form porosity superlattices. First, etch parameters like the current density or the light power density can be changed periodically during the etch process. These structures will be called type I superlattices [13]. Secondly, the layer structure can be predetermined by using a substrate with e.g. differently doped layers while keeping the formation conditions constant (type II superlattices).

Figure 5 shows the dependance of the porosity and the etch rate on the current density for differently doped substrates. For a given doping concentration of the substrate the porosity shows a linear dependence on the current density. Switching the current density between two values therefore results in the formation of layers with alternating porosity. To understand the formation of the so called type I superlattices it must be kept in mind that the electrochemical dissolution process takes place only at the bottom of the pores. One explanation for this is the quantum confinement effect which increases the bandgap of porous silicon and therefore creates a potential barrier for the holes [14]. The thickness of the single layers of a superlattice can easily be calculated by the etch rate for a given current density and the time (Figure 5).

Type II superlattices are formed using a fixed current density but etching a substrate with differently doped layers. Therefore, the variation of the porosity is given by the dependence of the porosity on the doping level for a certain current density.

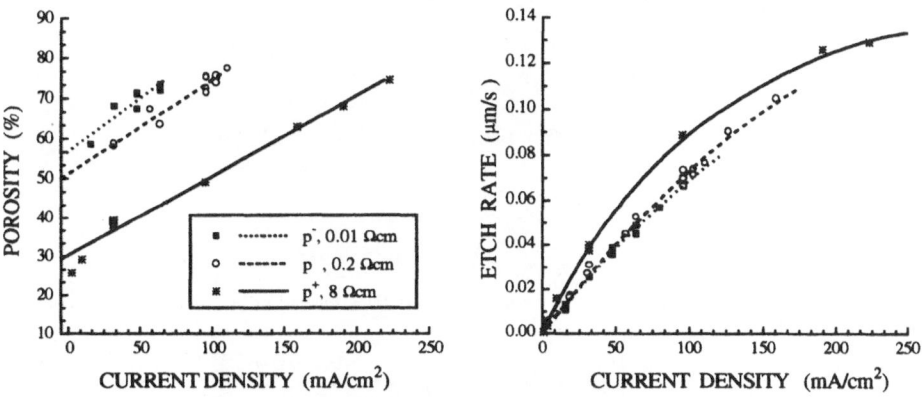

Figure 5: Porosity and etch rate as a function of current density and substrate doping level.

3.1 Experimental

Type I superlattices were formed on p-type doped substrates with resistivities of 0.01 and 0.2 Ωcm. As current source a Keithley 238 was used which allows a computer controlled etch process and a monitoring of the voltage.

For the type II superlattices periodically doped films were grown on Si (100) substrates (p-type, 0.01 Ωcm) by low pressure CVD. The doping levels were $1 \cdot 10^{17}$ and $1 \cdot 10^{19}$ cm^{-3}. Samples with thicknesses of the single epitaxial layers of 75 and 150 nm were investigated. The total number of periods was 10 and 5 respectively which results in a thickness of the epitaxial layers of 1.5 μm for all samples.

For the photoluminescence (PL) measurements the samples were excited using the 457 nm line of the Ar$^+$ ion laser. To avoid a photostimulated oxidation again the samples were measured in the UHV chamber. The power density for the PL measurements was 100 mW/cm^2. For spectral analysis a GaAs photomultiplier tube attached to the DILOR XY monochromator was used.

3.2 Properties of porosity superlattices

Figure 6 shows a TEM cross section of a type I porosity superlattice. Due to the higher density the low porosity layers correspond to the dark regions in the picture. The superlattice has been formed on 0.2 Ωcm substrate using a sequence of 19 mA/cm^2 for 1s and 175 mA/cm^2 for 2s which has been repeated 60 times. From these values the layer thicknesses can be calculated using the data in Figure 5. The layer thicknesses have been calculated to 218 nm for the high porosity and 16 nm for the low porosity layers. These values are in good agreement with the thicknesses estimated from the TEM picture which are 200 nm and 20 nm, respectively. The sharpness of the interface between layers of different porosity is very good on the microscopic scale. In the TEM picture a porosity gradient might be present. But it must be kept in mind that due to the finite thickness which is passed by the electron beam the gradient could also be an artifact of the TEM technique. Nevertheless, the interface is not perfect due to a roughness on the scale of some hundred nanometer.

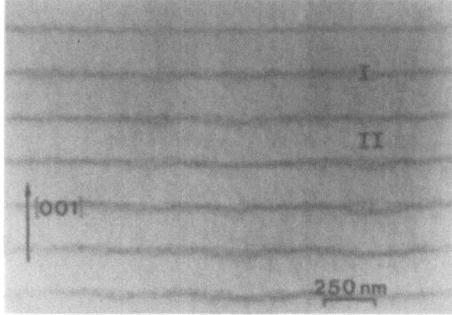

Figure 6: TEM cross section of a type I porosity superlattice formed on 0.2 Ωcm p-doped substrate. Layer I: 64%, 20 nm, Layer II: 84%, 200 nm.

Figure 7: TEM cross section of a type II superlattice. Etching was performed on a periodically p-doped substate (75 nm $1 \cdot 10^{17}$ cm^{-3}, 75 nm $1 \cdot 10^{19}$ cm^{-3}) using a current density of 50 mA/cm^2.

For the type II superlattices (Fig. 7) the interface is perfectly sharp because the microstructure follows the doping level of the epitaxial layers. From the epitaxial parameters only layers with two different doping levels and equal thickness were grown ($1 \cdot 10^{17}$ and $1 \cdot 10^{19}$ cm^{-3}) and therefore a two layer system is expected. However in the TEM picture three different structures are observed. Roughly in the middle of the lower doped layers the structure changes. This behaviour is presently not understood.

As already mentioned the most interesting property of porosity superlattices is their exciting optical behaviour. In Figure 8 the reflectance spectrum of a type I superlattice formed by a sequence of 19 mA/cm^2 for 3s and 207 mA/cm^2 for 6s with 20 repetitions on 0.2 Ωcm substrate is shown. These values result in a superlattice with layer thicknesses of 48 and 719 nm and porosities of 64 and 89%. In comparison with a single layer of porous silicon the reflectance of the superlattice shows strong modulation and reaches values up to 90% which strongly exceeds even the reflectance of crystalline silicon. In the near infrared a reflectance higher than gold has been achieved. The reason for the modulation are multiple reflections and interferences which occur in the superlattice due to the different refractive index of the layers. The strong modulation also influences the PL spectrum of porosity superlattices. This offers a very elegant way e.g. to narrow the broad luminescence band of porous silicon which typically has a FWHM of about 150 nm. Figure 9 shows the photoluminescence of a superlattice which was formed on 0.01 Ωcm substrate. The FWHM of the main peak is only 17 nm which is a reduction of nearly one order of magnitude compared to single layers of porous silicon.

It is even possible to form more complex types of structures with porosity superlattices. For example we have fabricated a Fabry Perot type of filter on top of a 10 μm thick porous layer. The filter consists of alternating layers with low and high porosity which act as high reflective mirrors and surround the cavity (see inset Fig. 10). The reflectance of this structure is shown in Figure 10. The sharp structures around 14000 cm^{-1} show the

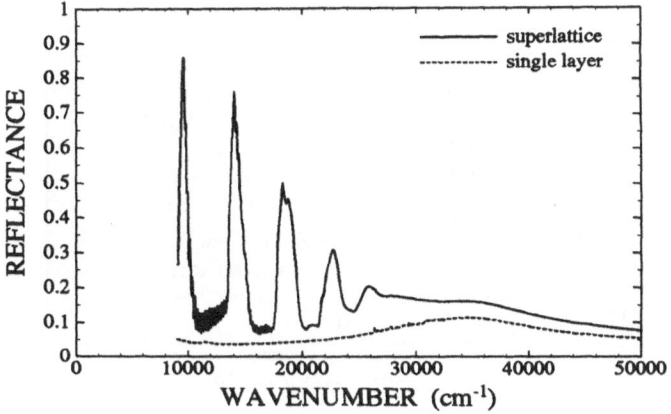

Figure 8: Reflectance spectrum of a type I porosity superlattice. Substrate: p-doped 0.2 Ωcm. Layers: 64% 48 nm, 89% 719 nm

effect of the filter. The transmission maximum of the filter corresponds to the dip in the reflectance at 13760 cm^{-1} and in spectral regions of high reflectivity the transmission is low. This can be visualized by the PL spectrum of this sample (inset in Fig. 10). It must be mentioned that this structure has not been optimized for a sharp filter characteristic so far and therefore could be improved in future.

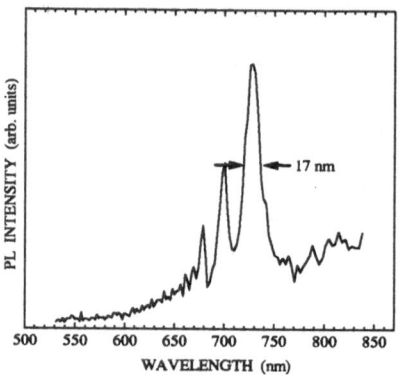

Figure 9: Photoluminescence spectrum of a type I superlattice formed on p-doped 0.01 Ωcm substrate. Layers: 34% 25 nm, 78% 264 nm

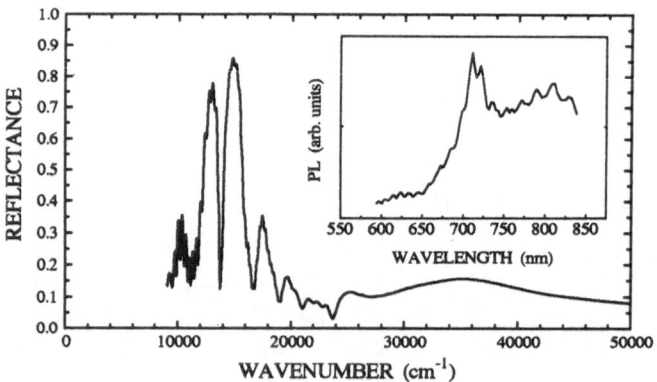

Figure 10: Reflectance spectrum of a Fabry Perot filter on top of a 10 μm thick 75% layer. The sample was formed on 0.2 Ωcm p-doped substrate. The inset shows the line narrowing of the photoluminescence by the filter.

4. SUMMARY

It has been demonstrated that Raman spectroscopy can be used to obtain informations about the microstructure of porous silicon. However using this technique the experimental conditions are important. Besides a low laser power density the ambient must be oxygen free in order to avoid a photostimulated oxidation during measurement. The crystal size distributions measured with Raman are in good agreement with TEM pictures but in addition reveal a more complex microscopic structure. Differences in the microscopic structure influence the effective dielectric function of the material.

Combining layers with different dielectric function can be done by the formation of porosity superlattices. We have shown different methods to obtain superlattice structures. For type I superlattices the layer structure is defined by changing the current density whereas for type II superlattices the layers are predefined by e.g. layers with different doping level grown by epitaxy. Superlattices can be used e.g. as optical filters or high reflective mirrors. We have demonstrated this by forming a Fabry Perot type of filter which is able to influence the broad luminescence spectrum of porous silicon. Studies for the long term stabilization of the optical peroperties of superlattices are in progress.

REFERENCES

1. R. Smith and S. Collins, J. Appl. Phys. **71**, R1 (1992).

2. L. Canham, Appl. Phys. Lett. **57**, 1046 (1990).

3. H. Richter, Z. Wang, and L. Ley, Solid State Commun. **39**, 625 (1981).

4. I. Campbell and P. Fauchet, Solid State Commun. **58**, 739 (1986).

5. Z. Iqbal, S. Veprek, A. Webb, and P. Capezzuto, Solid State Commun. **37**, 993 (1981).

6. U. Rossow, H. Münder, M. Thönissen, and W. Theiß, J. of Luminescence **57**, 205 (1993).

7. K. Tiong, P. Amirtharaj, F. Pollak, and D. Aspnes, Appl. Phys. Lett. **44**, 122 (1984).

8. H. Yoshida and T. Katoda, J. Appl. Phys. **67**, 7281 (1990).

9. Z. Sui, P. Leong, I. Herman, G. Higashi, and H. Temkin, Appl. Phys. Lett. **60**, 2086 (1992).

10. I. Reshina and E. Guk, Semiconductors **27**, 401 (1993).

11. H. Münder, M. Berger, H. Lüth, R. Herino, and M. Ligeon, Thin Solid Films **221**, 27 (1992).

12. H. Münder, M. Berger, S. Frohnhoff, M. Thönissen, and H. Lüth, J. of Luminescence **57**, 5 (1993).

13. H. Münder et al., Proc. ICFSI-4 Jülich (1993), in press.

14. V. Lehmann and U. Gösele, Appl. Phys. Lett. **58**, 856 (1991).

mpression : EUROPE MEDIA DUPLICATION S.A
F 53110 Lassay-les-Châteaux
N° 3666 - Dépôt légal : Janvier 1995